WEALTH FROM KNOWLEDGE

WEALTH FROM KNOWLEDGE

KNOWLEDGE

Studies of Innovation in Industry

J. LANGRISH
M. GIBBONS
W. G. EVANS
and
F. R. JEVONS

MACMILLAN

First published 1972 by
THE MACMILLAN PRESS LTD
London and Basingstoke
Associated companies in New York Toronto
Dublin Melbourne Johannesburg and Madras

SBN 333 12007 8

Printed in Great Britain by
R. & R. CLARK LTD
Edinburgh

Contents

Preface

Not much need be said here in justification of the theme to which this book is devoted. The process of technological innovation is so universally recognised to be of salient importance for the life of modern man, and yet so imperfect is our understanding of what makes it happen in the ways that it does, that there need be no undue academic reticence about bursting into print with something that may make a positive contribution, even though it does not have all the answers pat.

It may, however, be of interest if I briefly describe here how the book has come to be what it is. Its origin virtually coincides with that of the Department of Liberal Studies in Science, of which it is a product. The Department was set up in 1966 to devote itself to themes of the 'science-and-society' type: an aim the more unusual in that it was to be done in the context not of service teaching to students in other science departments but of an Honours School based on the Department. The importance of science for society has many facets, as I have emphasised elsewhere (*The Teaching of Science*, Allen & Unwin, 1969, chap. 3). Its role in helping to create wealth by underpinning technology and industry is by no means the only aspect that deserves more scholarly attention; but it is indisputably one of the most important ways in which science relates to the rest of society and, moreover, one which is of more than incidental interest in connection with practical policy considerations.

Much of the earlier work on technological innovation had been done by non-scientists and it seemed possible that a closer look from a more technical standpoint – as seemed appropriate for a department in a science faculty – might throw light on the ways in which science and technology are combined with other kinds of knowledge and skill in creating wealth. A grant was obtained from the Department of Education and Science on the recommendation of the Council for Scientific Policy and this made it possible for G. Williams to take up a Senior Research Fellowship at Manchester. It was he who had the idea of using the then newly instituted Queen's Award scheme to provide an independently selected sample of successful innovations as a source of case-study material. He was not, however, able to hold his Fellowship for long and in January 1967 the work was taken up by J. Langrish.

On the basis of his first few case studies, Langrish submitted an entry in a competition run by Shell and the British Association for the Advancement of Science for an essay on 'Human Attitudes and Institutions as Barriers to Technological Progress'. His entry shared a combined first and second prize and a revised form of the essay was presented at the 1967 annual meeting of the British Association at Leeds. Encouraged by this initial success, the scope of the work was enlarged into a team project to cover all the innovations that won Queen's Awards in 1966 and 1967. M. Gibbons and W. G. Evans, both Lecturers in the Department of Liberal Studies in Science, added their efforts to the case-study work. While Langrish continued with the innovations in the chemical and craft-based areas, Gibbons looked at those with a physics or electrical engineering bias and Evans at those based primarily on mechanical engineering. A selection from their case studies forms the bulk of this book. There will be many, I feel sure, who agree with me that they make fascinating reading. Indeed, a number of people, including some without technical qualifications, have already commented on them to that effect.

The full allocation of case studies is as follows (numbers refer to the list on pp. 85–9).

Langrish: 2, 3, 12, 13, 17, 19, 20, 21, 28, 30, 31, 33, 35, 39, 54, 56, 59, 60, 63, 67, 68, 73, 80, 81, 82.

Gibbons: 4, 10, 14, 15, 16, 37, 41, 44–52 inclusive, 55, 64, 69, 72, 79.

Evans: 1, 5, 6, 7, 8, 9, 18, 22, 23, 24, 25, 27, 34, 38, 40, 42, 43, 53, 57, 58, 61, 62, 65, 70, 71, 74, 75, 76, 77, 83, 84.

Nos. 32 and 78 were undertaken jointly by Evans and Gibbons. In the former case, that of the Smiths Industries aircraft automatic landing equipment, a fuller study has been done by J. E. H. Hartland in collaboration with them; the account that appears here has been prepared mainly by Hartland. The remaining five case studies – 11, 26, 29, 36 and 66 – were done by a research student, Miss V. Seal, under the guidance of Langrish.

My own contribution, apart from helping to get the project

off the ground and general editing of the book, has been confined to writing Part One, except for section (g) and parts of
sections (a) and (b) for which Gibbons provided the first drafts.
A glance at the general discussion which forms this part of the
book will suffice to show that most of the material in it derives
from my co-authors. Part Two is based on a report to the Council
for Scientific Policy which was prepared by Langrish. Many of
the ideas on which it is based are his own, though of course the
data that are presented in it derive from the whole range of case
studies; mostly they refer to all the 84 innovations that were
covered by the investigation.

In some sense, every part of the book is a product of our joint
efforts. It could hardly be otherwise, seeing that we were
working in almost adjacent rooms, with innumerable discussions of all degrees of informality.

Predominantly, the emphasis of our study has been not on
statistics of the head-counting type but on close-in assessments
of the situations in and around the innovating firms with
appreciation of the technical issues involved. The case-study
approach has continually rubbed our noses in the facts; our
book contains more of the complex realities of history 'as she
really happened' than of the abstractions and conceptualisa-
tions which achieve intellectual economy at the expense of
fidelity to the real world.

In some matters we have been able, by giving quantitative
data, to take the discussion beyond the stage of assertion backed
by anecdote, where it has so often finished in the past. For
instance, we are able to report figures for the relative importance in our sample of innovations of 'push' by scientific or
technological discoveries and of 'pull' by the needs of the market
or of management. We have also been able to estimate how
often outstanding individuals seem to have been important for
success, though those individuals have turned out to be less
often the 'lone inventors' beloved of romantic tradition than
'product champions' able to pilot their pet projects through the
unromantic bureaucracies of the modern industrial system.

Other favourite talking points in discussions of innovation,
however, seem to us to fade like mirages under attempts at close
scrutiny. Such, for instance, is the often repeated suggestion
that there is a decreasing time-lag between discovery and

exploitation. Such also is that policy-maker's wish-fulfilment dream, the idea of specifying optimum sizes for research teams. There are no slick quick tricks here to sharpen the research manager's job effectiveness: only chimeras to be dispelled by a substantial body of stubborn facts about real and recent experience.

We have paid particular attention to the relation of basic science to innovation. To our minds, our failure to find more than a small handful of direct connections is the more striking for the fact that we set out so deliberately to look for them. Our conclusions on this point have proved unpopular in some quarters. Some academic scientists find it difficult to accept what most businessmen already know: that the great bulk of basic science bears only tenuously if at all on the operations of industry. D. S. Greenberg has made some perceptive comments on this kind of situation in *The Politics of American Science* (Penguin Books, 1969, chap. 2). Science, he points out, is neither self-explanatory nor self-supporting and scientists tend to react to this predicament with a mixture of chauvinism, xenophobia and evangelism. Stirring professions of faith in basic science are coupled with dire prophecies that technology will die on the vine if it is starved of the rising sap of new ideas from undirected research. But in the long run, more solid reasons will be needed to ensure continuing public support for science on a large scale. In presenting our conclusions, we certainly do not intend to denigrate science; rather, we want to urge recognition of the fact that its value to industry is less direct and less overt than has been commonly supposed in the past. Perhaps science is not the father of technology but an anonymous well-wisher who sends it gifts through the post, as it were. If only the mechanism were more clearly understood there would be a better chance of increasing the benefits, and the justifications for public support of science would be strengthened.

We have many acknowledgements to make and we make them gladly. First and foremost, we are grateful to the Council for Scientific Policy, on whose recommendation we got the grants which paid for Langrish's Fellowship for three years as well as for the research expenses of the rest of us. Permission has been obtained to publish the results of research so financed.

Various members of the Council's secretariat have helped us by stimulating and fruitful discussions. Many people connected with the Award-winning firms have been generous with time and information; some but by no means all of them are mentioned in the case studies. Five of the 84 innovations were studied by Miss V. Seal, with guidance from Langrish; they are described in her M.Sc. thesis (University of Manchester, 1968). During the work she held a Science Research Council studentship. J. E. H. Hartland, who contributed the account of the Smiths automatic landing equipment, was supported in part by a grant from the Nuffield Foundation. J. S. Metcalfe has provided clear elucidation of some points of economics relevant to our investigation. We have benefited also from contact with our colleagues in the R. & D. Research Unit of the Manchester Business School, where Langrish is now working. Some undergraduates in the Department of Liberal Studies in Science helped with individual case studies. Last but not least, we much appreciate the willingness with which Miss M. Bruce and Mrs E. Foster have typed and re-typed the large volumes of material that it has been necessary to produce to make possible adequate consulting, checking, and correcting.

Department of Liberal Studies in Science   F. R. JEVONS
University of Manchester,
September 1970

Part One
General Discussion

(a) Wealth from knowledge

This book is about one of the outstanding features of the age in which we live: technological innovation. Progress in technology is one of the mainsprings of modern societies. Since the nineteenth century its effects have been obvious even to the most casual observers. In the twentieth century has come increasing awareness at government level of its importance for the well-being of nations; there has been no lack of exhortation from officials in high places about the central importance of technological innovation for industrial efficiency and economic growth. As Peck has put it, 'one persistent theme expounded by Britain's post-war leaders is that technology can be its major resource – the twentieth-century equivalent of nineteenth-century coal'.[1]

Economists, too, have turned their attention to this source of economic growth, and there seems to be agreement that the effect is big, though there are formidable difficulties in calculating exactly how big. Classical economic theory tended to ignore it and to ascribe increases in output per head to increases in capital per head – that is, economic growth was attributed to capital accumulation. More recently, however, it has become recognised that the effect of technical change is too big to be ignored. Many calculations indicate that it can outweigh the effects of increase in capital. In a pioneering paper, Solow[2] estimated a 'technical change' factor that contributed about four times more than capital accumulation to the growth of output per head in the United States private non-farm sector in the period 1909–49. The 'technical change' component is not due entirely to innovation in the sense of the introduction of new products and processes. Domar[3] identified within it the results of economies of scale as well as of improvements in the quality of labour, due to better education and health, and higher standards of management, which increase the efficiency with which resources are used. However, even after separating out economies of scale and some other smaller factors, Denison[4] is left with a substantial contribution from 'advances of knowledge'. During the period 1950–62, he estimates it to have been responsible for a growth in national income of 0·76 per cent per year in advanced countries.

It is, then, to a substantial extent through technological and associated managerial factors that man's intellectual powers create wealth and influence his material environment. The idea that this might be one of the major effects of human knowledge has been a dominant strain of Western thought since the seventeenth century. Francis Bacon saw the vision and proclaimed the message with unparalleled eloquence. 'Is truth barren ?' he asked. 'Shall we not thereby be able to produce worthy effects, and to endow the life of man with infinite commodities ?'[5]

Bacon's dream did not, however, become reality without further ado. We are still not as good as we might be at squeezing wealth from knowledge. One reason for this is the immense complexity of the processes involved. Truth does not of itself blossom in a spontaneous and straightforward way into worthy effects and infinite commodities. A multiplicity of other factors come into play, and there is usually a tangled jungle of complications before there are any tangible benefits, whether private, corporate or social. It is not surprising, therefore, that despite the acknowledged importance of the process of technological innovation, our understanding of its nature remains patchy. Modern scholarship has had only limited success so far in imposing the neat order of conceptual schemes on the chaos of raw observations.

Many managers and administrators, it is true, have not been loath to reflect in public on the experience they have won at the cutting-edge of innovation as it is put into practice in industry. It is not easy, however, to generalise from 'personal wisdom writings' based on individual experiences to principles which are in fact widely applicable. There are differences between firms in background, structure and objectives. There are also variations between sectors of industry; electronics is growing faster than cotton textiles, the aircraft industry is more research-intensive than shipbuilding. At the national level, too, diversity is marked. Different countries have different rates of economic growth, different frameworks of legal requirements and social pressures.

Uncertainty. In the circumstances, it is hardly surprising – perhaps even symptomatic – that some of the best-informed and most penetrating comment in recent years has focused on uncertainty as a key factor in understanding technological innovation. Charpie says that 'the process by which the idea, the men with energy and commitment, and the source of high-risk capital get together to produce consequential innovation has got to be one of the most haphazard and in a sense miraculous things that happens in this economy of ours'.[6]

Schon, too, selects uncertainty as a key factor in the understanding of technological innovation. He suggests that the innovative work of a corporation should be viewed as a process of trying to convert uncertainty to risk.

> Risk [he says] has its place in a calculus of probabilities. It applies to a specific course of action. . . . Uncertainty is quite another matter. A situation is uncertain when it resists analysis of risks. . . . [One] must invent what to do. [One] has no way of calculating with any precision the risks of action. [One] has only rough guidelines of skill and experience to help.[7]

In the course of a critical analysis of the assumptions on which much current thinking on technological innovation seems to be based, Gold[8] questions in particular the idea that the process is plannable by rational means. He identifies four hypotheses that underlie a great deal of recent enquiry in this area. First, there is the belief that technological innovation is intrinsically attractive to management – the confidence that innovations will increase efficiency and lower prices on the one hand and provide new products and open new markets on the other. Second, there is the view that technological innovation is plannable and controllable and that it occurs in areas chosen by management. Third, there is the assumption that major technological innovations are generated by a series of essentially rational decisions derived from well-defined organisational goals – that choices are based on close scrutiny of the techno-logical merits of ideas by scientists and engineers and that initial expectations are compared with final results so that significant deviations may in the future be used to provide better analysis. Fourth, there is the opinion that significant technological innovations are the result of industrial research and development (R. & D.), an opinion that derives strength from the common tendency to trace outstanding successes of selected prominent firms back to earlier R. & D. without seriously considering other possible influences.

None of these four assumptions emerges unscathed from Gold's critique. From an impressive survey of field work, he cites evidence which casts doubt on each of them. As a result, he makes a plea for a more critical study of the whole process of technological innovation.

Our investigation. It seemed to us that there is a need in this area of enquiry for more sheer information, more straight facts, so as to build up a fuller descriptive 'natural history' of the phenomena. As a basis for further analysis and higher-level generalisations, we wanted first to provide some rather detailed accounts of what actually happened in recent instances of technological innovation. Accordingly, we undertook the series of case studies on which we report in this book.

Rather than select our own sample, we used one that was selected independently of us, namely, the innovations which won Queen's Awards for technological innovation in 1966 and 1967. Whatever biases may have gone into the selection of these Award-winners (cf. p. 63), they were at least not put in by us, so that we could examine the histories without feeling that our conclusions might have been determined at the outset by preconceptions which we brought to the initial step of picking the cases. We felt, too, that concentrating on success stories would make it easier for us to get the kind of detailed informa-tion about particular cases that we wanted. Understandably, people are more often prepared to talk freely about successes than about failures. A selection from the case studies forms Part Three of this book.

Different emphases are possible in case-study work. Burns and Stalker,[9] for instance, concentrated on organisation structures, distinguishing between mechanical and organic types, suitable respectively for stable and for changing conditions. Bright[10] highlighted management processes, whereas the balance of the study by Carter and Williams[11] lay towards financial organisation and investment decisions. Allen[12] sought to relate scientific and technological advances to the social, economic and political developments that occurred over the period of the innovation. A British Association symposium[13] pointed to the difficulty of foreseeing the long-term consequences of technical advances.

In our study, we aimed in particular to relate the technological to the organisational and other aspects. We felt that, as Cahn[14] has pointed out recently in a review of case histories, too little attention has been paid to the scientific and engineering histories of individual innovations. Investigation of the technological background of each innovation was carried out in order to assess the type and magnitude of the technical breakthrough and to give indications of the types of institutions involved and the stage of development reached before the product or process was taken up by the innovating firm. Information on the industrial or social needs requiring the innovation was also gathered. The innovations were then studied within the organisational context of the firms in order to find out as much as possible about how they were taken up, developed and marketed. Familiarity with the technical background helped to provide a basis for informed discussion with people who were concerned with the innovations (see p. 82 for further description of the methodology).

By such means, we have been able to identify some factors that seem to have been important in promoting or delaying the innovations. These, together with some other generalised data, are given in tabular form in Part Two of this book. Most of these data refer to all the 84 innovations cited in the 66 Awards for technological innovation in 1966 and 1967 (cf. p. 62).

Our set of case studies brought us face to face with many of the unsolved problems and open questions relating to technological innovation and the ways in which wealth is created from knowledge. The rest of Part One of this book is devoted to discussions of some topics of this kind in the light of our case-history material.

(b) Sources of technological innovation

Much of our discussion is centred around the question of the sources from which innovation arises. In dealing with this complex topic, it is vital to be quite clear as to what is to be understood by the expression 'technological innovation'.

It is relevant here to consider the criteria used by the Queen's Award scheme. The report of the committee which drew up the scheme gives eight criteria for Awards.[15] Of these, the first six refer to Awards for export achievement, and it is the seventh and eighth which concern technological innovation. They are as follows:

> (vii) A significant advance in the application of advanced technology to a production or development process in British industry. Recognition should only be accorded under this head if greater efficiency results from the process.
>
> (viii) The production for sale of goods which incorporate new and advanced technological qualities.

These criteria give a rather embracing view of the process of technological innovation. Certainly it is one which goes far beyond mere invention. 'We do not recommend that inventors or inventions should be recognised as such. The scheme should concentrate on the practical application in industry of advanced technology whether in the form of processes or of products.'[16]

Invention does not in itself constitute innovation; rather, it forms a part, and only a part, of the overall process. In saying this, we want to do more than point to the fact, well known to industrialists and financiers, that certain steps have to follow after invention to turn it into successful innovation. We do not take the view that innovation is merely invention plus subsequent development and commercial exploitation. The conditions in which the inventive steps were taken must themselves be considered as essential ingredients of any proper understanding of the innovative process. As Gilfillan[17] pointed out, not only do inventions cause changes in the milieu but also, vice versa, changes in the milieu call forth inventions. With this in mind, we have concentrated a good deal of attention on the needs – the commercial, institutional, social and suchlike pressures – which form part of the environment in which inventions are made.

Here it is necessary to consider factors arising inside the innovating firms as well as external ones. Many of the inventive steps we identified were embedded within the innovation process, not made prior to it or at its beginning. An innovating organisation is as much a seedbed as a fruit of inventive activity. Invention and innovation are to be distinguished, but they are not separate. They are often inextricably interlinked because they stand mutually in a mixed causal relationship: each is part cause, part effect of the other.

Thus it may not be adequate, we believe, to accept the simple view taken by many economists, that innovation is a set of investment opportunities arising from invention. Correspondingly, success in our context is not judged solely by strict financial criteria such as profits in a firm's accounts or rates of return or present values. For some of the cases we have studied, the main financial returns may yet be in the future, but the innovations can be rated successful by the fact that they have led to 'worthy effects' and 'infinite commodities' – or, in more modern terms, that they have become efficient productive processes providing goods and services.

Multiple sources. Taking the inclusive view that we do of the innovation process, we are forced to the conclusion that the sources of innovation are multiple. A new productive process is the historical outcome not of some single point event but of the convergence of many strands of events. The plurality of sources of innovation is conceptually an important point. It prevents us from subscribing to models of innovation which view it as a linear process, as suggested, for instance, by the expression 'innovation chain' (see p. 72). And it makes us sceptical of attempts to define unique origins for particular innovations. Different workers select different 'origins' for given innovations. The point was forcibly brought home to us when we provided each other with given sets of facts and asked each other, 'What was the origin of this innovation?' or 'Where did the technical idea for this innovation come from?' Even among ourselves, different answers were forthcoming. In view of this uncertainty, the concept of a 'time-lag' between a discovery or invention and its exploitation in an innovation is a hazy one, and the debate about whether such time-lags are decreasing seems to us to rest on shaky foundations (see p. 35).

Our study has been carried out principally at firm level, a 'firm' in this context being any Award-winning unit, including research associations and public corporations. These are the managerial decision units primarily responsible for particular innovations. However, the innovation process does not respect legal, financial or organisational boundaries. For instance, technical novelty often comes from outside the innovating organisation. Mueller[18] analysed the origins of 25 important process and product innovations by Du Pont, an American firm noted for its technical progressiveness; only 10 of them were based on discoveries attributable to Du Pont's own employees.

In our study, we found a similar proportion of technical ideas arising within the Award-winning firm itself. Table 6 (p. 78) shows that 56 out of 158 important ideas arose internally. But in our data there are, on average, three 'useful' technical ideas per innovation. This shows how our approach differs from that of Mueller, who selected one idea as the specific origin of each innovation. Several of the technical ideas made use of by the firms in our sample had been known for a long time, and it was clear in many cases that the determining factor in bringing the innovation to fruition was not so much the availability of the ideas as the decision to use them or the changes in market or social conditions which prompted that decision.

It is often said that marketing does in fact, or should ideally, set the goals for research and development (R. & D.). We have examined this relationship of marketing to R. & D. in our case studies. Our predominant impression is one of considerable diversity. The highest common factor seems to be that, in one way or another, a need is brought together with a technical possibility for meeting that need (pp. 49–57). The data in Tables 3 and 8 show that the clear identification of a need that could be met was important more often than the realisation of the potential usefulness of a discovery except for the relatively small number of innovations involving large changes in technology or made in small firms.

There are marked differences of opinion, at least on the surface, about the part played by science in innovation. Some insist on its importance as a source of ideas on which technology is based. In the United States, the Office of Naval Research[19] sought to show that important technological developments derive from basic research, and more recently evidence pointing in the same direction was presented in a report prepared for the National Science Foundation.[20] However, in Project Hindsight,[21] an attempt was made to isolate critical events in the development of certain weapons systems and little relation to basic science was found. In a more academic context, Price[22] has compared the structures of publication in science and technology; he concludes that, just as science builds mostly on earlier science, so technology builds mostly on earlier technology, and direct interconnections between the two are relatively rare. If this is so, it becomes important to elucidate in more detail how it is that technology manages to build on technology. Some authors have maintained that moving individuals from one environment to another is more effective than circulating ideas on paper. Burns summarised this view by saying that 'the transfer of technology is one of agents not agencies, of moving people among establishments rather than of routing information through communication systems'.[23]

Among our innovations, we have found a number of instances where technological know-how was transferred 'on the hoof' (pp. 43–48). Table 7 (p. 79) shows that 'transfer via person joining the firm' is the most frequent single method of transfer of important ideas into the Award-winning firm from outside. The relation of science to the innovations in our sample is a matter we have looked into in some detail; our conclusion is that the input from science, though real, comes largely through indirect channels (pp. 33–42). The interactions between organisations that we have found among our cases are of the most various kinds, and are discussed on pp. 24–33.

Traditionally, the origin of an innovation has often been associated with the name of one person, though different countries have often selected different persons. Such individuals are usually thought of as outstanding inventors. We have not come across many such people in our studies. In relatively few of our cases have we found an individual of exceptional scientific or technological creativity to have been essential for success.

There are, however, other ways in which outstanding

individuals can be of vital importance. One is by helping to direct attention to suitable areas in which to commit R. & D. effort with good chances of success. It is often a person of some authority in a firm who is in a position to recognise that the state of technological knowledge is ripe or nearly ripe for an innovation that is viable in the existing social and economic climate. He combines a need of some kind with a technical possibility for meeting that need and thereby gets the innovation process under way.

Another function that may be fulfilled by a person in a position of authority is to overcome resistances to innovation within the innovating firm. It is sometimes assumed that technological innovation is inherently attractive to firms – that to innovate is obviously a good thing in itself. However, in every established organisation there exist pressures and forces making for stability, and these inevitably tend to hinder innovation. 'Always "dangerous"' is the entry against 'innovation' in Flaubert's *Dictionary of Accepted Ideas*.

Resistance may take various forms, active or passive. Bright[24] has listed some reasons why innovations are opposed. In particular, he identifies social status, custom, organisational structure, habit and personality as elements of passive resistance to rapid technological change. Similarly, Schon[25] points out that management attitudes to change are often ambivalent. Officially it may be recognised as essential but privately, because of the uncertainty it brings, it is liable to be viewed as a threat. Thus management may actually be taking positive steps to thwart innovation while believing themselves to be favourably inclined towards it. It is because resistances like these have to be overcome that successful innovation often owes much to individuals who commit themselves to pushing a project through. Various commentators have pointed to the importance of this factor, using names like 'entrepreneurs', 'innovators' or 'product champions' (e.g. Schon[26]). Stillerman,[27] speaking before a U.S. Senate subcommittee, gave it as his opinion that 'more often than not [modern inventions] originate with a persistence amounting to an obsession'.

If, despite a growing tendency towards teamwork, individuals still stand out in the histories of particular innovations, it seems to be less because of scientific or technological creativity than because of the ability to combine needs with technical possibilities and because of the drive and commitment shown by product champions. These two roles may, of course, be played by a single individual, and they are given together as 'top person' in Table 3 (p. 69). In all the groupings listed, this is the single factor most frequently identified as enabling the Award-winning firm to succeed.

Some of the issues raised above, together with some other related ones, are discussed in more detail in the sections which follow.

(c) Roles of individuals in innovation

One of the most important questions about research and development concerns the extent to which technological innovation can be forecast and planned. Managers, administrators and politicians have to ask themselves, is the provision of adequate resources enough to give a good chance of success ? Or does the process depend too much on random events ?

There are, of course, many kinds of factors whose fortuitous incidence makes innovation a process fraught with uncertainties. One kind which has given rise to some discussion is the emergence of outstanding individuals who play key roles in bringing innovation about. Genius is unpredictable. Little can be done to foretell or to influence where and when it will arise and which way it will turn. Commentators who emphasise the dependence of technological progress on key individuals tend towards a 'heroic' view of the process: it is the heroes of science and technology who exert the critical influences on the pace and direction of advance.

On the other hand, it is also possible to take the view that progress during any period is largely determined by the nature of the knowledge available at that time and the prevailing pattern of social needs. In that case, the broad managerial or policy problem concerns the efficiency with which the potentialities are uncovered and exploited. Waiting for key individuals to arise spontaneously and act freely may not be the best way.

This question has often been discussed. It, too, can be traced to the writings of Francis Bacon. The late Middle Ages had

produced three inventions – printing, the use of gunpowder in
firearms and the magnetic compass – whose revolutionary
impacts on learning, on war and on navigation particularly
impressed him. Yet these, he commented, were 'but stumbled
upon, and lighted on by chance'.[28] How much more might not
be achieved by conscious and planned effort! Speculating on the
way in which such effort might be organised, he envisaged
Salomon's House in the *New Atlantis*.[29] Here was a national
research institute in prototype, lavishly equipped by Bacon's
imagination with all the equipment and facilities he could
think of that might conceivably be of use. Among the thirty-six
fellows of the foundation, there was well-defined division of
labour and allocation of tasks. The programme was clearly
intended to be a corporate one. All the principal inventors were,
it is true, to be commemorated by statues – 'some of iron, some
of silver, some of gold'; but Bacon does seem to have placed his
trust more in his system than in exceptional individuals. The
lame man who keeps to the right road, he pointed out, out-
strips the runner who takes a wrong one. He even went on to
claim that 'the course I propose for the discovery of sciences is
such as leaves but little to the acuteness and strength of wits but
places all wits and understandings nearly on a level'.[30]

A more recent author who has taken a somewhat similar view
is Gilfillan. Impressed by the way in which changes in the social
environment 'cause' inventions as well as vice versa, he considers
invention 'by accident' to be rare. He is concerned less with the
problems of planning and organising research than with the
apparent inevitability with which inventions arise in the course
of history. 'A device can no longer remain unfound when the
time for it is ripe.' Given this view, it becomes readily understandable that duplicate invention is common. Social need
rather than individual inspiration is the determining factor.
'There is no indication that any individual's genius has been
necessary to any invention that has had any importance.'[31]

Independent inventors. In its modern form, the discussion owes
much to the important book on *The Sources of Invention* by
Jewkes, Sawers and Stillerman, first published in 1958. The
main conclusion from their study was 'that the sources of
invention were numerous, scattered and varied'.[32] In partic-

ular, despite the increase in corporate research, individual or independent inventors continue to make important contributions. The individual inventor is opposed in this analysis to the institutional one. In the extreme case, he 'chooses the field of ideas in which to work, employs his own resources or acquires them from others who exercise no control over his work, stands to gain or lose directly from his inventive success or failure, [and] works with limited resources and with colleagues subject to his guidance and leadership'.[33] Although few cases were quite as extreme as this, Jewkes *et al.* ranked more than half of their sample of twentieth-century inventions as individual ones (33 out of 61, increased to 38 out of 70 in the second edition, 1969).

The view of Jewkes *et al.* has been criticised by Freeman,[34] principally on the grounds of possible bias in the selection of the sample of inventions. Jewkes *et al.* reply that their sample is 'not in any sense a scientifically balanced one' and point out that it is difficult to see how any such sample could be constructed.[35]

The sample which we have used for the study reported in this book was selected by an independent mechanism external to the study, namely, the Queen's Award scheme. This does not, of course, necessarily free it of possible biases (see p. 63). We agree with Jewkes *et al.* that scientific or technological 'balance' is not easy to define, let alone achieve. However, because of the emphasis on the Queen's Award scheme on the *use* of new technology, the sample is possibly representative of the kind of innovations which make important contributions to national wealth.

We are in whole-hearted agreement with Jewkes *et al.* about the great variety and diversity of the circumstances from which technological progress arises. As regards the roles of individuals, however, we feel that it is necessary to make some careful qualifications.

Attention should be drawn, firstly, to an important difference between our study and that by Jewkes *et al.* The Jewkes study was concerned primarily with *invention*, as the title of the book indicates. 'Although when this work was started it was not intended to say anything in detail about the development of inventions, it subsequently became increasingly apparent that some comment on it was unavoidable.'[36] The discussion that was accordingly added 'is confined to that period, intimately associated with an invention, which ends when commercial utilisation appears at least to some people to be feasible'.[37] Our study, by contrast, in line with the relevant declared objectives of the Queen's Award scheme, is firmly directed at *innovation*. Awards for innovation are given for the application of advanced technology to a production or development process, or for the production for sale of goods incorporating new and advanced technological qualities (see p. 6). These activities may be very different from invention. Very substantial investment in development and in production facilities is usually required before a new process becomes established in production. Great novelty of invention within the innovating organisation is not, however, always a prerequisite; technical novelty may be brought in from outside, and there are many examples where this course has been adopted with successful results.

The context in which we discuss the roles played by individuals is, therefore, significantly different from that adopted by Jewkes *et al*. It does undeniably still make sense in the twentieth century to talk about 'the independent inventor', but 'the independent innovator' is almost a contradiction in terms. For technological innovation to occur, there must be some interaction between a set of ideas and an institution; the ideas must be interpreted in terms of a need of the institution and put into effect by it. Innovation is almost by definition a corporate and collaborative effort, and it is correspondingly difficult to disentangle the roles played by particular individuals.

The institutional environment. Jewkes *et al*. say of people like Midgley (given much of the credit for tetraethyl lead and for Freon refrigerants) and Carothers (nylon) that they 'apparently find it possible to work as employees in institutions, subject to some limit on the range of their activities'.[38] Even if one leaves aside the matter of the resources required for development and production, this gives only a one-sided picture of the institutional environment as a seedbed of technological advance. Some kinds of individual initiative would not be possible except within an institution. Organisations may exert an enabling as well as a limiting effect on the personal creativity of the people who work in them.

This may happen, for instance, as a complex engineering product undergoes continued evolution in design. 'Creative design may range from the ability to evolve the overall conception of novel systems or machines to the ability to conceive better design of elements of machines or structures forming part of larger complexes devised by others.'[39] An example from among our case studies concerns the machining of flats on diesel engine valve stems so as to achieve intermittent oil pressure pulses in the stem passage. The proposal for this came from Kryzwski, a development engineer at Mirrlees National (Hawker Siddeley). The problem to which it provided a solution had arisen within the context of a major programme of work in the firm on the improvement of large diesel engines. Improved design of the valve assemblies had brought valve stem temperatures down low enough to cause condensation of sulphuric acid from the exhaust gas, and corrosion of the valve stem would have been severe but for protection by the pulsating oil pressure.

The programmes carried out by chemical firms of testing large numbers of chemical compounds might appear to be strictly team rather than individual research. However, such apparently shotgun strategies may afford opportunity for individual initiative. For instance, Rattee, in the Wool Section of the Dyehouse at I.C.I. Dyestuffs, decided to try on cotton a compound originally intended as a dye for wool. The Procion dyes resulted. In a somewhat similar way, at the Beecham research laboratories, a discrepancy between the results of chemical and microbiological assays suggested to the scientists concerned that 6-APA, the penicillin 'nucleus', might be present. From this substance, the semi-synthetic penicillins were derived.

Managers and creative researchers. The roles played in innovation by key individuals fall into two broad types. On the one hand, there are those who act by virtue of occupying a position of some authority in the innovating organisation, as some kind of manager or perhaps as technical director or chairman. On the other hand, there are those who are 'creative' in the research and development sense. Such a person might be described by colleagues as a 'mechanical genius', or he may have 'green fingers' with organic compounds or with semiconductor materials, or he may have some special knowledge that he brings from education or from previous employment.

These two types are given as separate 'factors' – 'top person' and 'other person' – in Tables 3A, 3B and 8A (pp. 69 and 82). However, they are not entirely distinct. In small companies, in particular, the more and the less technical functions are of necessity telescoped and all may be carried out by one or a few people.

Thus Wood started Oxford Instruments by making magnets in his own house. Plasticisers, a family business owned by the Slack family, has been innovative largely because at least two of the Slack brothers 'like to chase new ideas'. A similar case of a family firm is that of H. S. Marsh, where Eric and Rex Edwards, second and third generation respectively in the firm, have themselves been closely involved with design.

Such situations are found even in firms of substantial size. Sir James Martin is both Managing Director and Chief Designer of Martin-Baker Aircraft. In 1968 this firm employed over a thousand people at the Denham site alone. Nevertheless, he retained a high degree of control over all the activities of the firm both as regards policy and in its technological developments. Each drawing and modification was personally approved by him before being sent to the production departments.

In a really large organisation, a technical idea is not likely, in the nature of the case, to come from near the top. A director may well, however, be the one to suggest a new area in which effort might be expended to good effect. Thus Sir John Toothill, as Managing Director, played an important part in turning the attention of Ferranti, Edinburgh, to numerical control equipment for machine tools. The development of semi-synthetic penicillins was the eventual outcome of the decision by Lazell, Chairman of the Beecham Group, to seek scientific advice about the prospects for research leading to new antibiotics.

A function performed by the 'top persons' in such cases as these is to assess marketing and technical situations in order to arrive at decisions about areas in which research should be prosecuted. Essentially, they associate particular needs with technical possibilities for meeting those needs. The question may be asked, is it necessary for single individuals to fulfil such functions? Just as the role of individual inventors has been supplemented by R. & D. teams, is it not possible for the role of single persons in management to be supplemented by teams of people performing corresponding functions? Organised effort within a large company could be directed towards combining technical opportunities with suitable market objectives. The problem of what research to do is, after all, a research problem itself and could be tackled like many other research problems on a collaborative basis by teams of workers. This would be more than, or rather different from, the sort of project evaluation which is already carried out by many firms. In project evaluation, the usual philosophy is that the R. & D. department acts as the source of a stream of ideas from which the best have to be filtered out and given priority rankings. What is mooted here, by contrast, is a mechanism to *initiate* research projects. It would act as an input into the R. & D. process, not merely to screen the output.

A few steps in this direction seem already to have been taken in British industry. However, the role of a 'top person' is not by any means confined to deciding on R. & D. goals in the light of formal or informal market forecasts together with an appreciation of what might be technically feasible. In many cases, he clearly acts rather – or in addition – as a *champion* of the project (cf. p. 10). Such a role is indicated when others apply to him terms like 'enthusiast' or 'driving force'. He may be effective by virtue of the authority which he himself holds, or he may, in a large organisation, have the determination to carve a path through it. He acts as a spokesman and provides impetus in favour of proceeding with the innovation. Maurice Laing seems to have done this in the case of Lytag. It seems doubtful whether the project would have been started without his support, since other people within the Laing organisation wanted money for alternative investments. White, Technical Service Director of I.C.I. Dyestuffs, provided a driving force in favour of work on reactive dyes, as is attested by both Stephen and Rattee, who were responsible for the discovery. Rose, the Division's Research Director, also gave enthusiastic support.

In White's words, 'the two of us were able to ginger the whole development machinery into an unusual activity'.

In summary, then, our emphasis on the importance of institutions in technological innovation should not be taken to mean that we belittle the importance of individuals. Individuals are important within institutions. Indeed, 'top person' alone is the most common single factor favouring success in all the groupings of Tables 3A, 3B and 8A. We have taken care over the evidence for listing this factor as important for any particular innovation. In no case have we merely accepted the uncorroborated testimony of the person concerned. As evidence we have used statements by others, in most cases from several independent sources.

As might be expected from the varied nature of the functions performed by key individuals, their educational backgrounds, too, are very varied. It is worth remarking, however, that even where the contribution has been predominantly technical, the person does not always hold high formal qualifications in science or technology. Naturally graduates are often encountered, but we have been struck by the frequent and real importance of people with the Higher National Certificate type of qualification. These are not 'qualified manpower' according to the categorisation adopted by the Committee on Manpower Resources for Science and Technology,[40] but there seems to be a good deal of justification for the view that 'yesterday's H.N.C.s are today's B.Sc.s'. Lowe, whose artistry in fashioning silicon contributed to the development of high-voltage transistors by Lucas, had no university training. Ransom of Short Bros has no formal engineering qualifications but as design draughtsman contributed no less than eight out of eighteen patents taken out on the Seacat missile system. Sir James Martin holds over a hundred patents but has no formal qualifications.

(d) The question of research group size

As mentioned above, it is widely believed that teamwork is coming to predominate over individual efforts in research. The complexities of modern research, its demands on background and knowledge and the range of skills required are said to make

collaboration necessary. On the other hand, large teams also suffer from disadvantages. They can easily become unwieldy; communication problems arise because person-to-person contacts cannot be effective between all members of large groups; and it may become difficult to maintain clarity and unity of purpose. Questions have accordingly been asked about the best sizes for research teams. The problem is of the greatest interest to those concerned with policies for research, whether at company, institute or national level.

Our studies unfortunately cannot provide clear-cut answers to such questions. Rather, they have convinced us that it is often virtually impossible to say what sizes the teams have actually been in the cases we have investigated, let alone to say what they should have been for greatest success. Research is not a sufficiently distinct component of the innovation process. The concept of research teams as discrete entities often turns out to be largely a myth.

One difficulty is that different parts of the innovation process are frequently carried out in different organisations. This type of phenomenon is discussed later (p. 25). Other types of difficulty arise from the lack of clear demarcation between research and other functions of a firm, from the fact that the numbers of people concerned with a given innovation may vary greatly over time during the course of its life history, and from the fact that individual workers may operate on a 'time-sharing' basis, concerning themselves with more than one project over any given period. These considerations are illustrated below.

Demarcation difficulties. Even where an innovation can reasonably be described as having originated within the organisation that brought it to fruition, there are problems of demarcation. Boundaries between the 'team' and the environment in which it operates, and between the research team's environment and the world at large, are commonly so fuzzy as to be virtually indistinguishable. Neither an 'inner circle' nor an 'outer circle' is at all sharply defined. At best, perhaps, one could speak of a 'field' in which intensity of involvement decreases with increasing distance from the 'centre' according to something like an inverse square law.

Sometimes, it is true, research is so separate as to be in a

B

separate research company within an industrial group. Thus the Beecham group formed Beecham Research Laboratories Ltd, and it was out of research there that semi-synthetic penicillins came. However, the work within a group that leads to innovation is not always that which is carried out in research companies or separate research laboratories. The Doulton group has had Doulton Research Ltd as a central research organisation since 1960, but English Translucent China had been developed by that year, the impetus coming mainly from three people in Doulton Fine China Ltd working part-time on it in addition to their other functions. Among the giants in the electrical industry there may be separate research laboratories, but they seem sometimes to be regarded not so much as a source of new products or processes of direct use in manufacturing but rather as a 'library' or consulting service to which either a long-term problem or a particularly complex one can be taken.

A research project may depend on certain facilities and services which are shared with other research projects and with non-research work. Such supporting services might be thought to be ancillary to the main research effort, but they can be vital to its success. One kind of example is the testing of chemical compounds for certain desirable properties, as mentioned above (p. 15) for reactive dyes and for semi-synthetic penicillins. In engineering firms there is likely to be similar sharing of drawing office facilities. Patent services, too, may be part of the research 'overheads' that are difficult to allocate.

Some of the demarcation difficulties that commonly arise may be illustrated by considering the I.C.I. Dyestuffs Division site at Blackley, Manchester. This site contains the headquarters of the Division, separated by a river from one of its six factories. In the headquarters, scientists and engineers are employed not only in the Research Department but also in technical service, patents, sales and so on. It is not at all clear how many of these should be considered to have contributed effectively to the 'environment' in which Procion dyes were discovered and developed. The research origin of this particular innovation can be defined more clearly than in most other cases: it was the synthesis by Stephen of a compound which was found by Rattee to react chemically with cotton. Stephen was working in the Research Department but Rattee belonged to the 'Dyehouse',

later (in 1965) combined with other departments into the Applications Research and Technical Service Department.

Similar demarcation difficulties arise in the engineering context. Complex engineering products consist of many units which are often changed one or a few at a time as a series of small problems is solved or better solutions are arrived at. A continuing process of organic evolution takes place as one component after another is modified or replaced or added. In such situations, close integration of research and development work with production is necessary to ensure that each change fits into the context of the complete product, which must at each stage remain a functional whole. Geographical proximity often helps such integration. In Mirrlees National (Hawker Siddeley), the Research Department is separated for convenience from production and testing, but it is in the next building on the same site and the engineers engaged on the programme of large diesel engine development used the Manufacturing Department's drawing office facilities.

Variations over time. One of the biggest difficulties in specifying team sizes is the fact that there are often gross variations over time in the number of people concerned with a given innovation. A wide range of activities is normally gone through before a new process becomes operational or a new product is sold.

Thus the Procion dyes arose out of a discovery by two people, Stephen and Rattee; but as a result of that discovery, a substantial development effort was set in motion (cf. above, p. 17) and it is estimated that more than a hundred people were involved at various times. Preparation of more dyes in the range was necessary, and so was work to stabilise the dyes, to develop manufacturing processes and textile printing techniques, to counter competitive patent activity and so on.

Similarly, in the case of semi-synthetic penicillins, the discovery of the penicillin 'nucleus', 6-APA, was only a starting point for a great deal of further work. The original team of eight graduates plus assistants was reinforced by the gradual transfer of workers from another programme. A commercial process for the production of 6-APA was established; work was carried out aimed at patent coverage; and new penicillins were made from 6-APA. This latter activity involved the chemical synthesis of a large number of compounds and testing them for their value as drugs, trying to combine the desirable properties of penicillinase resistance, stability to acid (necessary for activity when taken orally) and effectiveness against a broad spectrum of bacteria.

Work on the Smiths automatic landing system could be said to have started when Meredith wrote a company report in 1956 on a survey of the SEP-2 (autopilot) performance and of possible lines of development for the type 5 autopilot. By 1957, ten people were working on the project and two years later this figure had grown to 120 people at Cheltenham alone; eight other manufacturing sites as well as three other major participants (B.E.A., Hawker Siddeley and government groups such as the Air Registration Board and the Blind Landing Experimental Unit) were also involved in the development, production, testing and approving of the flight control system.

Time-sharing. Another difficulty in specifying team sizes is that it is common for individuals to operate on a 'time-sharing' basis; that is, instead of a given person working full time on a given project for a given period, he divides his time between several projects. Where a variety of component technologies goes into the making of a product, as is quite normal in many kinds of engineering, there may in a large firm be a population of up to 200 to 300 people on development work. At any one time there is present a set of groups, but these continually form and re-form in response to changing problems. When a particular stimulus comes, a team of, say, two to ten qualified people (with appropriate supporting staff) is formed and to some extent separated out from the 'pool'. These *ad hoc* groups tend, however, to overlap in personnel; one person may serve several groups as 'consultant' on some specialised aspect such as pumps or heat-resistant paints. The 'pool' remains fluid in the sense that the groupings in it undergo continuous restructuring. The responsibility for guiding this restructuring lies with someone like a chief engineer, who deploys and re-deploys the skills available to meet the particular objectives currently in view. It is a process analogous, one might say, to the way an engineer deploys items of hardware from the large selection of components listed in the catalogues piled high on his desk as he designs a new piece of equipment.

Thus, for example, as a result of the visit of a team from Delco (U.S.A.) to Joseph Lucas (Electrical) to enquire about transistorised ignition systems for cars, a team was hived off to do development work on high-voltage transistors, following up an earlier interest in this field which had, however, had different applications in mind. Similarly, in the case of G.E.C., when the need arose for reliable solid-state microwave transmitters of medium power and high frequency for overseas contracts, a group was formed which successfully met the specifications and then went back into the 'melting pot'.

In other cases, success leads to the formation of permanent new divisions. This happened, for instance, with Ferranti inertial platforms, Ferranti numerical control equipment and the English Electric satellite communication systems.

The myth of discreteness. Cases such as these show that research and development may be very closely linked with normal, 'bread-and-butter' engineering. The semantics of the word 'engineer' cover a range of activities stretching from the humdrum to the highly creative. This is one reason why, in the engineering context at least, a question about research group sizes could often be dismissed as being based on the 'myth of discreteness' – the false idea that a discrete group produces ideas or prototypes which are handed on to other groups for production.

Rather than attempt any such 'handover' process, many firms have consciously reorganised themselves so that projects can become the responsibility of the same group of people throughout the product's development from the research stage to production. Bragg, formerly Chief Scientist of Rolls-Royce, has called this 'taking the seedlings to the garden to grow'. The principle is that the originators of ideas should be responsible for the production of the hardware that embodies them. Smiths Industries, too, have adopted an organisational structure designed to facilitate the 'convection' of people through the company, carrying their ideas with them instead of requiring them to hand over the projects with which they have developed familiarity and to which they have committed enthusiasm.

Our overall conclusion, then, is that questions about optimum sizes of research teams cannot be given clear, crisp answers without ignoring or obscuring some of the very real complexities that surround the issue. Semantically, the concept of 'research' lacks clear definition. Organisationally, the research function is in most cases not sharply separated from other functions of a firm. Innovation is a process over time, not a point event, and at different stages in the genesis and evolution of a given innovation very different numbers of workers may be giving varying proportions of their time to it. Management has to decide in particular cases how much promise lies in given research topics and how much urgency attaches to them. This is a management problem that calls for skill, judgement and careful study by the organisation concerned.

(e) *Interactions between different organisations*

One of the reasons for the interest in research group sizes discussed in the previous section is the idea that certain minimum sizes are necessary for research teams to be viable. When this is coupled with the fact that much greater resources are usually necessary for turning a discovery or invention into a productive process than for making it in the first place, it leads to a line of argument about minimum sizes for firms to be successful in industries dependent on R. & D. Blackett[41] argued this way in 1968 in his memorandum to the Parliamentary Select Committee on Science and Technology. He presented calculations on 'the minimum size of firm which can manufacture successfully an advanced product requiring heavy R. & D.', based on the assumption that 'few firms can afford to spend more than 5 per cent of their turnover on R. & D.', so that sales must be twenty times greater than R. & D. expenditure.

In practice, it is difficult to define the effective size limits of an innovating organisation. We do not give in this book extensive data on the sizes of Award-winning firms because we believe they might well be more misleading than informative. One difficulty is that firms may be subsidiaries of other firms or members of a group of companies. Tables 8A and 8B (p. 82) are therefore based only on separating out a category of 'small' firms.

A major limitation of the line of argument about minimum firm sizes is that it assumes innovations to take place within the boundaries of single firms or groups. Our studies have, however, shown many cases where firms or organisations not linked to the Award-winning firm have contributed to the innovation concerned. Such interactions between different firms or between firms and other organisations may take any of a wide variety of forms.

The Queen's Award scheme recognises the existence of relations of this kind in that Awards have been given both for R. & D. which was subsequently used by a different organisation and for the utilisation of R. & D. done by a different organisation (cf. the criteria for Awards quoted on p. 6). Such situations may arise when work done in university or government laboratories is used commercially by firms. Government can in addition provide financial support in various forms for R. & D. within firms. Relations between firms – in the form, for instance, of licensing agreements, or of the provision of machinery, or of custom work – are often important. Customers, whether they are commercial firms or national corporations or government departments, may take part in innovation by specifying the requirements or by actively collaborating in defining them.

Often more than two organisations are found to have interacted in some of the above ways in the genesis of a given innovation. For instance, triangular relationships are not uncommon between an institution that does some of the early R. & D., a machinery or instrument manufacturer and an industrial customer who uses the machinery or instrument.

The following examples illustrate the diversity of the relationships that we have encountered.

University research. In the cases we have studied, we have found rather few direct and specific links in terms of personnel or organisation between universities and innovating firms. Perhaps the clearest instances are two instruments made by the Cambridge Instrument Co., Microscan and Stereoscan. Microscan, an X-ray scanning micro-analyser, was developed around 1956 by Duncumb, a research student working under Professor Cosslett in the electrical engineering laboratories at the University of Cambridge. It was used at Tube Investments to study various problems in metallurgy. As soon as Tube Investments began to publish papers reporting results obtained with the machine, other firms began to enquire if they could obtain one. Tube Investments were not really interested in becoming instrument makers and, through a personal contact between Tube Investment's Research Director and Cambridge Instrument's Managing Director, the latter firm acquired the manufacturing option on the Cosseley instrument. Little remained to be done other than development for production and the first instrument was sold within a year. A somewhat similar pattern was followed in the case of Stereoscan. A prototype scanning electron microscope was produced by Professor Oatley's group, also in the electrical engineering laboratories at Cambridge, as early as 1953. By the early 1960s, improvements in design had been made and the instrument was potentially a commercial proposition. Stewart, one of Oatley's students, abandoned his work for a Ph.D. thesis in 1963 and joined Cambridge Instruments. With his detailed knowledge of the Oatley instrument, he was put in charge of developing it for manufacture and the Stereoscan was ready within two years.

In the case of the Oxford Instrument Company's superconducting magnets, the connection with university research was rather less specific. Martin Wood, who formed the company in 1959, was at the time on the technical staff of the Clarendon Laboratory of Oxford University and working on the design of high-field magnets. He received encouragement from the senior staff in the Clarendon, but the technical idea on which his success was primarily based originated in the United States and he heard about it while attending a conference at the Massachusetts Institute of Technology in 1961.

Another type of connection with university research played a part in the development by Ferranti of numerical control equipment for machine tools. It seems that the selection in the early 1950s of this area as one of high growth potential was influenced by the awareness of Sir John Toothill, then Managing Director of Ferranti, Edinburgh, of the potential success of the Mark I computer being developed in the Manchester University electrical engineering laboratories under Professor F. C. Williams. Ferranti in the Manchester area associated themselves with this development and produced a commercial version in 1951.

Consultancy is a mechanism through which firms can tap the knowledge and expertise available in the academic world. Lazell, Chairman of the Beecham Group, made it a matter of policy to find and pay for the best advice available. In 1954, impressed by American sales figures for antibacterial compounds, he discussed the possibility of research leading to new antibiotics with Sir Charles Dodds, Professor of Biochemistry at the Courtauld Institute, London, who had been a Beecham consultant since before the Second World War. It was Dodds who recommended Chain as a source of advice. Chain, who was working in Rome at the time, suggested a line of research aimed at the production of semi-synthetic penicillins. Support for this suggestion came from another Beecham consultant, Sir Ian Heilbron, formerly Professor of Chemistry at Imperial College, London, and at that time Director of Research at the Brewing Industry Research Foundation.

Consultancy services were important to the Colchester Lathe Co. in the design of its hydrostatic 'Flowline' for the mass production of lathes. Professor Eastwood of Sheffield University advised on concrete foundations and Professor Loxham of the Cranfield Institute of Technology on hydrostatic bearings.

In addition to the rather direct and specific kinds of university–industry link in terms of personnel or organisation, such as those exemplified above, universities also, of course, contribute a great deal to the general pool of knowledge and ideas which is freely and internationally available. The effect of this is discussed later (p. 39).

Government research. 'Government' is taken here to include research associations and public corporations as well as government research stations. Using this definition, we found that government research was used directly by industry more often than was university research.

The ATOZ process, a new method for yarn production, was developed in the Linen Industry Research Association (L.I.R.A.). Two young Ulstermen, McCleery and L'Amie, saw ATOZ-produced yarns at the L.I.R.A.-Chemstrand Research and Development Exhibition in 1962 and became

interested in the possibilities. L.I.R.A. had intended to sell the
process to fibre manufacturers but these, since they could sell
all their output at the time, did not feel justified in investing
in new types of yarn. McCleery and L'Amie were therefore
encouraged to go ahead; they formed a limited company in
1963, worked out a licensing agreement, set to work to scale
up the ATOZ process and shared the Queen's Award with
L.I.R.A. in 1966.

The adoption of the Chorleywood Bread Process developed
by the British Baking Industries Research Association illustrates
clearly the triangular type of relationship between research
organisation, machinery maker and machinery user. The
process depends partly on the replacement of a three-hour
period of bulk fermentation of dough by a short period of
intensive, precisely measured working of the dough. No existing mixer was found to be right for this 'mechanical dough
development'. Attempts to interest the existing bread machinery
manufacturers who were associate members of the Research
Association met with little response. A small engineering
company, George Tweedy and Co., was, however, found to be
selling something along the right lines. Pickles, who ran the
firm, became enthusiastic and worked night and day on some
occasions to develop a mixer suitable for use in the new bread
process on a pilot scale. Working in collaboration with the staff
of the Research Association, he was eventually successful and
Tweedy sold a large number of mixers for the Chorleywood
process.

The system for automatic control of steel strip thickness, for
which Davy and United Instrument Co. (Davy-Ashmore) won
an Award, made use of earlier work in the British Iron and Steel
Research Association. Two of the people principally concerned
in that work, Sims and Briggs, transferred to Davy and United
to became Research Manager and Head of the Electronics
Section respectively. Their research team gradually solved the
problems of circuit and machine design that arose as the system
was prepared for commercial exploitation.

Innovations in the commercial use of gamma radiation for
sterilisation of medical equipment, for which both Ethicon and
H. S. Marsh won Awards, drew on work done in the United
Kingdom Atomic Energy Authority (U.K.A.E.A.). The

U.K.A.E.A. built a package irradiation plant, the first of its kind in the world, at Wantage. Development work by industry was required not so much to increase the scale of operation of such plants as to make their operation more economical. Ethicon used the Wantage plant until their own was ready; the latter was built for them by Nuclear Chemical Plant Ltd, using U.K.A.E.A. experience and licences, together with certain improvements. H. S. Marsh's design for irradiation plant was produced in collaboration with the U.K.A.E.A., for which the firm had earlier developed several pieces of *ad hoc* equipment.

The National Physical Laboratory (N.P.L.) has contributed to some of the innovations we have studied. There was close liaison between Sayce at the N.P.L. and the small team set up at Ferranti to study the application of numerical control to machine tools. The Ferranti workers needed large and accurate optical gratings, of a type similar to those which Sayce was designing for use in infra-red spectroscopy. Close co-operation resulted in a very accurate moiré fringe measuring system which Ferranti used on a range of numerical control equipment.

The aerodynamics division of the N.P.L. was closely involved with the genesis of two novel features in the design of the Severn suspension bridge, for which Freeman, Fox and Partners won an Award. They carried out tests in a large Ministry of Transport wind tunnel at Thurleigh, near Bedford. A scale model broke loose during testing and, to occupy the time before a new model could be built, it was suggested that the N.P.L. should build and test some simple alternative shapes in timber. These tests indicated the feasibility of the streamlined box section approach to design for aerodynamic stability. Further tests were carried out on a detailed sectional model in the industrial aerodynamics wind tunnel at the N.P.L. itself at Teddington. The aerodynamic characteristics seemed satisfactory for all conditions likely to be met except for a slight movement in a narrow range of wind speed around 15 m.p.h. It was in connection with this problem that Sir Gilbert Roberts of Freeman, Fox and Partners conceived the second major departure from previous practice, the replacement of vertical hanger ropes by inclined hangers in an inverted vee-formation.

Work at the Building Research Station provided both knowledge and people to be used by John Laing and Son Ltd, in developing Lytag, a new kind of aggregate for use in making lightweight concrete. Lytag is made from the pulverised fuel ash from power stations burning pulverised solid fuel, and the Central Electricity Generating Board, for which disposal of millions of tons of ash annually is a major problem, also played a major role in the innovation. The Board's precursor, the British Electricity Authority, commissioned some early work at the Building Research Station on the use of pulverised fuel ash in building applications. Its marketing organisation for pulverised fuel ash also gives publicity to various uses, including Lytag.

Government finance. Of the various ways in which government financial help can help a firm's research and development, contracts are one of the most obvious. The Photoplot radar navigational aid of Kelvin Hughes (Smiths Industries) grew out of work on the photographic display of radar data done for the Royal Radar Establishment acting on behalf of the Ministry of Supply. This work in the context of early warning defence systems led to application in the field of marine instrumentation, in which Kelvin Hughes are specialists.

The interest of Ferranti, Edinburgh, in components for inertial navigation systems was stimulated in a similar way by contract work for the Ministry of Supply, acting in this case via the Royal Aircraft Establishment. A team was built up doing work on gyroscopes in connection with the 'Blue Streak' guided missile programme. The termination of this programme in 1960 was the immediate occasion for implementing an earlier decision 'to go into the complete systems business' rather than making components only. Within a few years, lightweight inertial platforms had been developed. The situation in which the group found themselves helps to explain so short a development time for so complex a piece of equipment. A team of highly trained staff with much expertise was in existence but faced with a tight budget and the possibility that the group might be disbanded if no commercial project emerged.

Licensing between firms. Relations between different firms are often vital factors in successful innovation. One common form for them to take is the licensing agreement.

Sometimes a small firm initiates innovation and production is taken up by a bigger one. A notable example of a small innovative firm is Plasticisers Ltd, a family business owned by the Slack family. Three Slack brothers control the firm, and two of them have the reputation of being highly inventive. Prevented by the patent situation from using the usual spinneret technique to make polypropylene fibre, they investigated fibrillation, the conversion of sheets of material into fibrous masses by mechanical treatment. A simple and cheap method for making polypropylene twine was developed. Considered in isolation, Plasticisers might be though too small to justify their development of machinery for fibrillation, but they succeeded in licensing the process and supplying machinery to British Ropes Ltd, claimed to be the largest rope and wire manufacturers in the world.

The success of Joseph Lucas (Electrical) Ltd with high-voltage transistors owes a good deal to the relationship with the American firm, Delco, which has resulted in one of the relatively few solid-state component developments to originate in Europe and be licensed in the United States. A visiting Delco team enquired in 1959 about the possibility of a transistorised ignition system. Some work had already been done at the Lucas Group Research Centre on high-voltage diodes for aircraft alternators. On the basis of this earlier experience, the Lucas workers were able within six months to produce sample transistors capable of handling the high voltages required. Arrangements for the licence were accordingly drawn up. Collaboration between the two firms continued in further development to reduce the size of the device and make it suitable for mass production.

Acquisition of the right licence at the right moment may be a vital ingredient of an innovation. This was the case with English Electric's electricity generating equipment for aircraft. It was felt within English Electric that pressure from aircraft manufacturers for increase in total installed capacity, coupled with the needs for lighter weight and greater reliability, would not be met by stretching the existing d.c. systems to higher voltages. The main difficulty in changing to a.c. generators arose from the fact that the generators themselves are driven by the aircraft engines, whose speed varies widely at different times. A constant-speed drive had therefore to be provided. The first such device to score a major commercial success was one produced by the Sunstrand Corporation at Rickford, Illinois. Accordingly English Electric negotiated a licence agreement with Sunstrand.

Role of customers. Since an important part of industrial innovation is often the identification of customer needs, and indeed many firms commit substantial resources to market research to find out about them, it seems fair to consider that a customer who states his requirement clearly is thereby playing an active part in the innovation process. Often, in such cases, there is also collaboration in further defining the requirements or in meeting them. Vosper Ltd (which has since been bought by the David Brown Corporation) got clear specifications from the Admiralty about the type of fast patrol boat required, including top speed under given conditions, slow speed and size of crew. Vosper's work was carried out on behalf of and in conjunction with the Directorate of Naval Construction of the Admiralty. Several Admiralty establishments were involved in the evolution and testing of hull, power plant, transmission and propeller design.

Similarly, the Seacat guided missile of Short Bros and Harland was matched to the market requirements through the joint efforts of the project leader, Armour, and three naval liaison officers.

A particularly interesting relationship obtained in the case of John Holroyd and Co. (Renold Ltd), who developed milling machines for making helical rotors for compressors. They were asked by the technical head of Svenska Rotor Maskiner AB if they could produce a machine tool for making rotary compressors based on a new design. The Swedish firm were concerned with the marketing of licences to build these compressors, but difficulties in producing the rotors had hindered the exploitation of the patents. F. T. Stott, the Technical Director of Holroyd, found a way to adapt standard worm-gear milling machines to copy the desired profile without needing to resort to the complex calculations necessary to determine the blade shape and path in cutting the rotor. As a consequence, Svenska Rotor Maskiner were able to sell licences and Holroyd sold both milling machines and rotors to licensees.

Overlap of objectives. A conclusion that can be drawn from cases like those mentioned above is that, for interaction between organisations to contribute effectively to innovation, it is usually necessary for there to be some degree of coincidence or overlap of the respective objectives of the organisations concerned. Rarely is it enough for an institution like a university or government laboratory merely to make results available in the hope that someone will use them. Workers in non-commercial environments often feel that industry is slow in taking advantage of the fruits of their labours; when a firm eventually does show enough interest in a topic to invest money in it on a substantial scale, it may appear to them as an act of unusual foresight. From the point of view of the firm, however, the foresight lies in seeing a market or other need that can be met; that is the main objective, not making sure that research is put to use. The case of sterilisation by gamma radiation seems to be a case of this type (p. 246).

Lytag's interest in the possibility of using pulverised fuel ash for making a lightweight aggregate was helped by the fact that this objective overlapped with the Central Electricity Generating Board's interest in finding outlets for a material available in huge and embarrassing quantities as a waste product. Licensing between firms presupposes a measure of coincidence of objectives between the parties to the licensing agreement, and so does active participation by customers in identifying requirements and refining definition of them.

The need for some overlap of objectives helps to explain a feature of Table 6 (p. 78). It can be seen there that, of the important ideas made use of by the Award-winning firm, most come from industry and fewer from 'government' organisations. Universities are a relatively infrequent source. It seems, therefore, that most help comes from similar types of institution. Like contributes most to like.

(f) Roles of science in innovation

Discussion of the roles of science in innovation is bedevilled by difficulties of demarcation and definition. Under terms like 'pure', 'fundamental', 'basic', 'non-mission-oriented' and 'curiosity-oriented', different people in different contexts understand rather different things, and different boundaries are drawn between this kind of science on the one hand and applied research and development on the other. Partly for this reason, the categorisation adopted in Table 6 (p. 78) is based not on judgements as to the type of science but on the type of institution where the ideas originated – industry, 'government' or university. Some of the university contributions came from engineering departments, and there is room for difference of opinion as to how 'basic' these are.

The issue of the relation of basic science to innovation is so germane to our main theme of the creation of wealth from knowledge that it calls for further discussion here in a less rigid and restricted framework. There is a practical policy sense in which the question is important, for the generation of economic benefits is one of the main arguments used to justify support from public funds for education and research in science.

'Basic research provides most of the original discoveries from which all other progress flows.' This claim was made by the Council for Scientific Policy in its *Second Report on Science Policy*[42] in 1967. The view is widely held that science is indeed one of the mainsprings of economic growth. Many people regard it as almost self-evident. So far, however, there has been relatively little detailed study of it, and there are many questions about it that remain to be answered. Not much is known about the magnitude of the effect or about the mechanisms through which it is exerted.

Support for the existence of the effect has come in a report to the National Science Foundation in the United States on *Technology in Retrospect and Critical Events in Science* ('TRACES'). In this report are documented the events that were considered to be crucial in leading to five innovations (magnetic ferrites, video tape recorder, oral contraceptive pill, electron microscope and matrix isolation). Approximately 70 per cent of the key events were classified as non-mission-oriented, 20 per cent as mission-oriented and 10 per cent as development and application. The general conclusion was that 'innovations for the next generations depend on today's non-mission research'.[43]

The idea that wealth-producing applications are derived from specific scientific discoveries has led to discussions of the time-lags between discovery and exploitation. These time-lags are sometimes said to have been decreasing in recent times. In attempts to demonstrate this, lists of intervals between discovery and exploitation have been given, for instance, by Baker,[44] by Hafstad[45] and by Newth.[46]

Such evidence has been criticised on several grounds by one of us.[47] First, there is the difficulty of defining the scientific discovery on which an application is based. Different time intervals are in fact assigned by different authors for given innovations. Second, listing of examples is methodologically suspect in the absence of checks on possible biases in selection. It is in fact not difficult to produce sets of examples which show time-lags *increasing* substantially during the last hundred years or so. Third, it is inherently impossible to observe anything other than a short time-lag for a recent discovery. Discoveries made in the last few years may be exploited in the future, and such cases are of necessity excluded from consideration. Fourth, there are many cases of 'negative time-lags', industrial advances that come *before* the scientific advances that help to make them understandable. For instance, it was not until after the first synthetic rubbers and plastics had been made that polymer science developed. The authors of the 'TRACES' report noted 'cases in which mission-oriented research or development efforts elicited later non-mission research which often was found to be crucial to the ultimate innovation'.[48] This emphasises the point that the relation between science and technology is a matter of two-way interaction, not of unilateral dependence.

The Byatt–Cohen method tested. Byatt and Cohen[49] have used the idea that wealth is produced by application of specific scientific discoveries in a potentially more practical way. They have made it the basis of a suggested method for quantifying the economic benefits of curiosity-oriented research. The principle of the procedure they propose is to estimate the economic effects of notional marginal delays in the timing of scientific discoveries which have later been profitably applied; that is, to estimate how much less wealth, suitably discounted to a common year, would have been generated from an innovation if, because of a smaller research effort, a key scientific discovery underlying it had been made later than it actually was.

We have used our set of case studies in an attempt to assess the feasibility of this approach. The results lead us to believe that the process of transition from scientific knowledge to commercial application is more complex than is assumed in the proposed method.

In discussing this matter, rigour in categorising types of research is called for. Byatt and Cohen use the term 'curiosity-oriented', which is clearer than 'pure' or 'basic', indicating as it does research justified by curiosity in fields where no application is apparent. According to Byatt and Cohen (benefit F of their paper, paragraph 2), ideas discovered during curiosity-oriented research, which 'tend to arise unpredictably', are sometimes applied later in industry. This is contrasted with 'mission-oriented' research, defined by Byatt and Cohen (benefit C) as 'research in fields whose application is evident'. Thus work cannot properly be described as curiosity-oriented if it can be shown that some application was envisaged; though naturally this does not imply that scientists necessarily lose interest just because there is a practical goal in view. The term 'basic' is commonly used in a wider sense to include fundamental investigations in areas whose application is quite obvious.

These distinctions are not just semantic subtleties, for they represent a possible policy problem. Should basic research be funded regardless of whether the field is one where social or economic benefits are relatively likely? Or should one try to identify such general fields and selectively support work in them, opting out partly or completely from other fields? British science policy in recent times seems to have been sympathetic to the former course. The view of the Council for Scientific Policy in 1966 was that 'scientific knowledge is necessarily interrelated and interdependent in all its aspects, and . . . there are therefore dangers in attempting to opt out of particular major fields without damage to the whole'.[50]

In preparing their paper, Byatt and Cohen drew up a list of eleven innovations which they thought might be worth examining to see if their model was applicable. Two of these, the Chorleywood Bread Process and Pilkington's float glass process, were included in our list of Queen's Award winners, and so we examined them from the particular point of view of the suggested method.

The major feature of the Chorleywood process is the replacement of a lengthy period of dough 'development' involving fermentation by a combination of a controlled amount of intense mechanical work and the use of certain chemicals known as 'improvers'. The effect of mechanical work was described by 'mission-oriented' researchers in the United States in 1926, thirty-five years before the Chorleywood process was announced by the British Baking Industries Research Association. No commercial application resulted until J. C. Baker in the United States was able to combine the effects of mechanical work with the use of improvers. Baker carried out careful systematic studies of a 'mission-oriented' nature and no large input of 'curiosity-oriented' research was involved.

Baker's 'Do-Maker' process was not suitable for the needs of British bakers or housewives, but the Chorleywood process is now used by over two-thirds of the British baking industry. The Chorleywood process uses ascorbic acid (vitamin C) as a fast-acting improver in addition to conventional improvers which were discovered in the 1920s by accident and by empirical research. The commercial availability of ascorbic acid depends on research in organic chemistry in the 1930s. However, had ascorbic acid not been available, other fast-acting improvers could have been used.

The technical knowledge forming the background to the research which produced the Chorleywood process was thus available in the 1930s, but it could not have been used in Britain before the 1950s when chemical additives in flour and bread were allowed following extensive studies by the Research Association. It seems fair to conclude that, while the allocation of greater resources to the Research Association might conceivably have brought about the development of the Chorleywood process at a slightly earlier date, there is nothing to suggest that a greater research effort in the curiosity-oriented sector would have had any accelerating effect.

The float glass process seems to be a typical 'technological discovery' in that it cannot be described in terms of the application of a specific discovery made by academic or other 'curiosity-oriented' scientists. The concept of using molten tin for flat glass manufacture was patented in America in 1902; Pilkington had to spend seven years and £4 million on development work, however, before they announced their new process in 1959.

After the process had been developed empirically to the stage of producing satisfactory glass, scientists were used to increase understanding of the process and to solve technological problems; for instance, the cause of a surface bloom was traced to a layer of tin dissolved in the glass. Such work made use of certain fundamental concepts and analytical techniques which may have derived from curiosity-oriented research; but it does not seem possible to apply the concept of marginal delay to these science–technology connections, which occurred after the initial development of the process.

Besides the Chorleywood and float glass processes, we also considered the other innovations included in our study to see whether they might lend themselves to Byatt–Cohen analysis. None seemed suitable. Table 5 (p. 76) shows 'DS' entries indicating 'push' by scientific discoveries for two innovations. Both these, however, are dual entries. One case is that of the Award to I.C.I. for the development of titanium alloys for the aircraft industry by Imperial Metal Industries (I.M.I.). Useful titanium alloys were first developed by American industry. When I.C.I. started manufacturing titanium, it also produced two titanium alloys using American know-how, which can be described as a technological discovery. However, the improved alloys developed by I.M.I. made use of previous academic work on titanium alloys carried out by A. D. McQuillan and Mrs M. K. McQuillan who, after working in Australia, became respectively a professor at Birmingham University and the Research Manager leading I.M.I.'s titanium alloy research. The I.M.I. case does not lend itself to Byatt–Cohen analysis because there is no single curiosity-oriented discovery of which the new alloys can be said to be applications.[51] Also, the timing of I.M.I.'s new alloys was dependent on several factors other than the timing of the academic research.

The other case is that of the Award to Sanders and Forster for innovation in structural steelwork. This made use of a theory of plastic flow developed at Bristol and Cambridge Universities by J. F. Baker. There was also, however, a strong management need: the price of steel and its shortage in the 1950s put a premium on finding ways of using steel more economically. Sanders and Forster realised that the plastic flow theory offered a way of saving steel. The timing of the innovation was influenced more by the constructional opportunities available in the 1950s and 1960s than by the origin of the plastic flow theory, which Baker had started to develop in the 1930s. It should also be noted that Baker's work was not purely curiosity-oriented. He carried out research on steel structures at the Building Research Station before becoming Professor of Engineering at Bristol, and his early research was partially financed by industry through the British Steelwork Association.

Thus it has not been possible in any of our cases to pinpoint specific curiosity-oriented discoveries whose timing was crucial for the timing of innovations using them. On the basis of this and some other evidence, we therefore conclude that Byatt–Cohen-type innovations are in practice difficult to find, let alone investigate (see pp. 120, 248).

Outside our list of Queen's Award winners we have come across two innovations that seem to us to be genuinely based on specific discoveries of curiosity-oriented research: nuclear power and silicones. The concept of notional marginal delay would be difficult to apply in either of these two cases because of the important influence of war-time development work, which obscures the effects of possible delays in discoveries. The timing of the innovations, in other words, may have been determined more by the Second World War than by the timing of the curiosity-oriented discoveries. The massive effort put into the development of an atomic bomb is well known. Silicones could be said to depend on curiosity-oriented research on organo-silicon compounds by Kipping in 1904–8; industrial preparation of silicones, however, rests not on reactions discovered by Kipping but on a synthesis developed by Rochow working for General Electric in 1940, and silicones were first produced commercially in 1943 by the Dow-Corning Corporation to satisfy military needs.[52]

Routes for economic benefits from science. Despite all the foregoing, we do not wish to suggest that curiosity-oriented research is useless in economic terms. Our conclusion is only that the transition from 'pure' knowledge to wealth is less simple and direct than is commonly supposed. Science probably *does* work economic miracles, but it acts in rather mysterious ways its wonders to perform. Further study is clearly needed to elucidate these ways, but we believe we can identify three broad classes of effects.

First, curiosity-oriented science, practised largely in academic institutions, provides techniques of investigation. Second, it also provides people trained in using those techniques as well as in scientific ways of thought in general. One example of the use of science to tackle an industrial problem is given above in connection with the float glass process. Another type of example is provided by the need to synthesise large numbers of organic compounds in some kinds of industrial research, for instance in the development of Procion dyes by I.C.I. Dyestuffs Division and of semi-synthetic penicillins by Beecham; this calls for people who have a knowledge of chemical facts and concepts and are skilled in chemical techniques. Such inputs from science into innovation are less obvious and less spectacular than the major discoveries leading to application on which attention has usually been focused. However, they may be many in number. It seems likely that sufficiently detailed examination of any technological innovation could break down the science input into a host of individual contributions of this kind.

Third, science enters innovation already embodied in technological form. It may be relatively rare for a piece of curiosity-oriented research to generate a piece of new technology, but once this process has occurred, the technology can be used over and over again and developed into more advanced technology. There seems to be much justification for the view of Price that technology builds largely on earlier technology:

> The naïve picture of technology as applied science simply will not fit all the facts. Inventions do not hang like fruits on a scientific tree. In those parts of the history of technology where one feels some confidence, it is quite apparent that most technological advances derive immediately from those that precede them.[53]

The technology that embodies earlier science may take the form of something that appears as part of the final product, such as a hardware component or a new material, or it may be a service technology which, though it is not physically present in the final product, is used in making it. Examples of these types are provided by transistors, crystalline polymers and electron beam welding respectively. Curiosity-oriented research played a part in the development of solid-state devices such as transistors,[54] but it is the devices as such that are incorporated in many innovations. Sheer 'transistorisation' makes up much of the technical novelty in some of our innovations, such as Ferranti's summation metering equipment and G.E.C.'s telecommunication equipment. Stereospecific or 'crystalline' polymers came from chemical work which, though its strictly 'curiosity-oriented' status is not beyond question,[55] (see p. 394) was basic enough to earn Ziegler and Natta a Nobel Prize. Such materials were used by Plasticisers for fibrillation, but the innovation depends on mechanical properties of the material, not the chemical considerations underlying them. Electron beam welding derives from physics but is now an established technique and was used as such, for instance, to weld on 'Stellite' valve seats for diesel engines by Mirrlees National (Hawker Siddeley) (see also p. 110).

With considerations such as these taken into account, the conflict between our conclusions and those of the 'TRACES' study turns out to be perhaps more apparent than real. We fail to find much direct input of basic science into innovation but believe there is a substantial contribution in various latent forms.

Justifications for basic science. The qualifications we make seem important for achieving a realistic basis for some policy decisions. For instance, because the relation between science and innovation is largely indirect, so are the economic benefits that spring from basic research. They may not accrue in the country where the basic research was done. It was noted by B. R. Williams[56] that there is poor correlation between different countries' R. & D. expenditures and their economic growth rates. If this applies to figures for total R. & D., most of which is accounted for by applied and development work, then one would expect even less correlation between basic research and economic growth. It is difficult to believe that much *national* economic benefit arises from the *knowledge* output of a nation's own basic research.

The justifications for national expenditure on basic science therefore need looking at carefully. It may not matter much that the nation is backing the right horse for the wrong reasons; but it would become serious if the reasons came to be so widely recognised as false that the nation loses the will to back the

the horse because of the apparent lack of adequate justification.

There are, of course, non-economic reasons for pursuing basic science – culture and prestige, for instance – but in most circumstances these are not felt to justify expenditure of really large sums on a continuing basis. Support on anything like the present scale must, therefore, rest on prospects of economic returns of some kind. We can identify three main ways in which science can bring economic benefits. Scientific discoveries occasionally lead to applications in the form of new technology; this is rare, but the effects may be multiplied indefinitely as technology builds on technology. Science also provides techniques which make it possible or easier to tackle industrial problems successfully. Finally, basic research is an element contributing to the output of highly qualified men and women educated in science and its methods. Of these three factors, the manpower benefit may be the most important when the justifications for basic science are considered in the national context, partly because discoveries and techniques cross international boundaries more easily than men.

(g) Technology transfer

From the preceding discussion it would appear that the role of science in technological innovation is an indirect one, its most obvious contribution being the supply of trained manpower. If the ideas on which modern technical products are based do not often come directly from science, where do they come from and how do they arise ? According to a view advanced by Price, they come largely from earlier technology. Science and technology form two parallel structures; just as science builds mostly on science, so technology builds mostly on technology.[57]

Development of technology. To shift the onus for the origin of new ideas from science to technology, however, still leaves unanswered the basic questions. Where do the ideas in technology come from ? How do *they* arise ? The answers to these questions will, of course, depend partly on what one understands by technology. Some prefer an essentially 'hardware-based' definition. Thus, for Schon, 'technology [is] . . . any tool or technique, any product or process, any physical equipment or method of doing or making, by which human capability is extended'.[58] If one uses this formulation as a basis, then, when technology builds on technology, it is likely to be either by the extension of a technique or by the conjunction of various techniques, products and processes in ever more complex configurations. Technological development may then be seen as a process with a dynamic of its own, moving usually from more elementary to more complex systems. This view of technological development has had many proponents and has recently been taken up and developed by Jantsch[59] as a framework within which to describe the state of the art in technological forecasting. For Jantsch, the whole process of technological development is to be viewed as technology on the move to higher complexity and wider applicability, and this is what he describes as 'technology transfer'.

But a broader definition of technology than Schon's is possible if, in addition to hardware, one also includes explicitly the concepts and knowledge which are embodied in the hardware. In this case, technology transfer may be reduced simply to the movement of ideas and information from one technology to another or from one context to another. Ideas and information may be conveyed either through books, magazines and technical periodicals or by an individual as he moves from one area of employment to another. These modes of transfer are often referred to as technology transfer 'on paper' and 'on the hoof', respectively.

There is now a considerable body of evidence about the operation and relative importance of these transfer processes. It seems that the spread of technical information from one milieu to another is a significant element in the development of technology. Moreover, the movement of individuals from one job to another appears to be a more efficient mechanism of transfer than pushing ideas through communication networks of the kind traditionally recognised. Transfer of a technologist with his background and expertise seems often to work better than dissemination of information through printed matter, conferences and so on. The importance of this mechanism has been put very forcibly by Schon, who argues that the whole process is 'so to speak "packaged" and optimised by the movement of a technologist from one milieu to another'.[60] In the cases we have studied, the most frequent single mode of technology transfer was by a person joining a firm, as shown in Table 7. This result is in agreement with the general conclusion reached by Burns that the transfer of technology is the work of 'agents not agencies'.[61] We have also found, however, that commercial agreements, visits to conferences and industrial colleagues, personal contact and consultancy all played a part in the effective transfer of technology. Of course, not all innovations involve the transfer of a technological concept from outside the firm. The data shown in Table 6 indicate that, of 158 important technical ideas, 56 arose within the firm (p. 78).

Modes of technology transfer. An example of technological development as a consequence of a new person joining the firm is to be found in English Electric's development of fuses for the protection of semiconductor devices. E. Jacks, then Chief Engineer in the Fusegear Division, had identified the area of printed circuit technology as an area which might provide an answer to the manufacture of the fuse elements. Progress was, however, held up because no one in the development team possessed enough skill in the use of photofabrication techniques. This need was overcome when, 'by sheer luck', Jacks was interviewing an electrical engineer for a job and the applicant happened to mention casually that he had some skill in industrial photography which he had developed as a hobby. Photofabrication techniques were applied with great success and resulted in a completely novel process in the manufacture of fuse elements. It would be incorrect to assume that photofabrication was easily applied in the new situation; the change in manufacturing principles was radical and involved considerable development in the original ideas of the process. Technological progress here involved the process of taking a technology out of one context and discovering its applicability in quite a different one. This technology was transferred by an engineer changing jobs.

In this case the relevant technology was part of the new employee's background, but it was only a chance remark which indicated that some of his skills were of particular interest to English Electric at that time. It is much more common to seek the services of someone whose skills are realised to be needed. For example, when Stewart gave up his doctoral studies at

Cambridge University and went to Cambridge Instruments, it was realised that he possessed the skills required to turn the scanning electron microscope into a commercial proposition. Similarly, Dr W. C. Brown was recruited to Freeman Fox and Partners because of his specialised skill in design of structures. He joined the company from Imperial College after completing his doctoral dissertation on the design of stiffened plates. Another example is that of Sims and Briggs, who were recruited to join Davy and United to prepare for commercial exploitation the system for automatic control of steel strip thickness on which they had already worked in the British Iron and Steel Research Association. Sometimes the transfer of technology involves a transfer of something rather less specific than a particular process, gadget or idea. In the development of a shuttleless loom for narrow fabric weaving, Bonas Bros employed an engineer who was able to transfer the knowledge of the methods and materials of high-precision engineering which he had gained in his previous employment to a technology where such high-precision engineering had not been used.

Technology can be transferred by bringing previous experiences to bear on a new problem. In 1928, I.C.I. became a major shareholder in Thorium Ltd. This had an important effect after the Second World War. Thorium Ltd then had three I.C.I. nominees as directors. One of them, the Technical Director, A. M. Roberts, had worked on uranium purification using solvent extraction during the war and was aware of the possible development of new outlets for thorium and the rare earths. It was he who suggested in 1950 that solvent systems should be examined for the extraction of thorium and the rare earths. It was not obvious at that time that solvent extraction would turn out to be better than ion-exchange and other extraction methods, but Thorium Ltd has since been very successful in exploiting solvent extraction techniques.

Technology transfer may be encouraged by proximity to other technologies. This happened in the case of Ferranti Summation Meters. Ferranti were considering the redesign of a range of their electricity measurement meters which were based on a mechanical technology. As semiconductor devices had become reliable, Maurice Done asked the engineers who were building one of the Pegasus computers in an adjacent laboratory to design some circuits to count units of electric energy. Their circuits demonstrated the feasibility of the idea, but they were much too complex and would have been too expensive to install. Subsequently, Done acquired the services of a recent graduate, N. Mascarenhas from Manchester University, who designed most of the electronics for the meters. As a result of the combination of electrical measurement technology possessed by Done, and the new electronics technology learned by Mascarenhas while at university, Ferranti were able to market a new series of meters operationally so like the mechanically based meters that, in many cases, both types could be serviced by the same technicians without any need for retraining. The close but informal relations with the computer laboratory were certainly an important factor in stimulating interest in the application of computer technology to electricity measurement.

Sometimes technology can be transferred most effectively by sending individuals from one environment to another to learn new technology. This was the situation with both Plasticisers Ltd and Joseph Lucas (Electrical) Ltd. Plasticisers' technology was transferred to British Ropes Ltd when R. Cairns sent a team to learn how to use fibrillation machinery for synthetic materials. High-voltage transistor technology was acquired by Delco when they visited Joseph Lucas (Electrical) – but, by the same token, Lucas learned something of mass-production techniques when they sent a team to Delco to help with the design of the production line for the transistors.

It is not often that technology is transferred through personal contacts, but opportunities may be presented in these ways which subsequently alter the technological base of the company. Two such examples are to be found in the Cambridge Instruments and H. S. Marsh case studies. When Cambridge Instruments were trying to diversify into more modern technology, Mr T. Hughes, Research Director of Tube Investments Ltd, offered Mr H. C. Pritchard, Managing Director of Cambridge Instruments, an X-ray scanning microanalyser (later called Microscan) which Tube Investments had obtained from Cambridge University and did not wish to market. Through this contact, Cambridge Instruments was able to launch out into a new technology – electron tube technology. Again, it was through a chance meeting at a cocktail party that Eric Edwards was able to interest the Wantage Research Laboratory in the possibility of H. S. Marsh Ltd solving the problem of producing an irradiation plant of low capital cost. Similarly, when Drury and Brown of Ferranti's Inertial Navigation Group made an extended visit to the United States to assess the state of the art with regard to gyroscopes and inertial platforms and renew their contacts with Kearfott Co. and Norden Ltd, they were able to identify the seminal ideas behind what was to be their range of lightweight inertial platforms.

The transfer of technology through papers and trade magazines is harder to establish because it is often difficult for a scientist or engineer to recall a starting point of this kind for an idea to which he later devoted a lot of work. R. J. Callow of Thorium Ltd

> during 1950–51 . . . talked about solvent extraction, read the literature and selected methyl isobutyl ketone (MIBK) and n-tributyl phosphate (TBP) as being potentially the most useful solvents for thorium. . . . Two years of laboratory work with TBP between 1951 and 1953 showed the commercial promise of a TBP-based system for the production of pure thorium nitrate and a semi-pilot plant was set up (p. 464).

Similarly, when Joseph Lucas began work in semiconductor physics to solve a problem concerned with alternators for aircraft, the work was guided by some published research by J. A. Ditzenberger of Bell Laboratories on the diffusion of impurity elements in germanium and silicon, and this led Lucas to experiment with aluminium diffusants. Subsequently, this work was used to design the high-voltage diodes for automobile ignition systems requested by the American firm Delco.

Conferences are an obvious source of information exchange. Mr Wood of Oxford Instruments was attending a conference in the United States at a time when information about the manufacture of superconducting wire capable of withstanding large magnetic fields began to be reported and discussed. The last day of the conference had been allotted to superconductivity in the expectation that there was not much of major interest going on in this area. In fact, it turned out to be the most exciting day of the conference, with the conference hall full to overflowing. Wood, while he was there, purchased some of the superconducting wire and, when he returned to England, began to turn his skill as a specialist magnet maker to the winding of superconducting magnets to be made widely available.

Practical considerations. From a practical point of view there seem to be some ways in which technology transfer could be improved. For example, the sort of innovation that depends on a person having experience in another technology obviously requires movement of people. Perhaps changes in industrial or university pension schemes would help to lessen the barriers to mobility between employers.

The fact that technology is transferred most frequently by movement of people might imply that research workers are not looking widely enough for ideas. Many workers in an area of technology try to keep in touch with what is happening in their own technology as well as those areas of science which seem most relevant. Less often do they look for another area of technology that might be experiencing similar problems. For example, a firm attempting to produce large plastic panels for use in building structures might maintain contacts with chemical manufacturers, machinery manufacturers, the Building Research Station and possibly a university. If, however, this imaginary firm defined in a generic way what it was attempting to do as 'the economical production of large strong panels in new materials', it might look for another technology that faces similar problems and it would soon emerge that car body manufacturers have been carrying out just this sort of activity for some time. In the present industrial situation, it seems that the transfer of ideas from an existing technology like automotive technology to another one like building technology takes place mainly by chance. This situation could be altered if research and development personnel considered other technologies as possible sources of useful ideas. What seem at first sight to be the most unlikely pairs of technologies often have much in common. Bread baking and plastic foam manufacture, for example, are both concerned with the expansion and hardening of paste-like materials into solids. Do plastic foam technologists ever attend the meetings of baking technologists to see if they might get any new ideas?

Another way of getting useful ideas that is often ignored is bringing old ideas up to date. The patent literature is full of novel processes that were not operated successfully. There are, for example, large numbers of patents on shuttleless looms going back over one hundred years, but it is only recently that advances in precision engineering have made the shuttleless loom a commercial proposition. It might be worth while for technologists more often to carry out patent surveys to see if an old idea can now be used in the light of advances in other technologies or of altered market circumstances. As an example of the latter, one can quote the Taywood silent piledriver (Taylor Woodrow) which pushes piles into the ground rather than hammering them. The concept of 'pushing' piles can be found in several patents, but it was not until there was an increase in the amount of building in city centres combined with an increase in night work on building sites that the market was large enough to justify the expenditure of money on the development of this concept in a piledriving machine.

It might be that detecting the *change* in market circumstances is more important than the production of the idea. Table 7 (p. 79) shows that, of 102 technical ideas made use of, 24 could be described as common knowledge (15 entered the firm as a component of industrial experience and 9 via the educational process). In other words, at least 24 of the ideas were old ideas and could, in principle, have been used by anyone who had the ability to see the market potential for them or for developed versions of them.

(h) *Relations of research to marketing*

In the nature of the case, it is likely that any innovation which turns out to be successful will have involved some kind of awareness of need or market-consciousness. Many industrialists tend to take a 'need comes first' attitude, acting on the principle that one should make what one can sell rather than try to sell what one can make. However, there is also the alternative view that in general the main impetus and driving force for innovation is not need but the making of new discoveries and inventions. To give some indication of the relative importance of these two factors, we have analysed our cases in terms of the 'discovery push' and 'need pull' views. The results, further subdivided in both cases, are given in Tables 5A–5D (p. 76).

The previous two sections have discussed some aspects of the ways in which discoveries came about. This section deals with some of the ways in which market needs have come to influence R. & D.

The importance of market orientation was emphasised particularly by the report on *Technological Innovation in Britain* issued by the Central Advisory Council for Science and Technology in 1968. This report brought a new degree of economic awareness to government-level discussion of the deployment of money and men for R. & D. Among the factors which make for success in technological innovation, it listed 'the framing of planned programmes of innovation in relation to the assessment of opportunities revealed by a sophisticated analysis of market situations'.[62]

In our study, we have indeed found a number of cases where there were systematic attempts to define needs by formal market research or similar means, followed by sustained R. & D. to bring about innovation to meet those needs. But usually there were many complications to blur this simple picture. Clear definition of need plus efficient planning fails to account satisfactorily for the majority of innovations. It is not exceptional for the need eventually met to be different from that foreseen and aimed at earlier on. Innovation is a process over time, not a point event, and continuing awareness of changing patterns of needs and possibilities has often paid dividends. Some innovations proceed, in their early stages at least, in a market environment which is indifferent at best; brought eventually to fruition by some combination of personal enthusiasms, organisational pressures and sheer historical accident, they lead rather than follow the market.

Market-oriented R. & D. Market-directed R. & D. aimed at objectives clearly specified in advance has turned up in our study most often in the engineering areas. The properties of machines are in some ways more predictable than the properties of substances, and accordingly it seems to be rather more feasible for engineers to produce machines to fill given specifications than for chemists to tailor-make substances exhibiting specific new properties.

The developments in heavy diesel engine design by Mirrlees National (Hawker Siddeley) were clearly and explicitly market-oriented. Pope, as Technical Director and Director of Research, called on the Economic Appraisal Department of the Hawker Siddeley Group to carry out a world-wide study of the competitive position of diesel engines for base-load power generation. This showed that, to achieve a significant increase in the company's share of the power-generating market, it would be necessary to raise power output by 50 per cent without an increase in overall engine size. At the same time, it was important to design the engines to be capable of burning residual heavy fuel, since the cost of fuel is usually between 70 and 80 per cent of the total cost of power generation and the price of heavy fuel is often one-quarter to one-third lower than that of diesel fuel. Pope accordingly sanctioned and directed a series of tests on thermal and mechanical performance to determine the optimum configurations for various components. Improvements resulted particularly in the pistons and in the valve assemblies.

A similar effort to relate research and development targets to market requirements was made by Petters, which deals with the small engine interests of the Hawker Siddeley Group. J. Smith, who became Technical Director in 1958, was faced with the task of developing a rationalised range of small engines to counter increasing competition from abroad and from light-weight petrol engines. The mainstay of the then current Petters range was a 3-b.h.p. diesel engine weighing 236 lb. To determine the design requirements of the new generation of engines, data were collated from visits by company engineers to trade fairs, etc., from reports by marketing executives covering many parts of the world, from trend analysis by the Applications Department and from market surveys by the Economic Appraisal Department of the Hawker Siddeley Group. The synthesis of these requirements showed that the new engine should have a power output of 3 b.h.p., a weight of 85 lb, height 20 in., length 14 in. and width $12\frac{1}{2}$ in. The engine went through three phases of development and came to approximate closely to these specifications with overall dimensions of $17\frac{1}{8} \times 17\frac{3}{8} \times 12\frac{7}{8}$ in. and a total weight of 90 lb. It sold well

C

largely because it was able to penetrate a market previously served adequately only by lightweight petrol engines.

Market research was important, too, in the development of English Translucent China by Doulton Fine China. The new type of china depends basically on a new formula which gives a product that is non-porous and translucent under the glaze. Its introduction was the culmination of a series of changes in manufacturing techniques which bore in mind the results of design research in the North American market and increasing competition from European and Japanese non-porous porcelain. Technical developments, design effort and market research proceeded side by side in Doulton Fine China. The *Pottery Gazette* in 1960 described English Translucent China as one of the greatest achievements of market research.

A routine market survey enabled Metals Research to change the nature of the image-analysing device they intended to produce. The original plan was for a device costing about £500 for electronic scanning of photographs. The response of possible buyers to this proposition was discouraging but some of them said they would buy several even at a cost of thousands of pounds if the instrument could look directly through a microscope without needing an intermediate photographic step. Such an instrument was then developed.

The decision of the Beecham Group to diversify into ethical pharmaceuticals was guided by market considerations. The management were confident that demand in this field would increase but there was not unanimity as to whether research was necessary for the diversification. Lazell, the Chairman of the Group, was in favour of research. In 1954, impressed by American sales figures for antibacterial compounds, he obtained high-level scientific advice. The consultants agreed that there was room for further research on antibiotics and research facilities were extended to enable work to be carried out. An unexpected discovery in the course of that work led to successful semi-synthetic penicillins.

Reservations about market orientation. In the above cases, market information undoubtedly played a valuable part in directing R. & D. into useful channels. Some reservations must, however, now be entered.

One such reservation concerns the value of formal market research carried out by independent organisations. We have encountered some cynicism about this. Thus the decision to go ahead with the manufacture of Lytag was not made until after a market survey had been carried out by consultants; but Hobbs, who was in charge of the R. & D. that developed the technique for making Lytag, is sceptical about the real value of this market survey. He claims that the result was predictable from a knowledge of the total market for aggregates and the claimed cost and properties of the new material which only he and his colleagues in Laing could define. Predictably, therefore, it produced results that were in line with forecasts that he himself had made earlier, and subject to the same errors. However, the fact of being presented by an outside organisation may have helped by giving these forecasts an air of greater authority.

A major limitation of the idea that innovation should be planned in advance to meet clearly specified market objectives is the observed fact that successful innovation often emerges from activities whose principal objectives at an earlier stage were not those that in the end were met. Thus some early work on high-voltage diodes at the Lucas Group Research Centre was done with aircraft alternators in mind. This work was subsequently applied to the design of high-voltage transistors for automobile ignition systems as the result of an enquiry from Delco (U.S.A.) about such a system for American cars. There was no pressing need in Britain for car ignition systems incorporating high-voltage transistors, but the Delco request provided a stimulus for further efforts and resulted in a licence agreement and some joint development work. Lucas later put effort into finding British markets other than its traditional customers in the vehicle industry. New outlets appeared in television cameras, X-ray systems and voltage regulators.

Again, the first reactive dye made in I.C.I., forerunner of the range later known as Procions, was sent for testing in the wool section, but Rattee, having found that it did not give good results with wool, tested it on cotton, with which it reacted chemically. High standards of wash-fastness were already available over a wide range of shades in dyes for both wool and cotton, and Rattee seems at the time to have seen the advantages of chemically reactive dyes as lying not so much in increasing wash-fastness as in the solution of certain technical problems in continuous (as opposed to batch) dyeing. In the event, however, reactive dyes owed much of their success to the fact that wet-fastness and dyeing behaviour became properties independent of each other, so that dyeing properties could be improved without reference to wash-fastness. New brilliant shades thereby became possible. These new shades were a factor in the development of brighter fashions, including the pop-art trend, which in turn may actually have helped to reduce the amount of continuous dyeing carried out in this country, since there was less demand for thousands of yards of fabric all dyed in the same shade.

The Chorleywood Bread Process is another case where the desire to make a process continuous acted as a goal which turned out to be partly illusory. Mechanical development of dough was taken up for investigation at the British Baking Industries Research Association in 1958–9 in connection with continuous mixing, on which work had been done previously. The Chorleywood process turned out, however, to be suitable for batchwise production of bread and offers advantages to small bakers for whom continuous mixing is not feasible.

J. and S. Pumps were set up to exploit a very specific market, namely, leak-free 'canned' pumps for use in the chemical industry where dangerous chemicals are processed. They used their existing base of technical expertise to diversify into an apparently quite unrelated market. Among the firm's customers was British Petroleum. Through this contact, Somlo, the Managing Director, learned of a problem that B.P. faced of providing electrical power supplies on off-shore wellheads in the Arabian Gulf. Solutions that had been proposed included fuel cells and bacteriological cells. Somlo proposed a solution which virtually amounted to running one of their pumps 'backwards'. By using the energy in the fast-flowing gas and oil mixture emerging from the well to drive the impeller as a turbine, with permanent magnets attached to the rotor, electricity can be provided at zero energy cost.

The management of Thorium Ltd, aware in the post-war years of a possible increase in the demand for thorium if it came to be used as a nuclear fuel, sanctioned an expansion of the research effort and a solvent extraction system for purification was worked out. However, the expected requirement for thorium did not develop and the so-called pilot plant was able to produce all the thorium nitrate that the company required. On the other hand, similar solvent extraction systems were used for separating individual compounds of the so-called 'rare earth' metals. There was no real promise of large new markets for rare earths and it was really an act of faith in the future that led Thorium Ltd to develop extraction methods for them. The firm carried out little applications research (i.e. research to find applications) but concentrated on making available samples of compounds for a wide range of industrial research in the hope that unpredictable new uses would develop. As research workers looking for novel effects included rare earths in their investigations, new markets began to appear in the 1960s. For instance, yttrium compounds are used in a new phosphor for colour television and in garnets for microwave devices and lasers; and praseodymium oxide is used in making a clear brilliant yellow ceramic stain.

In this instance, then, innovation in rare earth purification proceeded so as to lead rather than follow the market. This is by no means the only example we have come across of innovation in the face of an indifferent market.

The Photoplot navigational aid developed by the Kelvin Hughes Division of Smiths Industries represents a diversification that was carried out through the enthusiasm of the technical personnel involved rather than because of encouragement from the marketing side. Early experience with photographic display techniques had been gained by the firm in connection with radar early-warning systems. In 1959, some thought was given to the possibility of using such techniques in the field of marine instrumentation, in which the firm specialises. The idea was pursued vigorously by the people who had been principally concerned in developing the rapid photographic process that lies at the heart of the system – Townley, Parsons and Embling. The Sales Department, however, was frankly pessimistic and management was, in general, reluctant to commit any private venture capital until there was some guarantee of at least a small market. In spite of this lukewarm reception, Townley, Parsons and Embling assembled a demonstration unit at Southend and installed another in a small ship. Captain Washer of the P. and O. line expressed an interest in a Photoplot for the new ship *Canberra* and an Admiralty contract followed in 1960.

Ferranti's innovation in electronic summation metering equipment was also carried out without much stimulus from the market. The range of equipment being sold in the late 1950s for the summation metering of electricity consumption enjoyed a good reputation with customers. Other suppliers did not appear to be attempting to introduce new designs. Nevertheless, it was felt that the existing range, which was based on electro-mechanical principles, was nearing the end of its useful life. Transistor circuitry became attractive as a possible development path in the early 1960s when, with the adoption of silicon instead of germanium transistors, reliability improved and prices dropped. The necessary electronics was developed within a short time in 1962–3, helped by the fact that it was possible to use circuits which were standard in computer technology. Energetic marketing effort was then put into creating customer acceptance of the new equipment. Lectures and courses at various technical levels were provided for potential users.

The relations of R. & D. to marketing thus show great diversity. The examples referred to above by no means exhaust the range. In some situations, for instance, the efforts of firms to look outside themselves virtually fuse into a single process with the activities of their customers in defining their own requirements; this occurred with the Vosper fast patrol boats, the Short Seacat guided missile and the Holroyd rotor-milling machines, as mentioned earlier (p. 32). Sometimes the 'market' for R. & D. to satisfy is within the firm itself, if the objective is to improve a production technique. The Colchester Lathe Co. wanted to change from batch to mass production of its centre-lathes. Existing types of conveyor line would have been quite unsuitable for the purpose because of the requirement for an extremely level platform if the lathes are to be built to a sufficiently high degree of accuracy. Long, the Works Director, explored various possibilities and it was decided that a conveyor system of 'rafts' floating on high-pressure oil would be feasible. The assembly line enabled production to be raised from 50 to 200 lathes per week.

In the light of the diversity illustrated above, it would clearly be an oversimplification to suppose that sophisticated market research is a sure-fire path to success across the whole spectrum of industrial innovative activity. Perhaps the highest-level generalisation that it is safe to make about technological innovation is that it must involve synthesis of some kind of need with some kind of technical possibility. The ways in which this synthesis is effected and exploited take widely differing forms and depend not only on systematic planning and the 'state of the art' but also on individual motivations, organisational pressures and outside influences of political, social and economic kinds. Because the innovation process extends over time, it is important to retain continuous sensitivity to changes in these factors and the flexibility to perceive and respond to new opportunities.

References for Part One

1. M. J. Peck, in R. E. Caves *et al.*, *Britain's Economic Prospects* (Allen & Unwin London, 1968) p. 448.
2. R. Solow, 'Technical Change and the Aggregate Production Function', *Review of Economics and Statistics*, **39** 312 (1957); see also W. P. Hogan, 'Technical Progress and the Production Function', *Review of Economics and Statistics*, **40** 407 (1958).
3. E. D. Domar, 'On the Measurement of Technological Change', *Economic Journal* **71** 709 (1961).
4. E. F. Denison, in Caves *et al.*, *Britain's Economic Prospects*, esp. pp. 235, 261.
5. F. Bacon, 'In Praise of Learning', in *Francis Bacon*, ed. A. Johnston, (Batsford, London, 1965) p. 13.
6. R. A. Charpie, quoted in 'The Management of Technological Innovation', *Harvard Business Review*, May–June 1969, p. 162.
7. D. A. Schon, *Technology and Change* (Pergamon, Oxford, 1967) p. 21.
8. B. Gold, 'The Framework of Decision for Major Technological Innovation', in K. Baier and N. Rescher (eds), *Values and the Future* (Free Press, New York, 1969) p. 396.
9. T. Burns and G. M. Stalker, *The Management of Innovation* (Tavistock Pubns, London, 1961).
10. J. Bright, *Research, Development and Technological Innovation* (R. D. Irwin Inc., Homewood, Ill., 1964).
11. C. F. Carter and B. R. Williams, *Industry and Technical Progress*, (Oxford U.P., 1957).
12. J. A. Allen, *Studies in Innovation in the Steel and Chemical Industries* (Manchester U.P., 1967).
13. I. Maddock *et al.*, 'Inventiveness and Innovation in Industry', *Advancement of Science*, Dec 1969, p. 107.
14. R. W. Cahn, 'Case Histories of Innovations', *Nature*, **225** 693 (1970).

15. *The Queen's Award to Industry*, Report by a Committee under the Chairmanship of H.R.H. the Duke of Edinburgh (H.M.S.O., London, 1965) S.O. Code No. 51–382, p. 3.
16. Ibid., p. 4.
17. S. C. Gilfillan, *The Sociology of Invention* (Follett Pub. Co., Chicago, 1935) p. 7.
18. W. F. Mueller, 'The Origins of the Basic Inventions Underlying Du Pont's Major Product and Process Innovations, 1920 to 1950', in National Bureau of Economic Research, *The Rate and Direction of Inventive Activity* (Princeton U.P., 1962) p. 323.
19. Naval Research Advisory Committee, *Basic Research in the Navy*, vol. 1 (June 1959).
20. *Technology in Retrospect and Critical Events in Science* ('TRACES'), Prepared for the National Science Foundation by the Illinois Institute of Technology Research Institute, vol. 1 (1968).
21. R. S. Isenson, 'Project Hindsight: An Empirical Study of the Sources of Ideas Utilized in Operational Weapons Systems', in W. H. Gruber and D. G. Marquis (eds), *Factors in the Transfer of Technology* (M.I.T. Press, Cambridge, Mass, 1969) pp. 155–76.
22. D. J. de Solla Price, ibid., p. 97.
23. T. Burns, ibid., p. 12.
24. Bright, *Research, Development and Technological Innovation*.
25. Schon, *Technology and Change*, pp. 42–74.
26. D. A. Schon, 'Champions for Radical New Inventions', *Harvard Business Review*, Mar–Apr 1963, p. 77.
27. R. Stillerman, Hearings before the Senate Subcommittee on Antitrust and Monopoly of the Committee on the Judiciary, 'Concentration, Invention and Innovation', Part 3, 89th Congress, 1st session (Government Printing Office, Washington, D.C., 1965).
28. Bacon, in *Francis Bacon*, p. 15.
29. Bacon, *New Atlantis*, in *Francis Bacon*, pp. 161–81.
30. Bacon, *Novum Organum*, book 1, aphorism 61; in *Francis Bacon*, p. 93. For a brief discussion of Bacon's vision of science organised, see F. R. Jevons, *The Teaching of Science* (Allen & Unwin, London, 1969) pp. 65–73.
31. Gilfillan, *The Sociology of Invention*, p. 10.
32. J. Jewkes, D. Sawers and R. Stillerman, *The Sources of Invention*, 2nd ed. (Macmillan, London, 1969) p. 204.
33. Ibid., p. 72.
34. C. Freeman, 'Science and the Economy at the National Level', in *Problems in Science Policy* (O.E.C.D., Paris, 1967).
35. Jewkes *et al.*, *The Sources of Invention*, p. 208.
36. Ibid., p. 152; see also p. 212.
37. Ibid., p. 154.
38. Ibid., p. 82.
39. *Chartered Mechanical Engineer*, **15** 225 (1968).
40. Committee on Manpower Resources for Science and Technology, *Report on the 1965 Triennial Manpower Survey of Engineers, Technologists, Scientists and Technical Supporting Staff*, Cmnd 3103 (H.M.S.O., London, 1966) pp. 46–8.
41. P. M. S. Blackett, *Nature*, **219** 1107 (1968).
42. Council for Scientific Policy, *Second Report on Science Policy*, Cmnd 3420 (H.M.S.O., London, 1967) para. 45.
43. 'TRACES', op. cit., p. v.
44. W. O. Baker, 'The Dynamism of Technology', in E. Ginzberg (ed.), *Technology and Social Change* (Columbia U.P., New York, 1964).

45. L. R. Hafstad, 'The Role of Industrial Research', *Science Journal*, Sep 1966, p. 79.
46. F. R. Newth, in the 'Swann Report', Committee on Manpower Resources for Science and Technology, *The Flow into Employment of Scientists, Engineers and Technologists*, Cmnd 3760 (H.M.S.O., London, 1968) Annex D.
47. J. Langrish, 'Does Industry Need Science ?', *Science Journal*, Dec 1969, p. 81.
48. 'TRACES', op. cit., p. v.
49. I. R. C. Byatt and A. V. Cohen, *An Attempt to Quantify the Economic Benefits of Scientific Research*, Department of Education and Science, Science Policy Studies No. 4 (H.M.S.O., London, 1969).
50. Council for Scientific Policy, *Report on Science Policy*, Cmnd 3007 (H.M.S.O., London, 1966) para. 15.
51. M. Gibbons, J. R. Greer, F. R. Jevons, J. Langrish and D. S. Watkins, 'Value of Curiosity-oriented Research', *Nature*, **225** 1005 (1970).
52. E. G. Rochow, *Chemistry of the Silicones*, 2nd ed (Chapman & Hall, London, 1950).
53. Price, in Gruber and Marquis (eds), *Factors in the Transfer of Technology*, p. 97.
54. M. Gibbons and C. Johnson, 'Relationship between Science and Technology', *Nature*, **227** 125 (1970).
55. Allen, *Studies in Innovation in the Steel and Chemical Industries*, p. 32.
56. B. R. Williams, 'Research and Economic Growth', *Minerva*, **3** 57 (1964).
57. Price, in Gruber and Marquis (eds), *Factors in the Transfer of Technology*, p. 97.
58. Schon, *Technology and Change*, p. 1.
59. E. Jantsch, *Technological Forecasting in Perspective* (O.E.C.D., Paris, 1964).
60. Schon, reported by Burns in Gruber and Marquis (eds), *Factors in the Transfer of Technology*, p. 12.
61. Burns, ibid., p. 12.
62. Central Advisory Council for Science and Technology, *Technological Innovation in Britain* (H.M.S.O., London, 1968) para 8.

Part Two
Some Quantitative Results

The study of innovation is just beginning to pass from the 'wisdom-writing' stage to the collection of quantitative data. Many problems remain to be overcome before such data become accepted as being truly descriptive of reality. For example, attempts to measure the efficiency of research and development in terms of the patents produced per qualified worker have obvious limitations. Without such attempts, however, 'research into research' is not likely to progress.

This part of the book describes some quantitative results of our study. We are aware of criticisms that could be made of such results but we feel that work along these lines is necessary if there are to be significant advances in the present understanding of the process whereby knowledge is converted into wealth. It is usually possible to find specific examples to support any particular view of innovation or any particular factor claimed to be of importance. The quantitative results given here are an attempt to measure the relative importance of some factors which have in the past been described and argued over in qualitative terms.

Except for Tables 6 and 7, the data refer to all the 84 recent technological innovations in Britain which have been included in this investigation. They concern the types of innovation studied, factors found to have been of importance for success, factors found to have delayed innovation, the kinds of innovation models which might be applicable, the sources and methods of transfer of technical ideas made use of and the sizes of the firms involved. There is also a note on methodology and a list of the innovations studied is given.

The word 'firm' is used here to mean any Award-winning unit, including research associations and public corporations.

(a) The Queen's Award Scheme

The scheme is based on the report of a committee under the chairmanship of the Duke of Edinburgh.[1] The purpose of the scheme is to recognise outstanding achievement by industry either in increasing exports or in technological innovation. The definitions of this latter given in the report should be noted as they explain why it is possible for firms to gain Awards even though they are not the originators of the innovation involved. The emphasis of the Award is on *use* of new technology rather than discovery or invention. (The patent system already provides a sort of Queen's Award for invention.) The two quotations from the report on p. 6 above illustrate this point.

Awards have been made annually since 1966. They are given for export achievement or technological innovation or both. Unsuccessful applicants are not named and the Office of the Queen's Award to Industry cannot state what proportion of the unsuccessful applications were for technological innovation, but it is known that the total number of applicants has been of the order of a thousand in each of the first four years of the scheme's existence.

The number of Awards given has been as follows:[2]

Award for:	1966	1967	1968	1969
Export achievement	86	48	61	69
Technological innovation	11	28	17	24
Both exports and innovation	18	9	8	6
Total Awards	115	85	86	99
Total Awards involving innovation	29	37	25	30

There were 66 Awards involving innovation in 1966 and 1967 but some involve more than one innovation. For example, the Award to I.C.I. in 1966 was for innovation in reactive dyestuffs by the Dyestuffs Division and innovation in gas production by the Agricultural Division. Similarly, the 1967 Award to Ferranti Ltd cites four different innovations involving different parts of the company. Thus the figure of 66 Awards involving innovation is increased to a figure of 84 innovations. These formed the basis for the study and are listed in Section (i), pp. 85-9.

The sort of conclusion that can be drawn from a collection of case studies of this type is that any factor which seems to be of importance in several of the studies is at least worthy of further investigation. It is therefore necessary to ask the question, 'Is there any factor which may crop up in several studies as a result of bias in the selection of studies ?'

In most investigations using case studies, there is the possibility that the preconceptions of the investigators may bias the selection of cases. This is not possible in the present study as the cases were selected by an external mechanism, the Queen's Award scheme. However, the scheme itself may be a source of bias in certain ways. For example, many of the Awards are applied for by the sales organisations of the firms concerned and it might be that Award-winning firms have sales organisations which are more on the look-out for new publicity angles than firms which do not apply for Awards. In other words, case studies of this type could not produce significant information on the importance of good sales organisations to successful innovation.

Another possible bias is connected with the dual nature of the Award scheme. Many of the firms receiving Awards for innovation also applied at the same time on the grounds of export performance. It might be predicted that Award-winning innovative firms as a group have an export performance better than average, but this could be a product of a bias in the selection of case studies.

The factors which are discussed below as being possibly important for innovation are not concerned with sales organisations or with export performance and there seems to be no reason to suspect that they are products of bias in the Award system itself.

(b) What kinds of innovations?

The innovations studied cover a wide range of industries. Table 1 gives the distribution of innovations among various sectors of employment. For comparison, the total numbers of qualified engineers, technologists and scientists employed on research and development in these sectors is given for 1965, the year that the Queen's Award scheme was instituted.

It will be seen from Table 1 that the innovations cover a wide range of British industry and that in the mechanical engineering, construction and vehicle industries the proportion of Awards is more than twice the proportion of qualified manpower engaged on research and development.

Table 1

	Number of Award-winning innovations, 1966 and 1967		Qualified engineers, technologists and scientists employed on research and development, 1965*	
MANUFACTURING INDUSTRY		(%)		(%)
Food, drink and tobacco	1	(1·2)	1,300	(2·4)
Chemical and allied industries	7	(8·3)	8,300	(15·4)
Metal manufacture	1	(1·2)	2,200	(4·1)
Mechanical engineering†	25	(29·8)	5,200	(9·6)
Electrical engineering	3	(3·6)	3,500	(6·5)
Electronics	16	(19·0)	7,100	(13·2)
Motor and other vehicles	5	(6·0)	1,000	(1·9)
Aircraft	7	(8·3)	4,200	(7·8)
Textiles, clothing, etc.	2	(2·4)	1,500	(2·8)
Other manufacturing‡	10	(11·9)	2,800	(5·2)
TOTAL MANUFACTURING			37,100	
Construction industry	3	(3·6)	300	(0·6)
Industrial research associations	2	(2·4)	2,000	(3·7)
Nationalised industries and public corporations	1	(1·2)	2,400	(4·5)
Central government	0		8,700	(16·1)
Atomic Energy Authority	1	(1·2)	3,200	(5·9)
Local authorities	0		200	(0·4)
TOTAL (excluding education)	84		53,900	

* *Report on the 1965 Triennial Manpower Survey*, Cmnd 3103 (H.M.S.O., London, 1966) Tables 4 and 10.
† Includes marine engineering and precision instruments.
‡ Includes ceramics, building materials, printing, rubber and other manufacturing.

It has been found convenient to categorise the innovations into four 'areas of innovation', viz. chemical, mechanical engineering, electrical and craft-based. The chemical area includes such innovations as a new process for the manufacture of acetic acid, 'Procion' dyes, semi-synthetic penicillins and also the sterilisation of sutures by radiation from a cobalt-60 source. The mechanical engineering area includes a wide variety of innovations such as machinery for dispensing plastic foam, zoom lenses, new aero engines, the hovercraft and the Moulton cycle. The electrical area consists mainly of electronics but also includes such innovations as equipment for producing very low temperatures. The craft-based area includes such innovations as a new method of baking bread, a new type of tableware and a new building material. The 'area of innovation' is not exactly synonymous with an industrial classification. For example, an innovation in electrical transmission has been classified as mechanical engineering and not electrical because the area of innovation involved the winding of cables. Similarly, an innovation in the automatic control of steel strip thickness has been classified as electrical because the area of innovation involved the development of electronic equipment. The distribution of the 84 innovations among these four 'areas of innovation' is given in Table 2 and is used in subsequent tables.

Innovations obviously vary in size of technology change and in extent of economic and commercial effects. These two are to a large extent independent of each other. Only the former has been assessed in this study because no consistent method of estimating the effect of the innovations was found practicable, the chief difficulties being the time dependence of any effect (the most important effects of some of the innovations may still be in the future) and the lack of comparability in such commercial data as were available.

The size of 'technology change' involved in the innovations was estimated on a five-point scale. For the purposes of this scale, a body of knowledge or industrial practice is considered to be 'a technology' if sufficiently developed to provide a university lecture course at M.Sc. or final-year B.Sc. level. The size of the change in a technology can be roughly estimated in terms of the amount of change required in a textbook of the sort that would be used for such a university course.

Five-point scale of size of change in technology

5 – Innovation leads to a new technology, i.e. a new textbook with a new title is required.

4 – Innovation makes several chapters of the standard book out of date.

3 – Innovation requires major change in one or two chapters of standard book or additions of new chapters. –

2 – Innovation requires alterations or additions of a few paragraphs.

1 – Innovation makes zero or very slight difference to the standard book.

The hovercraft is an example of a '5' on the above scale. 'Procion' dyes, the first commercial dyes to react chemically with cellulosic fibres, provide an example of a '4', making books on the technology or chemistry of dyeing out of date in several places. The development of solvent extraction systems for the production of rare earth compounds provides an example of a '3' in that, though other methods of extracting rare earths are used, solvent extraction is an important addition, making one or two chapters of a book on the rare earth industry out of date. The development of a particular system of structural steelwork for industrialised building provides an example of a '2' in that similar systems were available before the Award-winning innovation took place but certain important improvements were made which would require mention in a revised edition of a textbook on structural steelwork.

This scale obviously has limitations, but it has been used as it is important to examine any differences between innovations involving different sizes of technology change. Many of the current views of innovation derive from well-known type '5' innovations such as the development of the transistor.

Table 2 gives the distribution of the 84 innovations amongst the four 'areas of innovation' and the sizes of technology change. It will be noted that only 11 of the innovations are '5' or '4' on the scale. Differences between these 11 and the rest are shown in Tables 3B, 4B and 5D. The distribution of different 'sizes' among the four areas of innovation is very similar with the exception of the craft-based area, which has a higher proportion of both '4' and '2' than the others.

(c) Factors enabling Award-winning firms to succeed

In most of the cases it was possible to identify a small number of specific factors of importance in the firms' success. These factors can be grouped together in seven categories as listed below. Less specific types of factors such as the all-round competence of a design team or sheer luck are harder to identify clearly and have been excluded.

Table 2

Area of Innovation

Size of change in technology	Chemical n = 12	Mech. Eng. n = 40	Electrical n = 23	'Craft' n = 9	All n = 84
	(%)	(%)	(%)	(%)	(%)
5	–	1 (2·5)	–	–	1 (1·2)
4	2 (16·7)	3 (7·5)	3 (13·0)	2 (22·2)	10 (11·9)
3	5 (41·7)	19 (47·5)	10 (43·5)	2 (22·2)	36 (42·9)
2	4 (33·3)	15 (37·5)	9 (39·1)	5 (55·6)	33 (39·3)
1	1 (8·3)	2 (5·0)	1 (4·4)	–	4 (4·8)

(i) *Top person: the presence of an outstanding person in a position of authority*, e.g. a manager, managing director, technical director or chairman who made a special contribution to the innovation. There were two main ways in which the 'top person' phenomenon was shown. Firstly, the starting of a successful project was in many cases the result of a top person identifying a useful area to work in. Secondly, the top person can generate enthusiasm for a project by making sure that resources are available, taking personal interest in the results of workers and in general being a spokesman for the project.

(ii) *Other person: some other type of outstanding individual.* In one sense, all innovations depend on individuals. However, in some cases, it seems likely that the absence of one specific person would have prevented or delayed success. Examples of this type of individual are the person who is described by his colleagues and others in the industry as a 'mechanical genius' and the person who possessed some unique area of knowledge that otherwise would not have been at the disposal of the firm.

(iii) *Clear identification of a need.* In some cases an important reason for the firm's success was simply the identification of a need that they could meet. This factor is, of course, present to some extent in most innovations, but in some cases it is possible to demonstrate that the identification of a need was a major reason why the Award-winning firm succeeded in the innovation instead of its competitors.

(iv) *The realisation of the potential usefulness of a discovery*. In some cases success was achieved primarily through exploitation of a discovery made earlier, either elsewhere or within the Award-winning organisation. For example, a new method of baking bread arose from the realisation that a discovery made in the laboratory of a research association could be made the basis of a new process.

(v) *Good co-operation*. This category includes both inter- and intra-firm co-operation. In some cases good co-operation stands out clearly as an important factor in the firm's success. For example, the development of new titanium alloys of use to the aircraft industry was clearly the result of a close collaboration between Imperial Metal Industries and Rolls-Royce. Similar firms in the United States do not have the same degree of co-operation. Another example is provided in the development of new forms of penicillin by Beecham Research. The fact that chemists and biologists had a close relationship with each other led to an important discovery being made in what was then a small research organisation.

(vi) *Availability of resources*, either of men or of money. All innovations of course require men and money but in some cases the availability of resources stands out as being of special importance. For example, the float glass process is based on a concept first patented in 1902. The development of this concept to a stage where it could make satisfactory glass cost £4 million, far more than was thought necessary when the development began. Clearly, success would not have been achieved without the availability of resources. Another way in which availability of resources can be important is shown in cases where factories or specialised research teams had been working on products becoming obsolete or projects drawing to a close and the firm has succeeded in diversifying so as to avoid closing the factory or disbanding the research team.

(vii) *Help from government sources*. Such sources have been taken to include research associations and public corporations. For example, the success of John Laing in developing a new material for lightweight concrete depended to a considerable extent on the supply of knowledge and men by the Building

Research Station and also on assistance from the Central Electricity Generating Board.

Tables 3A and 3B give the relative importance of the above factors. Where more than one factor of importance was identified, fractional scores were used. Thus, where three factors were important to a firm's success each scored a third. In some cases no specific factors have been identified. These are represented in Table 3 as 'not classified'. Table 3A gives the distribution of factors among the four areas of innovation and Table 3B the distribution for the 11 'high technology change' innovations. It will be noted that the 'top person' factor, i.e. the presence of an outstanding person in a position of authority, is the major factor in all groupings. Table 3B shows that 'large change' innovations were more dependent on discoveries and less dependent on identification of needs than the other innovations, an important finding in view of the fact that much that has been written on innovation has been based on the comparatively rare 'large change' innovations.

Table 3A

Relative occurrence of factors (%)

Factors of importance in success of firm	Chemical $n = 12$	Mech. Eng. $n = 40$	Electrical $n = 23$	'Craft' $n = 9$	All $n = 84$
Top person	22·2	27·1	25·7	18·5	25·1
Other person	5·6	14·4	21·4	11·1	14·7
Clear identification of a need	19·4	18·8	14·5	9·3	16·7
Realisation of potential usefulness of a discovery	2·8	7·5	6·5	3·7	6·2
Good co-operation	8·3	3·1	5·1	7·4	4·9
Availability of resources	8·3	12·1	2·5	5·6	8·2
Help from government sources	8·3	4·6	2·5	11·1	5·3
Not classified	25·0	12·5	21·7	33·3	19·0

(d) Factors causing delay in innovation

Several factors causing delay were identified and these have been grouped together in the categories given below and in Table 4. These categories refer to delays between an innovation being apparently possible and the Award-winning firm achieving success. Some delays can be attributed to the firm concerned but other delays were outside its control.

Table 3B

	Relative occurrence of factors (%)	
	Large technology change	Smaller technology change
Factors of importance in success of firm	(5 or 4 in Table 2)	(3, 2 or 1 in Table 2)
	n = 11	n = 73
Top person	26·5	24·9
Other person	25	13·1
Clear identification of a need	6·1	18·3
Realisation of potential usefulness of a discovery	14·4	5
Good co-operation	6·1	4·7
Availability of resources	4·5	8·7
Help from government sources	8·3	4·8
Not classified	9·1	20·5

(i) *Some other technology not sufficiently developed.* In many cases an innovation was possible but could not be successfully developed until another technology had 'caught up'. For example, the concept of replacing the shuttle in a loom for narrow fabric weaving has been patented in a variety of ways since the last century. This concept did not become part of a successful innovation, however, until modern materials of construction, lightweight alloys, low-friction bearings, etc., made it possible to develop a shuttleless loom that worked at a high enough speed to be economical.

(ii) *No market or need.* The patent literature is full of inventions for which there is no use. In some of the cases studied, the reason for delay between an idea being technically possible and its successful development was simply that there was no commercial justification for developing the idea until certain economic, social or technical changes created a use for it.

(iii) *Potential not recognised by management.* In some cases it was found that both a technical possibility and its potential use existed for some time before active steps were taken to develop the technical possibility. The reason for this delay was in some cases the lack of recognition by management, i.e. by people responsible for allocating resources that could have developed the potential innovation at an earlier stage. The proportion of such cases was much higher in the chemical area of innovation than in others (Table 4A). This may be because there are so many courses of action open in the chemical area (the number of chemical compounds and chemical reactions is astronomical) that it is easy to miss something that could be of use.

(iv) *Resistance to new ideas* (or over-attachment to old ideas). This category includes such factors as clinging to an old way of designing a product and the known resistance to new ideas on the part of customers which reduces the commercial incentive to innovate. The mechanical engineering area contained the highest proportion of this category (see Table 4A).

Table 4A

Relative occurrence of factors (%)

Factors causing delay in innovation	Chemical $n = 12$	Mech. Eng. $n = 40$	Electrical $n = 23$	'Craft' $n = 9$	All $n = 84$
Some other technology not sufficiently developed	8·3	30·2	50·0	30·6	32·5
No market or need	37·5	25·4	8·7	25·0	22·5
Potential not recognised by management	29·2	4·7	2·2	5·5	7·6
Resistance to new ideas (or over-attachment to old ideas)	4·2	16	4·3	2·8	9·8
Shortage of resources (manpower or capital)	0	13·1	8·7	25·0	11·3
Poor co-operation or communication	4·2	5·6	4·3	0	4·4
Not classified	16·7	5·0	21·8	11·1	11·9

(v) *Shortage of resources* of manpower or capital. Table 4B shows that this factor was more important in the 11 large-change innovations than in the others.

(vi) *Poor co-operation or communication.* This category includes both inter- and intra-firm activity. For example, poor communication between the marketing and research functions can delay innovations just as much as poor co-operation between the firm and its customers or suppliers.

Table 4B

| | Relative occurrence of factors (%) | |
Factors causing delay in innovation	Large technology change (5 or 4 in Table 2) n = 11	Smaller technology change (3, 2 or 1 in Table 2) n = 73
Some other technology not sufficiently developed	16	35
No market or need	16	23·5
Potential not recognised by by management	9	7
Resistance to new ideas (or over-attachment to old ideas)	11·5	9·5
Shortage of resources (manpower or capital)	20·5	10
Poor co-operation or communication	9	4
Not classified	18	11

In some cases no specific factors have been identified. These are represented in Table 4 as 'not classified'. Where more than one factor of importance in delaying an innovation was identified, fractional scores were used.

(e) Linear models of innovation

Models are not just of academic interest; they are also important in practice since some policies seem to rest on assumptions which can be described as models. This section draws attention to some limitations of some current ones.

Most writers on innovation have either clearly stated or implicitly assumed that the innovation process consists of a linear sequence of events. These linear models of innovation can be divided into two types: those in which the start of the process is a discovery, and those in which the start of the process is some form of need.

An example of the first type, which can be called the 'discovery push' model, is provided by Blackett, who states:

In a simplified schematic form, successful technological innovation can be envisaged as consisting of a sequence of related steps: pure science, applied science, invention, development, prototype construction, production, marketing, sales and profit. Clearly the first steps .... cost money and only the later stages ... make money.[3]

An example of the second type, which can be called the 'need pull' model, is provided by Holloman, who claims:

> The sequence – perceived need, invention, innovation (limited by political, social or economic forces) and diffusion or adaptation (determined by the organisational character and incentives of industry) – is the one most often met in the regular civilian economy.[4]

These two types of model can each be further divided into two subdivisions, producing four models of the innovation process as follows. (The symbols DS, DT, NC and NX are used in Table 5 to refer to these models.)

1. DISCOVERY PUSH
DS. The *'science discovers, technology applies'* model.

Blackett's description given above falls into this category, in which innovation is seen as the process whereby scientific discoveries are turned into commercial products. Attempts to measure time-lags between scientific discoveries and their applications (see p. 35) assume this model to be a valid description, as do comments about foreign industry being better at applying the results of Britain's science.

DT. The *'technological discovery'* model.

Many innovations are not clearly based on any scientific discovery but can be described as being based on an invention or technological discovery. For example, Pilkington's float glass process can be regarded as being based on a technological discovery (see p. 37). Pippard[5] has listed the delays between some technological discoveries and their applications.

2. NEED PULL
NC. The *'customer need'* model.

Innovation can be considered as a process which starts with the realisation of a market need. Market research or a direct request for a new product from a customer can be the start of research and development activity leading to successful innovation.

NX. The '*management by objective*' model.

Some innovations can be described in terms of the start of the process being a need identified by the management where this need is not a customer need. For example, the need to reduce the costs of a manufacturing process can lead to resources being allocated to research and development which may produce a new and cheaper process. Another kind of example is the case of a firm producing a new product to avoid a take-over possibility.

When the Queen's Award innovations are examined, *very few of them fit any one of the above models* in a clear and unambiguous manner. The reason for this is quite simple. It is extremely difficult to describe the majority of the cases in terms of a linear sequence with a clearly defined starting point. For example, in the case of Lytag Ltd, a subsidiary of John Laing, the Award was given for the development and sale of a new lightweight aggregate used in the production of low-density concrete. If the process leading to the manufacture for sale of this new product is to be described in linear terms, what is the start of the process ? Is it the Romans' use of pumice as a lightweight aggregate, the manufacture of an artificial lightweight aggregate during the First World War, the need of the Central Electricity Generating Board to dispose of the pulverised fuel ash used in the Lytag process, or the work carried out at the Building Research Station which resulted in the supply of both ideas and people to the John Laing organisation ? The Lytag innovation can also be considered to start with Maurice Laing's desire to venture into new areas of activity, but he would not have gone ahead with Lytag without a belief in a potential market and, from this point of view, the Lytag innovation starts with changes in society creating a need for a material with good insulating properties.

Clearly, any of the four models given above can be made to fit the Lytag case, though the 'science discovers, technology applies' model can be brought in only in very indirect ways, such as by considering the manufacture of an aggregate from a waste product of electricity generation as an application following from the scientific discovery of electricity.

The complexity of the Lytag case is by no means unique. Innovation is a complex process involving the interaction of many factors. This complexity, however, can be simplified somewhat by restricting the process to the Award-winning firm. The question then becomes, '*What stimulated the firm into the activity that led to the successful innovation?*' (See pp. 135, 246.)

From the point of view of the Award-winning firm, it becomes possible to categorise some innovations as being clearly 'discovery push' or 'need pull'. If a sales manager realises that a product needs a particular new property and then persuades the firm to develop a product with this new property, then the innovation is of the 'need pull' type. If, on the other hand, a research department discovers a material with new properties and the firm attempts to find if the new properties have any commercial value, then it is an example of the 'discovery push' type. Similarly, if the managing director of a firm becomes fascinated with a new technological discovery and spends money on its development with no clear indication of any specific market potential, then it is also an example of the 'discovery push' type.

Even when the complex process of innovation is simplified by concentrating on what stirred the Award-winning firm into action, it is still very difficult in a large number of cases to state clearly that the innovation is of one type or another. However, if the above models are regarded as complementary rather than mutually exclusive, the innovations can sometimes be better described as a combination of two of the above models. Tables 5A and 5B show the numbers of innovations which fit the particular models. Table 5C is a combination of 5A and 5B with each 'dual' type scoring 2. It can be seen that, numerically, 'need pull' is more important than 'discovery push'. However, Table 5D shows that the larger technological changes tend to be of the 'discovery push' type. (See above for explanation of symbols.)

Table 5A

Area of innovation	Type of innovation from point of view of firm						
	Not known	*DS*	*DT*	*NC*	*NX*	*Dual*	*Total*
Chemical	1	0	1	1	1	8	12
Mech. Eng.	4	0	3	13	7	13	40
Electrical	2	0	1	3	7	10	23
Craft-based	1	0	0	1	3	4	9
TOTAL	8	0	5	18	18	35	84

Table 5B

Analysis of the 35 dual cases

	DS/DT	*DS/NX*	*DT/NX*	*DT/NC*	*NC/NX*	*Total*
Chemical	1	0	2	4	1	8
Mech. Eng.	0	0	3	9	1	13
Electrical	0	0	3	5	2	10
Craft-based	0	1	1	2	0	4
TOTAL	1	1	9	20	4	35

Table 5C

	Total occurrences in dual and single types				
	'Discovery push'		'Need pull'		*Total*
	DS	*DT*	*NC*	*NX*	
Chemical	1	8	6	4	19
Mech. Eng.	0	15	23	11	49
Electrical	0	9	10	12	31
Craft-based	1	3	3	5	12
	2	35	42	32	111
TOTAL	37		74		

Table 5D

Comparison of 'large technology change' innovations with the others

Innovation model	Large technology change (5 or 4 in Table 2)	Smaller technology change (3, 2 or 1 in Table 2)
Not known	1	7
DT	3	2
NC	1	17
NX	2	16
Dual	4*	31
TOTAL	11	73

* The four 'dual' cases consisted of three DT/NC and one DT/NX.

(f) Sources of ideas

Fifty-one of the innovations have been analysed in terms of major technical ideas or concepts made use of. The sources of these ideas from the point of view of the Award-winning firm are listed in Table 6. 'Government' refers to such organisations as the Atomic Energy Authority, the armed forces, research associations, the B.B.C., etc.

Of the 158 important ideas made use of in 51 innovations, 102 came from outside the firm and Table 7 lists the routes by which these ideas reached the firm.

Using the Lytag case as an example, four important technical ideas were identified. These were:

(i) the idea of manufacturing an artificial aggregate for use in lightweight concrete;
(ii) the idea of using pulverised fuel ash as the source material;
(iii) the idea of making pellets out of pulverised fuel ash;
(iv) the idea of sintering the pellets by a method developed in the treatment of iron ore.

From the point of view of the Laing group, the sources of these four ideas are, respectively, Sweden, the Building Research Station (two) and the iron and steel industry. The ideas reached the Laing group via a director visiting Sweden, the recruitment of staff from the Building Research Station and by attendance at a symposium organised by the British Iron and Steel Research Association.

Table 6

Source of 158 important ideas made use of in 51 innovations

Within firm itself 56

U.K.

Industry (different firm)	18	
Industry (different division or subsidiary)	5	
University	7	
Government	17	
Second World War effort	2	
	—	49

U.S.A.

Industry	26	
Government	2	
Second World War effort	2	
	—	30

Europe

German industry	2	
German Second World War effort	3	
Italian industry	2	
Italian Government	1	
French industry	1	
French university	1	
Dutch university	1	
Dutch industry	1	
Swedish industry	4	
Swiss industry...	1	
	—	17

Other

Australian university...	1	
Canadian Government	1	
Japanese industry	1	
Mexican industry	1	
Common knowledge of uncertain origin	2	
	—	6
		——
		102

Total external sources	102	
Total ideas		158

When innovation is considered as a process encompassing different companies, it is very difficult to answer questions about the origin of ideas. The idea of manufacturing an artificial aggregate for use in lightweight concrete, for example, has cropped up many times, e.g. in the United States during the First World War. However, from the point of view of the Award-winning firm, it is much easier to identify sources of ideas.

Some of the categories in Table 7 are not mutually exclusive. For example, the organisation of a symposium by the British Iron and Steel Research Association scores $\frac{1}{2}$ for the category 'passed on by government organisation' and $\frac{1}{2}$ for the category 'conference in U.K.'.

Table 7

Method of transfer of 102 important ideas from outside the Award-winning firms

Transfer via person joining the firm	$20\frac{1}{2}$
Common knowledge via { industrial experience	15
{ education	9
Commercial agreement (incl. take-over and sale of know-how)	$10\frac{1}{2}$
Literature (technical, scientific and patent)	$9\frac{1}{2}$
Personal contact in U.K.	$8\frac{1}{2}$
Collaboration with { supplier	7
{ customer	5
Visit overseas	$6\frac{1}{2}$
Passed on by government organisation	6
Conference in U.K.	$2\frac{1}{2}$
Consultancy	2
	102

The distinction between a 'government' organisation in Table 6 as the source of an idea and the passing on of information by a 'government' organisation in Table 7 needs clarification. In the Lytag case, the Laing research organisation was gathering information about methods of sintering. It was known that the iron and steel industry used sintering and so a conference organised by the British Iron and Steel Research Association was attended. The source of the idea was therefore the industry, but the method of transfer was 'passed on by government organisation'.

It will be noted from Table 6 that the most important external source of ideas was U.S. industry, followed by U.K. industry and U.K. 'government' organisations. (Research associations are included as 'government' as they are partially funded through the Ministry of Technology.) University sources are small in comparison and some of these relate to important ideas produced by academic science many years ago.

Table 7 shows that the most important route for the transfer of ideas was the recruitment of a person into the firm. This agrees with the view that the transfer of technology is largely a matter of 'agents not agencies' (see pp. 9, 44).

It can be seen from Table 6 that, from the point of view of the Award-winning firm, only 10 of the important ideas from external sources came from universities (7 from U.K. universities and 3 from foreign universities). This can be contrasted with the 59 important ideas that came from elsewhere in industry (23 from the U.K. and 36 from abroad). Of the 10 ideas from university sources, 4 were from engineering, 3 from chemistry, 2 from electrical engineering and 1 from physics departments. The methods of transfer of these ideas consist of 5 by transfer of person from university to the Award-winning firm, 2 classified as common knowledge made available by a university education, 1 from personal contact, 1 from a consultancy and 1 from a conference. None came from the scientific literature. In four of the five cases of transfer by a person going from university to industry, the person concerned was recruited by the firm because of specialist knowledge in an area which was becoming important to the firm. The cases involving the 10 ideas from university sources are not unusual in other respects. For example, a higher level of technological change is not involved (two were classified as 4 on the scale of technology change, three as 3 and five as 2). These results support the view expressed by Price (see p. 40) that technological advances derive more from earlier technology than from the application of basic science.

(g) Sizes of firms

Attempts were made to collect data about sizes of research teams and of firms, but it was found that the complex nature of the process of innovation makes it very difficult to compare such data in a meaningful manner. For example, in many of the cases the number of people carrying out research or development concerned with the innovation varied considerably with time (see p. 21). Similarly, the overall number of research workers and the sizes of the firm concerned varied during the time that innovation was proceeding.

Questions of the type, 'What was the size of the Award-winning firm at the start of the innovation ?' could not be answered meaningfully because, as already stated, different observers can pick on different points in time as the 'start' of an innovation. In many of the cases studied there were no clear points in time at which the innovation could be said to have started or been successful.

A further problem is the difficulty in defining the size limits of an innovating organisation. Several of the Award-winning firms are subsidiaries of other firms or members of a group of companies. In some cases, the firm was taken over by a larger organisation during the time that the innovation was proceeding.

However, it has been found possible to pick out some Award-winning firms that can be described as 'small'. These firms have had less than 1,000 employees at all stages of the innovation and were not in any sense subsidiaries of other companies that would bring the total number of employees over 1,000. This definition excludes some firms that would normally be considered as small. Thorium Ltd, for example, had 110 employees in 1966 but, since its formation in 1914, it has been owned by larger organisations. At one time I.C.I. was a major shareholder and the early research that led to Thorium's success made use of I.C.I.'s experience, equipment and capital. Such firms as Thorium Ltd can therefore be reasonably excluded from the group of small firms.

There were 18 innovations which occurred in small firms as defined above. Eleven of these were in the mechanical engineering area, 5 in the electrical area, 2 in the craft-based area and none in the chemical area.

Table 8A compares the distribution of factors making for success in these cases with the overall and mechanical engineering distributions. There is a greater tendency for the smaller firms' innovations to depend on a discovery and a smaller tendency for the identification of a need being important. The 'top person' phenomenon is more prevalent in the small firms.

Table 8B gives the distribution of factors delaying the

innovation. Shortage of resources is slightly more important
in the case of the small firms but not outstandingly so.

Table 8A

Factors of importance in success of firm	Relative occurrence of factors (%)		
	'Small' n = 18 (11 are Mech. Eng.)	'Mech. Eng.' n = 40	All n = 84
Top person	37	27·1	25·1
Other person	11·1	14·4	14·7
Clear identification of a need	9·3	18·8	16·7
Realisation of potential usefulness of a discovery	13	7·5	6·2
Good co-operation	4·6	3·1	4·9
Availability of resources	12	12·1	8·2
Help from government sources	7·4	4·6	5·3
Not classified	5·6	12·5	19·0

Table 8B

Factors causing delay in innovation	Relative occurrence of factors (%)		
	'Small' n = 18	'Mech. Eng.' n = 40	All n = 84
Some other technology not sufficiently developed	31·0	30·2	32·5
No market or need	18·1	25·4	22·5
Potential not recognised by management	4·6	4·7	7·6
Resistance to new ideas (or over-attachment to old ideas)	12·5	16·0	9·8
Shortage of resources (manpower or capital)	19·9	13·1	11·3
Poor co-operation or communication	2·8	5·6	4·4
Not classified	11·1	5·0	11·9

(h) *Methodology*

1. GENERAL METHOD OF OBTAINING INFORMATION FOR CASE
   STUDIES

Stage I: Initial approach. Starting with the name of the firm
receiving the Queen's Award in the official list supplied by the

Office of the Queen's Award to Industry, a suitable address was obtained from a trade directory, buyers' guide or similar publication and a request for information was sent to 'The Information Officer', 'The Publicity Officer' or some other person but *not* the person likely to have been concerned with the innovation.

The response varied from a Press handout to copies of a large number of publications.

Stage II consisted of a *literature survey* using relevant scientific, technical, commercial and patent literature. The aims of this part of the study were:

(i) to know enough about the innovation and the industry to be able to discuss the innovation in an informed way with the people concerned;

(ii) to trace the history of the technical idea or ideas utilised in the innovation;

(iii) to trace the change in the commercial situation related to the need for the innovation;

(iv) to obtain information about the firm, its organisation, its personnel and the points in time at which major decisions concerned with the innovation took place;

(v) to find out what competitors were doing and if any other organisations played a part in the innovation that resulted in an Award to a specific company.

Stage III: Second approach. In many cases, Stages I and II produced a considerable amount of information in which important gaps were apparent. It was then possible to write either to a member of the Award-winning organisation or to some other body which was concerned with the innovation. A few specific questions relating to the particular innovation were asked with the suggestion that they might best be answered during a discussion. As the questions were specific and relevant to the interests of the person written to, invitations to visit resulted in most cases.

Stage IV: Visits. Visits varied from a two-hour interview with a senior member of the organisation to a full day in the organisation with opportunities to meet staff at all levels. No formal questionnaire was used but there was a check list of points to be covered. The check list was specific to the case

D

being studied but was based on a continually growing list of factors found to be of importance in previous studies. In general, more open-ended questions were used in the earlier parts of interviews.

Stage V: Other sources. Information provided by the Award-winning unit was not considered sufficient in itself, but was tested against independent sources of information. Stage II often provided these, but in some cases it was necessary to contact competitors, firms who co-operated with the Award-winning firm, people who had left the firm or independent bodies who had specialist knowledge of the innovation.

The process described under Stages III and IV was repeated with different sources until sufficient information had been obtained. The information obtained was then written up as a case study complete in itself.

2. COMPILATION OF DATA

The data were gathered from the case studies by asking questions of the people who carried out the case studies rather than the firms concerned. Comparability was tested for by giving the same six completed case studies to different people and seeing if they gave the same answers. Cases where different answers were given initially helped to give a useful under-standing of the innovation process. For example, questions about the origin of an innovation led to quite different answers and this led to the realisation that in many cases, even from the point of view of one firm, innovation is not a simple linear process with a clearly defined starting point. The questions and answers were modified after discussion until agreement was reached.

Methodologically, it might be objected that case studies of success are of little value unless accompanied by case studies of failure as controls. However, it seems to us that the study of innovation is at a stage where a straightforward 'natural history' approach is appropriate. So little information is available about how innovation actually takes place that simple identification of factors of importance – whether in producing success or in delaying success – is of interest in itself. We do not claim that such factors always lead to the results observed in these studies. For example, there are several cases where it was important for success that one man at management level took a decision to innovate and then forced action on his decision. Obviously, this sort of situation might often lead to disaster instead of success.

Attempts that have been made to compare successful innovations with failures have met with problems. Success and failure are only the extremes of a continuum of possible outcomes. Lack of financial success for a particular firm may be accompanied by economic benefits to the nation as a whole in the form, for instance, of a saving in the balance of payments, or of reduced costs of goods or services to consumers or to other firms, or of new jobs created in regions of high unemployment. Furthermore, outcomes are strongly time-dependent; an innovation classified as a failure at one time may unexpectedly become a success at a later time (p. 467).

(i) *List of innovations*

Listed below are the 84 innovations which resulted in the granting of Queen's Awards to Industry in 1966 and 1967 for technological innovation or for innovation and export achievement combined.

	Firm	*Innovation in:*
	Awards in 1966	
1.	Automotive Products Associated Ltd	Motor-vehicle transmissions (Automotive Products Co. Ltd, Borg and Beck Co. Ltd, and Lockheed Hydraulic Brake Co. Ltd)
2.	Beecham Group Ltd	Antibiotics (Pharmaceutical Division)
3.	British Baking Industries Research Association	Breadmaking processes
4.	British Insulated Callender's Cables Ltd	Electrical transmission (Central Research and Engineering and Power Cables Divisions and British Insulated Callender's Construction Co. Ltd)
5.	British Motor Corporation Ltd	Motor-vehicle design and manufacture (Engineering Division)

6. David Brown Corporation Ltd	Boat building and design (Vosper Ltd)
7. Charles Churchill & Co. Ltd*	Automatic chucking lathe
8. Charles Churchill & Co. Ltd*	Numerically controlled profiling lathe
9. Charles Churchill & Co. Ltd*	Gear hobbing machine
10. Derritron Electronic Vibrators Ltd	Environmental test equipment
11. Distillers Co. Ltd	Yeast production (Yeast and Food Division)
12. Distillers Co. Ltd	Acetic acid production (Research and Development Division)
13. Doulton Fine China Ltd	Ceramic tableware
14. Elliott-Automation Ltd	Automation (Elliott-Automation Computers Ltd)
15. Elliott-Automation Ltd	Automation (Elliott Process Automation Ltd)
16. English Electric Co. Ltd	Aircraft generating systems (Aircraft Equipment Division)
17. Ethicon Ltd	Sterilisation of surgical materials by irradiation
18. Gullick Ltd	Coal-mining equipment
19. Imperial Chemical Industries Ltd	Reactive dyestuffs (Dyestuffs Division)
20. Imperial Chemical Industries Ltd	Gas production (Agricultural Division)
21. Lytag Ltd	Building materials
22. Martin-Baker Aircraft Co. Ltd	Aircraft escape equipment
23. Molins Machine Co. Ltd	Cigarette-making machinery
24. Monotype Corporation Ltd	Filmsetting and typesetting machinery
25. Newman and Guardia Ltd	Film-processing machinery
26. Nuclear Enterprises (G.B.) Ltd	Scientific instruments
27. Petters Ltd	Diesel-engine development
28. Pilkington Bros Ltd	Glass production
29. Rank Organisation Ltd	Optical engineering (Rank Taylor Hobson Division, Leicester Unit)
30. St Anne's Board Mill Co. Ltd	Manufacture of paperboard
31. Sanders and Forster Ltd	Structural steelwork
32. Smiths Industries Ltd	Automatic aircraft landing equipment (Aviation Division)
33. Taylor Woodrow Ltd	Piledriving equipment (Taylor Woodrow Construction Ltd)

* One citation 'for technological innovation in machine tools' (Applied Research and Development Division, Churchill Gear Machines Ltd and Churchill-Redman Ltd) has been counted as three innovations.

34. Westland Aircraft Ltd	Hovercraft development (Hovercraft Unit)

Awards in 1967

35. Bonas Bros Weavematic Looms (England) Ltd	Needle loom
36. British Drug Houses Ltd	Organic chemical processing (Pharmaceutical Division)
37. Cambridge Instrument Co. Ltd	Scanning electron microscope
38. Colchester Lathe Co. Ltd	Methods of manufacturing lathes
39. Concrete Ltd	Industrialised building
40. Coventry Gauge and Tool Co. Ltd	Screw-thread grinding machines (Machine Tool Division)
41. Crosfield Electronics Ltd	Electronic equipment for use in printing
42. Davy-Ashmore Ltd	Automatic control of the thickness of steel strip production in rolling mills (Davy and United Instruments Ltd)
43. Dowty Group Ltd	Hydraulics for mining equipment (Dowty Mining Equipment Ltd and Dowty Electrics Ltd)
44. Edwards High Vacuum International Ltd	High-vacuum engineering (that company and Edwards Vacuum Components Ltd, Edwards Instruments Ltd and Edwards High Vacuum Ltd)
45. English Electric Co. Ltd	Electrical fuse element construction (Fusegear Division)
46. English Electric Co. Ltd	Colour television camera (Marconi Co. Ltd)
47. English Electric Co. Ltd	Airborne automatic direction finder (Marconi Co. Ltd)
48. English Electric Co. Ltd	Overseas earth station for satellite communications (Marconi Co. Ltd)
49. Ferranti Ltd	Electronic summation metering equipment (Instrument Department)
50. Ferranti Ltd	Numerical control equipment (Numerical Control Division)
51. Ferranti Ltd	Lightweight inertial platform (Inertial Systems Department)
52. Ferranti Ltd	Digital microcircuits and digital computers (Electronics Department and Automation Systems Department)
53. Freeman Fox and Partners	Suspension bridge design
54. Gas Council	Town gas manufacture

55. General Electric Co. Ltd	Telecommunications (G.E.C. (Telecommunications) Ltd)
56. Harrison and Sons Ltd	Printing of multicolour postage stamps by photogravure (Harrison and Sons (High Wycombe) Ltd)
57. Hawker Siddeley Group Ltd	Design of a diesel engine of medium speed for industrial and marine use (Mirrlees National Ltd)
58. Holman Bros, Ltd	Rotary screw compressors
59. Imperial Chemical Industries Ltd	Manufacture of 'Fluon' (Plastics Division)
60. Imperial Chemical Industries Ltd	Titanium alloys for the aircraft industries (Imperial Metal Industries (Kynoch) Ltd)
61. J. and S. Engineers Ltd	Sealed motor pumps (J. and S. Pumps Ltd)
62. J. and S. Engineers Ltd	Turbine alternator units (J. and S. Pumps Ltd)
63. Linen Industry Research Association jointly with McCleery and L'Amie Ltd	Production of new knitting yarns from man-made fibres
64. Joseph Lucas (Industries) Ltd	High-voltage transistors (Joseph Lucas (Electrical) Ltd)
65. Joseph Lucas (Industries) Ltd	Diesel-engine components (C.A.V. Ltd)
66. Marley Tile Co.	Production of antistatic flooring tiles (that company and Marley Floor Tile Co. Ltd)
67. H. S. Marsh Ltd	Design and construction of plant for the sterilisation of medical equipment by gamma radiation
68. Metalastik Ltd, a subsidiary of Dunlop Co. Ltd	Suspension components for railway locomotives and rolling stock
69. Metals Research Ltd	Image-analysing computer for quality control
70. Moulton Developments Ltd	Hydrolastic suspension for motor vehicles
71. Moulton Developments Ltd	Bicycle of new design
72. Oxford Instrument Co. Ltd	Equipment for producing high magnetic fields and very low temperatures
73. Plasticisers Ltd	Synthetic materials for the production of cordage
74. Renold Ltd	Rotor milling machines (John Holroyd and Co. Ltd)

75.	Rolls-Royce Ltd*	Air cooling of components in aero engines
76.	Rolls-Royce Ltd*	The 'Bypass' design
77.	Rolls-Royce Ltd*	The V.T.O.L. aero engine
78.	Short Brothers and Harland Ltd	Guided weapon system (Precision Engineering Division)
79.	Smiths Industries Ltd	Marine radar (Kelvin Hughes Division)
80.	Thorium Ltd	Production of individual rare earth compounds
81.	United Kingdom Atomic Energy Authority	Production for sale of radio-isotopes for use in industry (Radiochemical Centre)
82.	Viking Engineering Co. Ltd	Machinery for production of rigid and flexible plastic foams
83.	W. Vinten Ltd	Television camera mountings
84.	W. Vinten Ltd	Design of aerial reconnaissance cameras

* One citation 'for technological innovation in aero engines' (Aero Engine Division) has been counted as three innovations.

References for Part Two

1. *The Queen's Award to Industry* (H.M.S.O., London, 1965).
2. Lists published in the *London Gazette*, 15 Apr 1966, 20 Apr 1967, 18 Apr 1968 and 15 Apr 1969.
3. P. M. S. Blackett, 'Memorandum to the Select Committee on Science and Technology', *Nature*, **219** 1107 (1968).
4. J. H. Holloman, in R. A. Tybout (ed.), *Economics of Research and Development* (Ohio State U.P., Columbus, 1965), p. 253.
5. A. B. Pippard, Annex D to the 'Swann Report', *The Flow into Employment of Scientists, Engineers and Technologists*, Cmnd 3760 (H.M.S.O., London, 1968).

Part Three
Case Studies

1 AUTOMOTIVE PRODUCTS LTD: AUTOMATIC TRANSMISSION SYSTEM

This case study is concerned primarily with the design and development of an automatic transmission system by the Automotive Products Group Ltd. Attempts to provide automatic transmission systems had been made for many years with varying degrees of success. The American motoring public had reacted well to the opportunities offered by automatic transmissions and by as early as 1950 such units formed a major segment of the American transmission system market. The European car market posed different problems for the transmission industry and provided a less receptive market for automatic transmission systems. The failure of semi-automatic transmission systems to gain public acceptance stimulated Automotive Products Ltd (A.P.) to embark on the development of a fully automatic transmission system. Such a system would not only provide a new product for the specialised company but would also consolidate its commanding position with respect to the supply of clutch units in Britain. Coincident with A. P.'s initial efforts to develop an automatic transmission system was the development in Northern Ireland of a simple automatic transmission system using bevel gears and inertial actuators. A. P. accepted the offer to develop this system and it provided the nucleus for their subsequent development efforts.

Features of significance in this case study are that it shows how a large firm responded to the predicted needs of a market, how individuals can radically affect the development process, and how innovation may consist of the synthesis of a series of well-known ideas to provide a new answer to a well-known problem.

(a) The Automotive Products Group

The history of the Group has been conveniently summarised in a recent report.[1] In 1920, a partnership trading under the name of Automotive Products Company was set up to import and sell American-made components. The partners soon realised that for real growth some manufacture in the United Kingdom was essential and they obtained licences to make Lockheed hydraulic brakes and later Borg and Beck clutches. By 1938 the millionth clutch had been made at Leamington.

Soon after the Second World War, A.P. reached the conclusion that the development of automatic transmissions in America was likely to be followed in Europe and that, as this would be at the expense of the conventional clutch, it should take an interest in these devices. A number of systems were investigated and rejected from 1948 onwards and about 1953 A.P. developed a 'manumatic transmission' of its own design. This was a semi-automatic system which dispensed with the clutch pedal (the clutch was actuated from a device on the gear-lever) and accordingly provided 'two-pedal motoring' but required the driver to change the gears manually.

The 'manumatic' semi-automatic transmission system represented a cautious advance in the 'state of the art' for A.P., an advance which reflected the doubts of a leading motor journal:

> Whether the benefits to be gained from a fully automatic transmission are sufficient to warrant the extra complexity and expense involved, is still a matter of debate. The general consensus of opinion is not in favour of such devices for use in this country. Here, the conditions are generally such that the increased driving simplicity would be more than offset by the reduction in overall efficiency that seems to be inescapable when an hydromatic transmission is employed. For the time being at least, there appears to be more potential applications for the A.P. semi-automatic transmissions.[2]

The 'manumatic' system consisted of a centrifugal clutch

(based on the Long design of 1936) and a series of vacuum-operated servo units and their controls and could be fitted to any normal synchromesh gearbox then in production.[3] Its initial reception was favourable:

> The introduction of the 'manumatic' must be regarded as a definite step in the right direction. It is less complicated than the full automatic transmission or pre-selector gearboxes of the epicyclic system, and, as the various pneumatic and electrical components are standardised, it is to be expected that car manufacturers will seriously consider the adoption of this system.[4]

Its success was not sustained, however. 'The transmission was sold commercially, on a limited scale, for a few years but it failed to find favour with the public and was withdrawn.'[1]

(b) Transmission systems

According to Chayne,[5] a Vice-President of General Motors in 1951,

> ever since the adoption of the internal-combustion engine for automotive use, industry has conducted a constant search for a suitable transmission to overcome the undesirable characteristics of the engine. A transmission with infinitely variable ratio and fully automatic operation of close to 100 per cent efficiency would be the optimum answer and until it has been evolved into a durable, compact, and low-cost unit which is easily maintained, the search will go on.

Indeed, in the United States, there were in 1951 nine major automatic transmission systems in production and their combined sales in that year amounted to 2,422,131 units. It was obvious that the concept of automatic transmission systems as being the norm for European cars would have to be studied closely by manufacturers such as Automotive Products, and it was equally obvious that the trend of developments in the United States would influence similar developments in Europe.

Four types of transmission system had been studied intensively during the preceding sixty years of the automobile industry's existence: infinitely variable systems, layshaft gear systems, epicyclic gear systems and fluid-drive transmission systems.

Various types of infinitely variable systems were developed,[6, 7] but owing to difficulties in production, cost and efficiency they failed to gain a general acceptance by the motoring world and the search for a practical automatic transmission system was concentrated on adapting the two established methods of obtaining different speed ratios in automobiles, viz. the layshaft gearbox and the epicyclic gearbox.

The layshaft type of gearbox had come into favour in 1889 when Daimler used two sets of gear clusters, one on the drive shaft to the rear wheels and one on an engine-driven countershaft, to obtain three speeds. Four years later the French Panhard and Levassor car used a similar system with a cone clutch to take the drive from the engine to the gearbox. By 1898 Panhard[8] had developed his gearbox design into what can be truly seen as the forerunner of the sophisticated layshaft gearboxes that automatic transmission systems were to displace over fifty years later. This gearbox had two parallel shafts running fore and aft along the car with two clusters of four gears that could be moved into mesh, each stage giving a different drive ratio. The design is all the more of interest in this study, for it illustrates how taking the drive from either of two diametrically opposed sides of a bevel gear can be used to give forward or reverse outputs – a feature that was to cause Reid (see below) to exclaim when he heard of this method being applied to his own design in 1957, 'I never thought of that'. Yet another Panhard development which was recorded in his patent dated 11 November 1899 was the separation of the reversing gear from the forward gears to provide one reverse ratio with four forward ratios. This design also separated the main gear cluster into two independently moved clusters, and so introduced the H-form selection mechanism which has become an industrial standard today. Refinements in the design of the layshaft gearbox followed with the introduction of the two principal types of synchromesh unit.[9, 10]

An epicyclic gear had been used by James Watt in his beam engine of 1788 to overcome the difficulty that the crank mechanism had already been patented. This form of gearing gradually dropped out of favour owing to difficulties in producing the internal gear forms accurately enough to be able to compete with external gears. It was not until the end of the nine-

teenth century that this production difficulty had been overcome and the gearing was installed in such vehicles as the Benz
tricycle of 1899, the De Dion 'Voiturette' of 1900, and a series
of cars produced by Dr F. W. Lanchester from that year on.[8]

The epicyclic gear gave the possibility of greatly simplifying
the procedure for gear changing and it was this form of gear
that was selected by Ford for his 'Model T' design of 1907,
when 'the then conventional type of gearbox which required
a fair measure of skill to operate was completely disregarded
in favour of an epicyclic form of gear train. Gear shifting was
accomplished by an easily learned procedure of depressing foot
pedals and was for all practical purposes foolproof in operation'.[10]

In Britain, Major W. G. Wilson designed a four-speed
gearbox which consisted of two epicyclic gearsets in series
giving four gear ratios to the Wilson-Pilcher car of 1904. 'This
gear with its four ratios was moderately successful but needed
considerable maintenance. Major Wilson decided that epicyclic gearboxes had to satisfy two conditions before they
could become established: they should have self-adjusting
brakes on the annuli and these brakes should be balanced'.
These features were subsequently developed by Wilson in his
designs for the drive units of tanks used in the 1914–18 war,
and for the passenger vehicles of the post-war years''. The
Lanchester 40-h.p. car of 1919 used a three-speed epicyclic
gearbox and conventional single-plate clutch but, because of the
inadequate brake surfaces provided, the system was expensive
to maintain and required frequent relining of the brakes.[12]
Wilson's developed system overcame these difficulties by utilising
two brake bands wrapped around the annuli to give balanced
and adequate braking. Its features were first offered on the
Armstrong Siddeley cars of 1924 and in 1930 the Daimler
Company introduced the Wilson epicyclic gear system into
their automobile range, but with the important additional feature of coupling it to the engine via a fluid flywheel.[13, 14] Thus two
elements that were to form the heart of virtually all subsequent
significant automatic transmission systems had been combined.

Fluid flywheel couplings and torque converters, originally
used mainly for marine drives of very high horsepower, were
developed for automobile use from the 1920s onwards. Up to
1957, twenty-two types of transmission system using fluid

couplings had been produced[15] following the pioneering work of Fottinger in 1905.

(c) Origins of A.P.'s automatic transmission

In 1955 it became evident to A.P. that the 'manumatic' semi-automatic transmission system had not succeeded in obtaining the desired market support and so negotiations were conducted with the Borg-Warner Corporation in the United States and with its subsidiary company, Borg-Warner Ltd, with a view to Automotive Products manufacturing part or all of Borg-Warner's automatic transmission system under licence. The British subsidiary began production of automatic transmission systems in Britain in 1956, but the negotiations between Automotive Products and Borg-Warner failed to reach a suitable agreement and so the decision was taken for Automotive Products Ltd to proceed with the development of its own automatic transmission.

Outlining the 'history and prototype development' of their automatic transmission system, the Chief Project Designer of Automotive Products Ltd, Mr F. E. Ellis, recorded that

> the earliest studies followed, in the main, the use of the spur gear layshaft type box with multiplate clutches interspersed to give power shifts from ratio to ratio, and the usual means of coupling these devices to the engine was via a torque converter. In parallel with these design studies, exploratory work was also being carried out on a three-element torque converter based on substantially traditional lines.[16]

The converter was, in fact, based on the Schneider converter coupling of 1940.

> Although the torque converter was eventually built, the earliest transmission schemes never progressed beyond the drawing board because during 1956 Mr Hugh C. Reid of Belfast approached the company with a bevel-gear device, designed in the first place for use as an overdrive, but which he suggested could form the basis for an automatic gearbox. We were impressed and decided to have a closer look at the system.[16]

In fact, Reid claimed that he designed the bevel-gear unit in the first instance as an automatic gearbox and that it was at a later date, when it appeared unsuitable for a full gearbox, that he decided to try and develop it as a more limited two-speed overdrive unit. 'My idea was a two-stage unit – not an overdrive unit as the Press have claimed. I wanted to make something very simple without hydraulics, clutch or springs. It failed because it wasn't sophisticated enough for the market'.[17] Reid had started his career as an apprentice in a firm owned by Harry Ferguson in Belfast and had become one of his assistants in the development of the Ferguson tractor. In the 1930s he became a car salesman with the firm and later rose to become Managing Director of the firm which was renamed Thompson-Reid Ltd. In his capacity as Managing Director of a large firm of motor distributors he was naturally in a position to monitor trends in development and, perhaps equally important, to pay for a substantial amount of initial experimentation.

Reid began his experiments with £100 worth of 'Meccano' construction kit and the advice of a lecturer at Queen's University, Belfast, Mr R. Jennings. Gradually the design evolved. Reid's idea consisted essentially of attaching masses to the planet wheels of an epicyclic gear so that the axis of rotation of the masses and the plane of rotation of the planet assembly were both at right-angles to the input shaft. By 9 June 1956 Reid had developed his concept to the stage where he was ready to apply for patent protection on his device, and in his application of that date he recorded that 'the mass of the weights and the radius at which they revolve are selected to predetermine by centrifugal force or control the desired characteristics of the unit so that the gears change down automatically and change up semi-automatically substantially under load and speed conditions at which gear changes would be effected by a manual control in a conventional gearbox'.[18] In his complete patent specification, Reid showed two two-speed units coupled together to give his basic three-speed gearbox, the overall ratios and speeds at which they would be in operation being determined by the dimensions of each unit. The gear arrangement that Reid proposed was similar to the conventional differential gear used in the back axles of most vehicles since the earliest days of motoring. It was by the introduction of the two masses on each planetary gear that

Reid had given the unit its patentable characteristics.

Once the design had been finalised, Reid offered his design to the Austin Motor Company – an entirely understandable move since he held the main distribution rights for their cars in Belfast. But Austin were not interested; 'they said in effect, "If we have to use an automatic transmission we will not make them ourselves – we will buy from specialists"'.[17] Reid's next move was to contact the Automotive Products Group at Leamington Spa and propose that they discuss his design. On 13 December 1956 Reid showed his design to Mr A. C. Burdon, Mr Simpson and Mr F. E. Ellis of Automotive Products. They expressed interest in the device and so, in association with the project to develop a torque converter, it was decided to build a two-speed bevel-gear drive unit corresponding to Reid's concept. This was first fitted behind a conventional Hillman Minx gearbox but the arrangement proved too bulky and so it was matched to the torque converter developed by Ellis and fitted to a Humber Hawk car for testing on 23 August 1957.

On 30 August 1957 the Deputy Managing Director of Automotive Products wrote to Reid referring to Reid's patent applications. He outlined an agreement whereby Automotive Products would have, for one year,

> an exclusive licence throughout the world . . . to manufacture and sell for all purposes, gear units coming within the scope of the claims of these patents and also of any further patents that you may file during the option period relating to this gearbox . . . the royalty to be passed by us to you in respect of the exercise of the option is not expected to be in excess of 5 per cent of the nett invoiced selling price of the gearbox, nor is it anticipated that it will be less than 2 per cent of this amount. In the marketing of this licence we would agree so far as we are able to associate your name with it. During the option period we will, at our own expense, design and manufacture a gear unit incorporating the principles described in the above patent applications, for the purpose of carrying out tests in a car in order to determine our future interest in the device. During the period we will keep you informed of the progress and consult you from time to time in

connection with design and other matters arising from our investigations. If, during the option period, we evolve some patentable invention relating to your gear unit, which inventions could not be used without infringing your own patent, then such invention will be the subject of a patent application in the joint names of yourself and the company.[19]

In the event of Automotive Products not exercising their option, all patents and data were to be supplied or assigned to Reid.

The Humber Hawk trials had shown the limitations of a two-speed unit and had also taken place without the use of a reverse gear – a problem that Reid had not solved. Automotive Products then proceeded to build two further units for installation in two modified Austin Cambridge 1·5-litre cars and these incorporated the first major improvement in Reid's design. In a patent application dated 30 December 1957, submitted by Ellis and Reid, it was stated that 'the object of the present invention is to make such modifications of the two-speed gear disclosed in the prior applications so as to provide only two forward speeds giving a smooth transfer from one speed to the other. The invention is characterised in that a reverse bevel wheel meshes with the inner planetary pinion on the side thereof opposite to the forward bevel wheel. . . .'[20] In effect, the method of obtaining reverse[8] that had been used in the Panhard car of 1898 was being used with the minor difference that in that case the reverse bevel was only driven when reverse gear was required.

Following rapidly on the development of the reverse gear capability came a further major breakthrough in March 1958 when Ellis was again named as inventor. 'The invention is characterised in that a friction or other clutch when operative serves to couple together the shafting of the input and output wheels so as to give a direct high-ratio drive, the clutch when inoperative enabling a low-ratio drive to become operative, the planet carrier being then held against backward rotation'.[21] This method provided an alternative means of obtaining the direct drive for which Reid had attached the masses to the planetary gears. Several previous transmission systems had, in fact, used this method of obtaining direct drive, including the Wilson, Borg-Warner, Z.F. Hydramedia and Lyshom-Smith transmission systems.[14] In another patent, Ellis outlined how either the Reid or direct clutch systems could be used, 'the gearing being completely automatic during forward running or during reverse running in adapting itself to the varying conditions of torque, speed, engine idling, over-running or braking as called for by adjustment of the accelerator pedal'.[22] A governor system was used to actuate the hydraulic control units of the brakes and clutch.

From tests carried out on the two-speed plus reverse gearbox, it was evident that a three-speed unit would be required to give a minimum acceptable performance in hilly country. To meet this need, a research engineer working with the Ellis group, Mr P. Standbridge, proposed in the autumn of 1958 that the reverse gear be used in addition as a sun gear so that when held stationary the overall output ratio would lie between direct drive and low speed.[37] This development is of interest, for though it was a relatively simple change, it represented a significant development in the gearbox's capability. A similar method of obtaining an additional gear had been employed in the Eaton two-speed axle of 1950 when a choice of two speed ratios was provided by engaging the sun-wheel with either the differential housing or the planet carrier.[23]

> The level of simplicity [of the two-speed gearbox] ended abruptly, however, with the conversion of the second two-speed box to three speeds. Not only did it include the third gear but it also embraced a free-wheel system to ensure smooth gear-changes and, as a result of this, an additional clutch for overrun braking. Further complications set in when it was decided that to match the high torque capacity rating of the system a cone-clutch would be more efficient and have less drag than the equivalent five-plate clutch. Invariably the idiosyncracies of this type of clutch made themselves evident and it was only through the perseverance of the Mintex engineers in trying all types of friction lining that a highly efficient and reliable gear-change finally resulted.[24]

(d) B.M.C. and the Mini

In the beginning of 1956, Alec Issigonis (knighted in 1969) rejoined the British Motor Corporation as Technical Director responsible to the then Managing Director, Sir Leonard Lord (later Lord Lambury). One of Issigonis's tasks was to develop a series of cars to compete on the European market. The first project undertaken by Issigonis and his team was code-named XC 9001 and in bodywork it 'went back some ten years to internal dimensions exactly as finalised for the Morris Minor', and had an 'all-aluminium 4-cylinder overhead camshaft $1\frac{1}{2}$-litre engine'. 'But in March 1957 it was decided not to go foward with either the XC 9001 or with a smaller version which had been projected at the end of 1956 and given the number XC 9002. Animated by the Suez crisis, Sir Leonard said that what he wanted was a small car that could go into production really soon'.[25] Issigonis had the option to 'use any sort of engine you like so long as we have it on our present production lines'. The choice fell on the A-series engine which had been designed in 1950 to power the A-30 saloon. This unit had originally been designed with a capacity of 803 c.c., had been increased to 950 c.c. after 556,000 units had been produced and was reduced to 848 c.c. when the decision was taken to proceed with the ADO-15 (Austin Drawing Office Project 15), as the project was called.

> An obvious way of saving space lengthwise is to put the engine sideways, but the experimental Morris Minor (which Issigonis had designed in 1951 prior to leaving B.M.C. to join the Alvis Motor Company) had shown that this produced grave problems in accommodating clutch, four cylinders and a gearbox within the biggest track that can be contemplated for a small car. The notion of shortening the power unit by putting the whole of the transmission beneath the crank and driving it by an intermediate gear seems, now that it has been done, so simple a solution that one cannot think why it was never done before. But it can rightly be compared with the centrifugal governor, the rear-mounted jets on the Caravelle, and the combination of an hydraulic coupling with an epicyclic gearbox, as an example of the simplicity which is rightly recognised as genius and summarised by da Vinci when he said *follow the briefest way possible*.[25]

In March 1957 nine people got to work to design the ADO 15. Just 120 days later there were mock-ups of the car and its principal components and by October two prototype units were on road test. In July 1958 the decision was made by Sir Leonard Lord to put the ADO 15 into production, and in the week ending 4 April 1959 the first two cars came off the production lines at the B.M.C. Longbridge factory. The cars were demonstrated to the motoring press on 18 and 19 August 1959 and made public on 26 August 1959.

A.P.'s reaction to the announcement of the Mini was a decision to hasten the development of their automatic gearbox, and so on 28 August 1959 they purchased, privately, a Mini to which they planned to fit the proposed gearbox. The basis of the Mini automatic gearbox design was to be the three-speed unit with torque converter that had been tested in the Cambridge test-car. It retained the bevel gearset that had been introduced into the A.P.'s project group thinking by Reid, but it no longer featured his inertial actuators. In a letter to Reid, his Patent Agent referred to

> our recent conversation in which you indicated that Automotive Products intended to manufacture your gear without the centrifugal weights but using a clutch. We must confirm the view we expressed during our conversation that the gear they intend to produce will be outside the scope of your various patents when granted since your claims in each case are limited in some way or another to the provision of centrifugal weights. This means, in our opinion, that anyone could produce this gear without infringing the claims of your various patents.[26]

The basic form of the proposed automatic gearbox for the B.M.C. Mini was that of the three-speed Cambridge experimental gearbox consisting of a torque converter, a brake- and clutch-controlled gearset with three forward ratios and reverse, and a double-cone clutch to take the drive to the crown wheel of the transmissions' differential gearset. Throughout 1960 and 1961, Ellis and his group were engaged in the development of the three-speed gearset. By November 1960 the design of the prototype had been settled and in the beginning of 1962 the first unit was on trial. In June of that year the Technical Director of A.P., Mr M. Cutler, demonstrated the three-speed prototype gearbox to Alec Issigonis at the B.M.C. Longbridge factory. He in turn demonstrated it to the B.M.C. Chairman, Sir

George Harriman. The outcome of the demonstration was a request from B.M.C. for a pre-production unit for extended trial and a discussion with A.P. to draw up a production agreement.

In August 1962 S. M. Parker, the Engineering Director of A.P., wrote to Reid and traced the progress that had been made in their developments to date.

Two B.M.C. 1½-litre cars were fitted with two-speed transmissions which gave inadequate performance. These cars were converted, with some limitation on detail, to three-speed transmissions whose performance was adequate but for which the commercial demand did not seem very promising.

This was followed by re-orientating our effort towards a three-speed transmission which would fit into the small B.M.C. transverse-engined car. This was running at the beginning of this year and was demonstrated to B.M.C. in June, when they indicated their interest.

In parallel with this last effort, and about six months behind it, we devised a four-speed transmission which goes in the same space and which has been simplified to a point where it is potentially cheaper than the three-speed.

The development of motorways and the call for a high top gear has made this potentially of even greater interest to B.M.C. than the three-speed edition.

From the beginning of our discussions with B.M.C., the development was beamed at an engine slightly bigger than 850 c.c. and it is only comparatively recently that B.M.C. have been able to provide such an engine to us.

My conversation with Alex Issigonis was somewhat in the following context:

I told him that design-wise we had been interested for some years in the development of automatic transmission in order to protect our clutch business. During the earlier stages of this work, you had approached us with a device which gave an automatic change without external control mechanism and this had intrigued us, especially in relation to a simple, cheap, two-speed box plus torque converter.

We had found, however, that two speeds were not enough,

that your idea in its simplest form required a reversal of torque to effect an upward change, and this and more detailed and sophisticated requirements had compelled us to move more in the direction of conventional governor control, which we had endeavoured to do in the simplest possible way.

I also reminded him that even now, the trend was to a greater number of gear ratios until such time as someone produced a really practically infinitely variable arrangement.

I think he was interested in the logical way in which our activity had developed, and of course B.M.C. as a whole are understandably attracted towards a proposition which leaves them making a large number of gearbox parts themselves and thus keeping fully occupied their quite considerable gear plant.[27]

The first of the four-speed gearboxes was designed by Standbridge and Ellis in 1961 and represented an important extension of the unit's capabilities.[38] By repeating the step which Standbridge had taken in 1958 when he used the reverse gear wheel as a sun gear to provide a third speed ratio, they provided a fourth speed by adding another gear wheel and brake over the reverse wheel which could act as another sun-wheel for the outer planet gear. 'In the first instance the four-speed unit retained the double-cone clutch but it was found that using the device in a small car with short stiff drive shafts and no lengthy transmission shaft to absorb the torsional stresses proved too great a problem and the unit was re-designed to incorporate hydraulically operated multi-plate clutches'.[24] As Ellis commented, 'with the three-speed Mini we seemed to reach the peak of the complexity curve for with the appearance of the four-speed unit the whole design seemed to sort itself out into a much clearer set of units'.[16]

(e) Product development [28, 34]

The 'sorting-out' process to which Ellis referred is one that is well known to any production engineer charged with taking a complex prototype unit and converting it into a simplified and highly reliable unit capable of being produced at the rate of several thousand per week. It is often not a spectacular process for it consists of a confirmation of characteristics which have already been generated. Almost by definition it is a process which generates friction for it frequently calls for the sacrifice of components which have been developed at great expense in terms of money and time, for the sake of the overall objectives. It is the task of the production engineer to match the design of each component to the overall design objective for the system so as to optimise its characteristics. The task of ensuring that this was done fell largely to Mr A. J. Atkins as Chief Product Engineer and Mr W. E. King as General Manager of the Transmission Division of A.P. Atkins had joined the Automotive Products Group in 1957 through the Borg and Beck Division on the initiative of a director of that Division, Mr Watkins. Watkins wished to ensure that someone in his Division was responsible for monitoring progress in the development of an automatic gearbox and Atkins was provided with a small team to proceed with the development of a layshaft type of gearbox. To this, and his subsequent work, he could bring the experience he had gained with both the Self-Changing Gear Company (founded by W. G. Wilson to manufacture his gearbox) and that gained with Rootes Ltd where he had been responsible for a comparative evaluation of the different gearboxes then available. King had been with A.P. since 1942 and had held a wide range of posts where he had been responsible for quality control, production planning and specialist vehicle design.

Typical of the decisions which had to be taken in the production-development phase were those concerned with the form and deployment of the clutch units, the location and type of oil pump to be used, and the improvement of the hydraulic control system to give an acceptable level of snatch-free performance when changing gear. Of these, the most important from the points of view of production and reliability was the development of the clutch unit. The reasons for introducing the double-cone clutch unit into the Cambridge prototype gearboxes are not too clear for they represented a marked departure from what Automotive Products Ltd had most experience of – single- and multi-plate clutches. There was available in the automotive industry a wide range of experience in the use of cone clutches and, among others, they were used in the Daimler 'Victoria' design[8] of 1901, the Wilson-Pilcher design of 1904, the 'Clyno' design[33] of 1927 and the General Motors Hydramatic[5] of 1952. Each of these firms subsequently abandoned this form of main clutch member, the normal reason being that although they provided a high torque capacity their rate of torque 'take-up' was too high for comfort. They had been extensively employed in marine drives because of their high torque capacity since Forest's reduction/reverse gear developments[20] of 1885, but in these cases the propeller was available to act as an efficient shock-absorber. The Russel-Newberry marine transmission system 'is an excellent example of the simplifications made possible for already existing designs, it frees the engineer from having to design mechanical controls that are complicated and difficult to build',[29] using as it did the internal pressurisation of cone clutches to obtain forward or direct drive. The cone-clutch units were proposed by Ellis and he was awarded a patent[39] on his method of using them in association with the developed three-speed gearbox proposed by Standbridge. Atkins advocated the retention of the control system used with the cone-clutch units but he proposed that the clutches themselves be modified from the cone type to multi-plate polar lever type. For understandable reasons this move was resisted, for the cone clutches represented almost two years of development effort by Mintex Ltd, one of A.P.'s six clutch-lining suppliers, apart from the effort that had been devoted to the problem of designing and developing them within A.P. Atkins, however, proceeded to design the necessary modifications and, within two weeks of obtaining sanction for the change at board level, the unit was running with multi-plate clutches of a novel design.[35]

For reverse drive, a single clutch unit must be engaged and, because of the low gear ratio used in reverse drive, the clutch must have a higher torque capacity than is needed when direct or top gear is engaged. For direct drive both the reverse drive and forward drive clutches are engaged to lock the elements of the transmission system together and these share the load transmitted. Atkins's novel design provided a greater force acting on the reverse drive clutch when it alone was in operation than when the two clutches were in operation for direct drive.

This meant that the torque capacity of the reverse drive clutch was automatically 'matched to the torque transmission requirements in both cases, and smoother engagement under the lower torque conditions is provided'.[35] Further efforts to smooth the engagement of different gear ratios resulted in a simple but novel and effective modification to the hydraulically operated brake-band control units. By interposing a set of springs of known characteristics between the piston of the control unit and the brake-band lever, Atkins was able to ensure that the load on the brake would increase smoothly and so ensure that the engagement of the selected gear ratio would be accomplished without snatching.[36]

Another significant modification was the removal of the ring-pump unit that had been proposed in Ellis's patent[22] of 1959 and its subsequent replacement by a camshaft-driven pump.

It is likely that with further development the power losses [of the ring-pump] could have been slightly reduced, but a problem with this type of pump, which was driven off the converter, and therefore fitted over its snout, is that, of necessity, it requires large-diameter gear rings. This involved not only high rubbing speeds but large contact areas compared with the camshaft-driven pump, which runs at half engine speed and has ground rotors which are considerably less in diameter with correspondingly reduced rubbing surfaces.[30]

Other detailed modifications were made to the original three-speed design including the repositioning of the bearing supports for the input bevel-gear drive, the introduction of controlled overlap on the hydraulic signals between gear shifts, the use of brazing (and subsequently electron beam welding) to permit an increase in the load capacity of the planet assemblies by 70 per cent over that of the original splined design which was 30 per cent under capacity, the redesign of the crankshaft extension to the torque converter, and the provision of overrun braking on all gear ratios. Each of these changes represented the elimination of difficulties that would have very severely restricted the commercial success of the system. Their importance should not, therefore, be underestimated.

By June 1964 twenty-five pre-production models had been made. The overall design had advanced sufficiently for Parker to write: 'We continue to make good progress, we have, in fact, found a way of avoiding the use of cone clutches, to my personal relief, as I have always thought them inconsistent devices'.[31] In October 1965 details of the transmission were made public, commercial deliveries began in January 1966 and in April of that year the Group received the Queen's Award to Industry for technological innovation.

To accommodate the production lines necessary for the new transmission system, the Filter Division of the Group was moved to Bolton and the production lines for the transmission system laid down in the space vacated. A production line capacity of 2,000 units per week was laid down. In 1968 a report stated:

> There were many who thought that the Leamington group had started at the wrong end when it chose to develop a system for the small car. Critics said that the market – and the motorist – was not yet ready for it. To some extent they have been proved right. The new plant built by A.P. two years ago to produce 2,000 automatic units a week for B.M.C. is working at less than half capacity. Hopes that French, German and Japanese car makers would follow B.M.C.'s lead have not materialised so far. But this costly venture into the small car field may well prove to be Automotive Products biggest selling point in its attempts to break into the medium-plus class.[32]

(f) Comments

(i) With any firm the ability to respond to changes in market needs is of vital importance if it is to retain its sector of the industry's sales. It was evident in the 1950s that the American trend towards incorporating automatic gearboxes in cars as standard equipment was a stable one, and that it would be only a matter of time before such a trend would be established in Europe. A.P.'s response to the challenge this trend implied for the traditional clutch unit industry was to design first a semi-automatic gear-changing mechanism. When this failed to gain full acceptance by the market they started work on designing a fully automatic gearbox aimed at the small car end of the market. What they could not have foreseen was the effect which the Suez crisis would have in giving that sector of the market an almost explosive growth because of the restricted availability of petrol. The design of the Mini range of cars at B.M.C. (as it then was) by Issigonis, and its ready acceptance by the market (albeit after a slightly hesitant start), put a premium on the space-saving characteristics of the automatic gearbox that A.P. were developing. It incorporated a bevel gearset which, with a diameter of little more than 6 in., could hardly have been more compact or suitable for fitting into the crankcase of the Mini engine. A.P.'s immediate market was not the public but rather the motor manufacturer, and it was thanks to the ready response of B.M.C. to the idea of providing an automatic gearbox option for the Mini that the design was given a market test.

The subsequent behaviour of the market has also been of interest. When the Mini had established itself in commanding an appreciable sector of the market, other transverse-engined cars of larger capacity were introduced. So the automatic gearbox introduced by A.P. has been uprated to deal with the higher loads to be transmitted and it now directly competes with the more conventional type of epicyclic gearbox (which in the meanwhile had been matched to the smaller end of the car range).

(ii) It is clear from an examination of the history of the automatic gearbox marketed by A.P. that it is not the outcome of a single clear-cut invention. Rather it is the product of an evolutionary process going far beyond A.P.'s first interest. It began as a layshaft-type gearbox, was modified to incorporate Reid's novel form of gear selector and only incidentally gained the compactness inherent in his design. When the need for Reid's patented feature was removed by incorporating a direct-drive clutch unit, the compactness of his proposal remained. Reverse gear was provided by taking a drive from the opposite side of the input bevel gear, a feature to be seen in most two-faced clocks and in virtually every differential drive system in vehicles. The reverse-drive bevel was used as a sun wheel to produce a third forward gear ratio, and a second application of this concept gave the gearbox its fourth gear ratio. Subsequent developments included the designing of novel clutch units, the modification of the hydraulic control system, and the simplification of the unit for manufacture.

This evolutionary process is the rule rather than the exception in the development of an engineering product from concept to marketable unit. A great number of possible lines of development of a product may exist at any one time; many of them will already have been tried and perhaps rejected, but a particular combination yields an integrated solution which meets a need at a given time.

(iii) Jewkes *et al.* have commented (cf. p. 13) on the role of the individual inventor in innovation, and Reid clearly falls into this category. His invention provided the stimulus for the redirection of A.P.'s thinking, but it was not itself used in their final design, and so his invention, instead of earning money, cost him over £1,500 to patent and develop.

(iv) The difficulties of estimating the benefits accruing from R. & D. are well known. They arise partly from the tendency for work on one project to spill over into others. Thus, though A.P. set out to develop a small car transmission system, enlarged versions have been produced for the larger end of the car range. Similarly, the benefits to be allocated to expenditure in such scientific fields as high-energy electron focusing devices become rather difficult to calculate when it is realised that these were incorporated in the electron beam welding units used to produce the bevel gearsets of the A.P. transmission system and gave the gearsets a 70 per cent increase in load capacity over the prototype units, which were 30 per cent undersize for the loads they had to carry. Thus, expenditure in science may produce benefits not directly in the service technologies (such as electron beam welding) which it influences, but indirectly in the industries influenced by these service technologies.

References

1. Sir Ashton Roskill, Monopolies Commission Report, *Clutch Mechanisms for Road Vehicles* (H.M.S.O., London, 1968).
2. 'Engines and Transmissions', *Autocar*, 27 Nov 1952.
3. 'The Manumatic Control', *Automobile Engineer*, Feb 1953, p. 70.
4. 'International Motor Show', *The Engineer*, 14 Nov 1952, p. 657.
5. C. C. Chayne, 'Automatic Transmissions in America', *Inst. of Mech. Eng. Auto. Div. Proc. 1952–3*, p. 9.
6. Ewen McEwen, 'Recent Developments in Automobile Transmissions', *Inst. of Mech. Eng. Auto. Div. Proc. 1947–8*, p. 97.
7. E. S. Moult, 'An Engine Designer's Scrapbook', *Inst. of Mech. Eng. Auto. Div. Proc. 1965–6*, p. 1.

8. E. B. Weston, *Theory and Design of Automatic Transmission Components* (Butterworth, London, 1967) p. 4.

9. G. T. Smith-Clarke, 'Chairman's Address to the Automobile Division', *Inst. of Mech. Eng. Auto. Div. Proc. 1947–8*, p. 1.

10. T. C. F. Stott, 'A Survey of Fifty Years of Transmission Development', *Inst. of Mech. Eng. Auto. Div. Proc. 1954–5*, p. 267.

11. 'Self-changing Gears RV 28 Gearbox', *Engineering*, 16 Aug 1963.

12. 'Motor Car Show at Olympia', *The Engineer*, 21 Nov 1919, p. 513.

13. 'Motor Show at Olympia', *Engineering*, 17 Oct 1930.

14. A. G. Wilson, 'Transmission Developments for Public Service and Heavy Goods Vehicles', *Inst. of Mech. Eng. Auto. Div. Proc. 1956–7*, p. 93.

15. J. G. Giles, 'A Review of Hydrokinetic Fluid Drives and their Possibilities for the British Motor Industry', *Inst. of Mech. Eng. Auto. Div. Proc. 1956–7*, p. 43.

16. F. E. Ellis, 'An Automatic Transmission for Small Cars – History and Prototype Development', *Proc. Inst. of Mech. Eng.*, **181**, pt. 2A (1966–7) 173.

17. H. C. Reid, interviewed by W. G. Evans, 14 May 1969.

18. H. C. Reid, 'Improvements in or relating to a Two-speed Epicyclic Gear for Use in Motor Vehicles'. British Patent No. 851,172. Applied for 9 June 1956, published 12 Oct 1960.

19. A. C. Burdon, letter to H. C. Reid dated 30 Aug 1957.

20. F. E. Ellis, 'Improvements in or relating to Variable Speed Gear Units'. British Patent No. 878,364. Applied for 30 Dec 1957, published 27 Sept 1961.

21. F. E. Ellis, 'Improvements in or relating to Variable Speed Transmission of Power'. British Patent 878,365. Applied for 26 Mar 1958, published 27 Sep 1961.

22. F. E. Ellis, 'Automatic Variable Speed Transmission System for Motor Vehicles'. British Patent No. 878,366. Applied for 1 May 1958, published 27 Sep 1961.

23. 'Eaton Two-speed Rear Axle', *The Engineer*, 10 Nov 1950, p. 440.

24. 'The Breakthrough', *Precision*, House Journal of A.P., Aug 1967, p. 13.

25. L. Pomeroy, *The Mini Story* (Temple Press, London, 1964) pp. 20, 29.

26. Letter to H. C. Reid from his Patent Agent, dated 15 Oct 1959.

27. Letter to H. C. Reid from S. M. Parker, Engineering Director, Automative Products Co. Ltd, dated 16 Aug 1962.

28. W. E. King and A. J. Atkins, interviewed by W. G. Evans, Leamington Spa, 26 Apr 1968.

29. 'Les Inverseurs de marche de marine', *Le Génie Civil*, **130**, no. 15 (1953) p. 309.

30. A. J. Atkins, 'Production Design of an Automatic Transmission', *Proc. Inst. of Mech. Eng.*, (1966–7) p. 191.

31. Letter to H. C. Reid from S. M. Parker, Engineering Director, Automotive Products Co. Ltd, dated 4 June 1964.

32. C. Webb, 'A.P.'s Challenge in Automatics', *The Times Business News*, 4 Sep 1968.

33. Address by the Chairman, A. G. Booth, M.B.E., *Inst. of Mech. Eng. Auto. Div. Proc. 1956–7*, p. 4.

34. A. J. Atkins, letter to W. G. Evans, 2 Dec 1969.

35. A. J. Atkins, 'Improvements in or relating to Power Transmission Systems'. British Patent No. 1,096,667. Applied for 6 Aug 1965, published 29 Dec 1967.

36. A. J. Atkins, 'Improvements in or relating to Hand Brakes for Use in Change-speed-gear Mechanism'. British Patent No. 1,101,941. Applied for 12 Aug 1965, published 7 Feb 1968.

37. F. E. Ellis, P. J. Standbridge and M. Taylor, 'Improvements in or relating to

Power Transmission System'. British Patent No. 910,003. Applied for 18 Feb 1960, published 7 Nov 1962.

38. F. E. Ellis and P. J. Standbridge, 'Improvements in or relating to Power Transmission Systems providing Automatic Changes of Gear Ratio'. British Patent No. 931,877. Applied for 14 Apr 1961, published 17 July 1963.

39. F. E. Ellis 'Improvements in or relating to Power Transmission Systems providing Automatic Changes of Gear Ratio'. British Patent No. 913,806. Applied for 27 June 1960, published 28 Dec 1962.

2 BEECHAM GROUP: NEW ANTIBIOTICS

The new antibiotics which earned Beechams the Queen's Award were semi-synthetic penicillins, discovered by Beecham Research Laboratories Ltd and marketed by the Beecham Group's Pharmaceutical Division.

Penicillin is not a single chemical substance; there are many chemicals that can be described as penicillins. They all contain the same complex ring structure but differ according to the nature of a chemical side chain which is attached to the ring structure. Beecham Research developed a chemical way of introducing this side chain and, as a result, were able to produce new compounds known as semi-synthetic penicillins. Some of the new compounds had better properties than previously available penicillins and have become important new antibiotics sold by Beecham in seventy-four countries.

(a) *The Beecham Group and the start of a research project* [1,2,3,4]

The Beecham Group has grown from Thomas Beecham's activities in selling his famous pills in 1842. In the 1930s, under Philip Hill, large-scale expansion through the acquisition of other companies took place. The company's products were aimed at chemists' shops and included toiletries and health beverages as well as proprietary medicines.

The Beecham Group first appeared on the penicillin scene when it applied for a licence to produce penicillin in 1943. They were refused a licence by the Ministry of Supply on the grounds of their lack of experience in ethical pharmaceuticals but in 1945 they managed to gain an Air Ministry contract for the production of penicillin pastilles. Also in 1945, Brockham Park in Surrey was purchased for a research laboratory and Beecham Research Laboratories Ltd was formed. The new laboratories were opened by Sir Alexander Fleming in 1947 and various projects were started including an attempt to stabilise penicillin G for oral use. The Beecham management were sure that the demand for ethical pharmaceuticals would be increasing and that they should diversify into this field but they were not unanimously in favour of research as being a necessary part of diversification.

The necessity for research was supported by H. G. Lazell, who became Chairman of the Beecham Group in 1952. Lazell had joined Macleans at the age of fourteen and studied accountancy in his spare time. He rose to prominence in Beecham Pills Ltd, which acquired Macleans in 1938. One of Lazell's methods was always to find and pay for the best advice available. This policy was vindicated when Lazell established the Beecham Toiletry Division in North America and Brylcreem became the best-selling hairdressing in the United States through the help of top-rank American marketing consultants.

In 1954, impressed by the American sales figures for antibacterial compounds, Lazell discussed the possibility of research leading to new antibiotics with Sir Charles Dodds, who had been a Beecham consultant since before the war. The then available antibacterial drugs, sulphas, tetracyclines, streptomycin and chloramphenicol, all suffered from limitations. Penicillins G and V were widely used but were effective against only a limited number of infection-producing bacteria. Dodds recommended Ernst Chain as the best person to give advice on research aimed at new antibiotics.

Antibiotics are the product of one micro-organism active against another. A massive world-wide search for soil cultures capable of inhibiting bacterial growth was being carried out, mainly by American industry, in the hope of finding new antibiotics. There did not seem much point in Beecham competing with this activity. Some work was being carried out aimed at the production of known antibiotics by chemical synthesis but the possibility of developing a chemical route that would be cheaper than the existing biosynthetic fermentation processes seemed remote.

Chain, however, was able to suggest a line of research that could be carried out by Beecham with some possibility of success. This was to attempt the production of semi-synthetic penicillins, different from the existing penicillins, in the hope that some new penicillins with improved properties would be obtained. Semi-synthetic antibiotics are obtained by chemical modification of a product of biosynthesis. Thus, streptomycin produced by fermentation can be converted chemically to dihydrostreptomycin, the first semi-synthetic antibiotic. The existing penicillins which had been obtained by adding chemical precursors to the fermentation flasks in which they were produced were not easy to modify chemically but Chain suggested that many new penicillins could be obtained from p-amino benzyl penicillin, a product of biosynthesis. The reactive amino group in this compund would make it possible for chemical modification to take place.

Another Beecham consultant, Sir Ian Heilbron, who had worked on penicillin during the war, supported Chain's suggestion and Lazell obtained approval for the extension of research facilities at Brockham Park to enable Chain's proposed research to be carried out. At that time, Chain was working in Rome at the Istituto Superiore di Sanità, where he had gone after the war because no one in Britain would provide him with the pilot plant fermentation facilities that he considered to be essential. With the approval of the Director of the Institute, Chain agreed to co-operate with Beecham Research and two new Beecham recruits, G. N. Rolinson, a microbiologist, and F. R. Batchelor, a biochemist, went to work with Chain whilst microbiological facilities were installed at Brockham Park.

In October 1956, within nine months of going to Rome, Rolinson, Batchelor and Italian co-workers succeeded in the biosynthesis of reasonable quantities of p-amino benzyl penicillin which were sent to Brockham Park for chemical modification. Rolinson and Batchelor returned to the expanded research laboratories at Brockham Park, where the techniques of fermentation, microbiology, biochemistry, chemistry, mycology and pharmacology became available.

One other line of research that might have produced new

penicillins was the attempt at total chemical synthesis. It was considered unlikely to be economically competitive with biosynthesis but there was the possibility of the production of a penicillin with such improved properties that extra cost would not be a barrier to its sale. This line of research was being financed in America by the Bristol-Myers Corporation.

(b) The isolation of the penicillin nucleus (6-amino penicillanic acid)

At Brockham, the team of young workers attempted to make use of *p*-amino benzyl penicillin. Penicillin fermentations in the absence of added side-chain precursors were used as control experiments in the course of work on the preparation of *p*-amino benzyl penicillin. In this work, penicillin contents were determined by two methods: a microbiological assay using cultures of bacteria and a chemical assay of the beta-lactam component of the penicillin molecule.[5] It had been observed before in other laboratories that these two methods did not always give comparable results. The discrepancy was discussed by the chemists Doyle and Naylor with Batchelor and Rolinson at one of their regular Monday morning meetings.[2] It seemed that there must be present a form of penicillin which responded to the chemical determination by reaction with hydroxylamine but had little or no antibacterial activity. Doyle suggested that the discrepancy was due to 6-amino penicillanic acid and he and Naylor then realised that such a compound could be readily detected by acylation.

The penicillins (formula I) can be regarded as derivatives of 6-amino-penicillanic acid (hereafter referred to as 6APA), the amine formed by the substitution of a hydrogen atom for the R.CO group in formula I.

$$
\begin{array}{c}
\mathrm{S} \\
\diagup \diagdown \\
\mathrm{R.CO.NH.CH\!-\!CH \quad C.Me_2} \\
\mid \qquad \mid \qquad \mid \\
\mathrm{CO\!-\!N \!-\!\!-\! CH.CO.OH}
\end{array}
$$

Formula I

The various penicillins differ in the nature of R and when no chemical precursor is used a mixture of penicillins is obtained

E

from fermentation. If a side-chain precursor is used, a majority of one particular R is obtained depending on the precursor. Penicillin G contains $PhCH_2$ as the R group in (I) and a mixture containing the suspected 6APA was found to yield penicillin G when acylated with phenylacetyl chloride.

Thus, in May 1957, strong evidence was obtained that 6APA was produced in fermentations lacking a side-chain precursor. The result was communicated to Dr J. Farquharson, the Director of Research at Brockham, and the p-amino benzyl penicillin project was stopped so that the possibilities could be fully explored. 6APA, if it could be produced at reasonable cost, would be a much better start for chemical modification than the p-amino benzyl penicillin. The 6APA was isolated from fermentation liquors in pure crystalline form and chemical evidence of its structure was obtained. The evidence for the identification of 6APA was published[6] in *Nature* in January 1959 but the details of its isolation were not published[7] until 1961. This latter paper contains the name of Chain among its five co-authors. Chain had returned to London as Professor of Biochemistry at Imperial College, where a fermentation pilot plant was installed. He continued to act as a consultant to Beecham.

In 1959 Beecham signed an agreement with Bristol-Myers and both companies started research into commercial methods for the production of 6APA. The American research was carried out by Bristol Laboratories Inc. who were advised by Sheenan of the Massachusetts Institute of Technology, famous for achieving the first chemical synthesis of penicillin[8] in 1957. The two companies were able to find different processes. Bristol carried out a detailed study of the fermentation that produced 6APA and were able to find optimum conditions for a satisfactory yield. Beecham noted a Japanese publication mentioned in *Chemical Abstracts* that showed the possibility of removing the penicillin side chain by enzymatic hydrolysis. The Brockham team were able to find a suitable enzyme which readily removed the side chain from penicillin V to give a good yield of 6APA.

The original research team of eight graduates plus assistants was reinforced by the gradual transfer of workers from an anti-tubercle programme and, within two years of the isolation of 6APA in 1957, the Beecham workers had established a commercial process for the production of 6APA, had carried out work aimed at patent coverage and had started on the search for new penicillins made from 6APA.

(c) New drugs from 6APA

In hospitals and similar environments where penicillin had been widely used since the Second World War, bacteria that were resistant to penicillin had a greater chance of survival than bacteria that were destroyed by penicillin. In particular, bacteria that destroyed the activity of penicillin through their production of penicillinase had become increasingly common and there was an urgent need for a penicillinase-resistant form of penicillin. This target had been present from the start of the project and by 1958 the Beecham chemists had shown that 6APA could be used to synthesise a form of penicillin highly resistant to destruction by penicillinase-catalysed hydrolysis. Unfortunately, there seemed to be toxicity problems with this new penicillin but the chemists were confident that they would produce an effective compromise between the requirements of penicillinase resistance and lack of toxicity.

In fact, the first new penicillin to be marketed by Beecham was not penicillinase-resistant. The first new semi-synthetic, 'Broxil', was developed by Bristol Laboratories Inc. and launched by Beecham and Bristol in September 1959. Broxil was a slightly improved form of Penicillin V and was regarded as rather an anticlimax after the publicity surrounding 6APA. Beecham had good reasons for starting with a product of Bristol research in spite of the fact that no great market for 'Broxil' was foreseen. In 1959 Beechams were not short of capital; there were no other projects competing with antibiotics and there were no take-over bids in the offing. In view of the possibility of a penicillinase-resistant drug becoming available, the Beecham management decided to invest $£1\frac{1}{2}$ million in the initial phase of an antibiotic factory capable of expansion when required. The design of the factory used Bristol know-how and the manufacture of 'Broxil' provided experience that was to be valuable when Beecham research produced new penicillins that were major advances.

The Beecham research involved the chemical synthesis of a

large number of compounds and the testing of many potential drugs. Clear commercial targets were known by the research workers who were able to produce three important new drugs, 'Celbenin', 'Orbenin' and 'Penbritin'. 'Celbenin' met the target of penicillinase resistance and was marketed in September 1960 when the *Lancet* described it as a 'major event in chemo-therapy'.[9] Since it is not absorbed from the gastrointestinal tract, 'Celbenin' had to be given by intramuscular injection. The next target to be met was a penicillinase-resistant drug that, unlike 'Celbenin', could be given by mouth. Beecham chemists had acquired a knowledge of the type of side chain that conferred penicillinase resistance. They also knew the type of side chain that conferred acid resistance apparently necessary for a drug taken orally. Many compounds combining these properties were prepared and as a result 'Orbenin' was marketed to meet the target of an oral penicillin that was penicillinase-resistant.

One more target remained to be met: a broad-spectrum penicillum effective against a greater number of bacteria. A convenient way of classifying bacteria is to divide them into Gram-positive and Gram-negative according to a staining test. The older penicillins were most effective against Gram-positive bacteria; other antibiotics, chloramphenicol and the tetracyclines, were used against both Gram-negative and Gram-positive bacteria but were more toxic than the penicillins. There was therefore a market for a drug combining the breadth of activity that was available from other antibiotics with the very low toxicity typical of penicillins. Beecham chemists followed up some earlier indications that the introduction of an amino group into the side chain conferred Gram-negative activity to penicillin and were able to combine a breadth of antibacterial activity with acid stability in 'Penbritin', marketed in 1962. 'Penbritin', unlike 'Orbenin' and 'Celbenin', does not possess penicillinase resistance. This was added to 'Penbritin' not by developing another penicillin but by marketing 'Ampiclox', a mixture of 'Penbritin', with 'Orbenin'. This completed the Beecham range of new penicillins which are now sold in seventy-four countries. 'Penbritin' in particular has been 'outstandingly successful'.[10] (Since this account of Beecham's research was written, another compound, alpha-carboxy benzyl penicillin, has been marketed as 'Pyopen'.)

Production of the semi-synthetics takes place at Worthing where the plant, including extensions, has cost at least £6 million in capital outlay. One side result of the antibiotic developments has been the setting-up of a veterinary department at Brockham. 'Orbenin' has proved useful in the control of mastitis, a disease causing considerable losses in milk yield.

(d) Comments

(i) *Profits from research.* The cost of Beecham's R. & D. from its modest start has increased steadily and by 1966 the total annual R. & D. cost was in excess of £1 million. As with most studies, data on the profitability of new products are not available. However, the Beecham Annual Report[10] gives separate figures for the sales of its three Divisions. The increase in sales value of pharmaceutical products can therefore be compared with the corresponding increases in sales values for the other Divisions as in the following table:

	Beecham Group world sales (£m.)			
Division	1959	1962	1964	1966
Food, drinks and confectionery	16·6	24·3	23·5	24·5
Toiletry products	13·1	20·4	21·4	27·4
Pharmaceutical products	10·7	14·0	16·2	25·1

The sales figures for pharmaceutical products include both ethical drugs and proprietary medicines. The contribution of proprietary medicines to the total sales is probably substantial as Beecham still continue the home remedy tradition. Photographs of new products in the 1966 Annual Report, for example, include 'Minards massage for the relief of aches and pains' and 'Theralax' laxative tablets. The significance of sales figures is further reduced by the fact that prices of ethical drugs tend to fall. The cost to the National Health Service of 250-mg tablets of penicillin G, for example, has fallen from about £5 per hundred in 1957 to £1 in 1967.

None the less, the sales figures suggest that Lazell's faith in the value of research was justified, though it is interesting to note that the increase in sales of toiletry products is of a similar order. Of the £14 million increase in toiletry sales from 1959 to 1966, only £2 million is due to increased home sales; the remainder is due mainly to successful market research and sales promotion in North America.

(ii) *Science and new drugs*. New drugs are sometimes pictured as one of the fruits of applying the discoveries of 'pure' science. On the other hand, research leading to new drugs is sometimes described as the production of large numbers of compounds either by moulds or by teams of chemists working like battery hens until, by chance, one of the new compounds is found to be useful.

Neither of these extreme views is true in the present case. Certainly a substantial number of chemical compounds were prepared by chemists and tested by other workers, but the choice of which compounds to prepare was made on a rational basis and their preparation involved a high degree of skill in preparative organic chemistry leading to the publication, between 1962 and 1965, of nine papers in the *Journal of the Chemical Society* and involving in some cases the development of new synthetic routes.

The preparation of new penicillins was made possible by a discovery. This discovery, the isolation and preparation of 6-APA, was scientific but it was not the product of 'pure' science. It was, in fact, the product of some scientists of high ability being employed to work on a project which had been selected on the grounds of possible commercial value to the Beecham Group.

What, then, was the role of 'pure science' or 'curiosity-oriented research' or 'academic science' (to use but three terms describing an activity financed mainly by non-industrial sources and carried out with the aims of satisfying curiosity or advancing a scientific reputation rather than promoting industrial advances) ?

The Beecham scientists made available to industry the results of previous 'academic science' as passed on to them through the educational function of universities. Most of the 'academic science' made use of by Beecham scientists in their work was of distant origin. In the work aimed at penicillinase resistance, for example, the chemists found that introducing steric hindrance into the molecule inhibited the hydrolysis that inactivates the penicillin. The concept of steric hindrance is nineteenth-century in origin. One particular form of steric hindrance is the reduction in chemical reactivity of benzoic acids caused by the presence of two adjacent chemical groups in the benzene ring, a phenomenon discovered and investigated by Victor Meyer in 1894. This basic piece of chemistry was made use of in 'Celbenin', the first commercially available penicillin protected from penicillinase-induced hydrolysis. (For 'Celbenin', the R in Formula I is 2:6 dimethoxyphenyl.) Another example is provided by 'Orbenin', which contains a sterically hindered isoxazole ring, a chemical structure investigated and described by Ludwig Claisen in the 1880s and 1890s.

Previous work on penicillin, of course, also contributed to the knowledge and understanding of Beecham workers. For example, inactivation of penicillin by penicillinase was first described by Chain in 1940. In 1943 penicillin became a strategic material of value to the war effort and much research was carried out by British and American scientists in a variety of institutions including five British industrial laboratories. This research was carried out under the joint sponsorship of the Office of Scientific Research and Development, Washington, and the Medical Research Council, London. A summary of some of the results of this war-time research, *The Chemistry of Penicillin*, was published in 1949 and is over 1,000 pages in length.

Although it was the work of Florey, Chain and others at Oxford that made possible the first human trials with a crude form of penicillin in 1941, it was not until 1949 that the chemical structure of penicillin was correctly ascertained. The background knowledge of penicillin that was made use of by the Beecham scientists is not, therefore, the product of 'curiosity-oriented academic research' but comes from a variety of sources working with a variety of motives, including, in the early days, the desire to improve the health of the fighting forces.

Whether semi-synthetic penicillins can be regarded as a product of science or not is a matter of how 'science' is defined. There is no doubt, however, that they were produced by scientists!

(iii) *The organisation of scientists in industry.* The successful Beecham research provides an example of the value of co-operation between chemists and property testers (biologists in this case). The testing of chemicals for desirable properties should be more than routine, otherwise discoveries can be missed. For example, the first acid-stable penicillin (penicillin V) was first prepared in 1948 but its acid stability and hence its commercial value was not noticed until 1954. Another important factor in testing is to make sure that the test really measures the required property; for example, oral absorption is a desired property of penicillins and acid stability is a test, but acid stability in itself does not ensure absorption. Communication between chemists and testers is therefore essential if chemists are to understand what they should be looking for.

Doyle, who helped to identify 6APA, is now Director of Research at Brockham Park. He attributes the successful development of new penicillins to good co-operation between a small team of enthusiastic young workers of different specialities. The discovery of 6APA, for example, came from discussion between chemists and testers. It was a far from obvious discovery, having been missed by previous workers.

Since then, Beecham Research Ltd has expanded considerably (employing about 300 graduate scientists in 1970) and Doyle has given much thought as to how the conditions of the successful small team could be repeated. He considers that the only practical way of doing this is to place the organisational emphasis on projects rather than on functional departments. A new Immunochemistry Department, for example, contains a variety of scientists ranging from chemists to a veterinary physician, the obvious aim being to recreate the conditions of the early Brockham penicillin team. Another attempt to recreate these successful conditions has been the expansion of Beecham Research Laboratories on to a separate site at Harlow in Essex, where a Medicinal Research Centre has been established.

(iv) *The management of research*. One of the factors making for success in new penicillins was the way in which the research effort was linked with overall management considerations. Although Brockham Park is geographically isolated from the manufacturing and selling functions of the Beecham Group, it is not an example of leaving scientists on their own in a country house in the hope that they will discover something. The antibiotic field was selected by Lazell as being an expanding one in which discoveries might be made. His policy of employing consultants enabled Chain to suggest the primary research project. The covering against other research being successful by an arrangement with Bristol-Myers and the decision to invest £1½ million in a factory before the really successful new penicillins had been obtained are further examples of the integration of scientific advice with overall management considerations.

References

1. D. W. F. Hardie and J. D. Platt, *A History of the Modern British Chemical Industry* (Pergamon, Oxford, 1966), p. 276.
2. F. P. Doyle, interview, 1967.
3. G. T. Stewart, *The Penicillin Group of Drugs* (Elsevier, London, 1965).
4. 'Beecham Research Laboratories', Beecham publicity booklet (1966).
5. 'The New Penicillins', Beecham publicity booklet (1965).
6. Batchelor, Doyle, Naylor and Rolinson, *Nature*, **183** 257 (1959).
7. Batchelor, Chain, Hardy, Mansford and Rolinson, *Proc. Roy. Soc. B*, **154** 498 (1961).
8. Sheehan and Henery-Logan, *J. Am. Chem. Soc.*, **79** 1262 (1957).
9. 'Penicillinase-resistant Penicillin', *Lancet*, 10 Sep 1960, p. 585.
10. Annual Report of Beecham Group, 1965–6.

3 BONAS BROS WEAVEMATIC LOOMS: SHUTTLELESS LOOMS

The Award to Bonas Bros recognises their achievement in developing commercially successful shuttleless looms for the weaving of narrow fabrics. For centuries, narrow fabrics – ribbon, tape, elastic web, etc. – have been woven on looms similar in operation to the looms used for broader fabrics with a shuttle carrying the weft between the warp. During the past hundred years, various attempts have been made to replace the shuttle in narrow fabric weaving by some other means of weft insertion. Prior to the Second World War, only the Clutsom loom, developed by Charles Clutsom in the Midlands during the 1930s, was even partially successful.

In 1954 Harry Bonas, the son of one of the original Bonas brothers, became interested in developing a better loom for the manufacture of zip-fastener tape as

a result of meeting Silberman, an American manufacturer who was using looms based on Clutsom's
invention. The Silberman looms were not successful
and in 1956 H. Bonas employed W. C. Arnold, a
qualified engineer (A.M.I.Mech.E.), to work on
loom development. Arnold was successful in redesigning the zip-tape loom which used a reciprocating needle instead of a shuttle and, encouraged by
Bonas, he developed the loom to produce a needle
loom for elastic webs. A range of needle looms was
built up which has been sold abroad successfully,
especially in America and Germany. Other manufacturers followed Bonas's example and today the
traditional slow-moving shuttle looms have been
replaced by fast needle looms in a large part of the
narrow fabric weaving industry.[1]

This study provides an example of how ideas from
a previous century can become commercially useful
in the light of changed requirements and the development of other technological advances which can be
used to bring the old idea up to date. It also provides
an example of how a person with little experience in a
particular field can produce an important innovation
through using experience gained in a different field of
technology.

(a) Early methods of weaving narrow fabrics

Even in the Neolithic period, man knew how to spin and weave.
Illustrated manuscripts show that woven girdles were in use in
the early Anglo-Saxon period and paintings from the Tudor
period show a profusion of woven tapes, etc. Narrow fabrics
were woven by the same method as broader fabrics with only
one tape on the loom.[2]

The first specifically narrow fabric innovation was the
development of looms that could weave four or more narrow
pieces at the same time. The so-called Dutch loom, a series of
single looms mounted in one frame and operated by a central
mechanism with toothed wheels for moving the shuttles, was
invented and suppressed several times. It is reputed[2] that a

multiple loom of this type was invented as early as 1579 in Danzig but the inventor was privately strangled by the burghers of Danzig. By the seventeenth century, tape and ribbon weaving was a highly organised industry in Holland and England. Despite opposition in the form of legislation and violence, the Dutch loom, described by Karl Marx as the harbinger of the industrial revolution, gradually spread throughout the weaving centres, reaching Manchester, for example, in about 1670.

For many years the narrow fabric weaving industry was an example of what would perhaps be described in modern parlance as advanced technology. Ribbon weaving and hat manufacture were both carried out in factories in the same centres. Kay and Stell's flying shuttle of 1745 led to the ribbon loom becoming automatic at least fifty years before an efficient powered broad loom was developed and factories using water power instead of foot treadle power gradually became common. [2] Developments in the nineteenth century were mainly concerned with increasing the size of the loom and this led to the use of steel instead of wood to give increased strength and hence length to the loom. By 1919, a conventional cotton-tape loom had two tiers with 100 shuttles in each tier operating at 160–180 picks per minute and requiring one operator to look after it. [3]

The advent of rubber gave narrow fabric weavers a new material. As early as 1841 there were four smallware looms in Manchester producing india-rubber web[4] and in 1832 the first 'two-way stretch' patent appeared. [5] The elastic web industry was to be the home of Clutsom's shuttleless loom.

(b) Early attempts at shuttleless looms

The first patents for shuttleless looms appeared in 1844 and 1866 in England[6] and America respectively. Between 1860 and 1866 there were seven English patents which claimed various methods of inserting the weft by needles and hooks instead of using a shuttle and there has been a steady stream of patents ever since. [7]

One of the problems associated with shuttleless methods of weft insertion is that a double length of weft is inserted into the warp, leaving a loop at the edge opposite to the weft-inserting mechanism. Various methods of dealing with these loops were attempted, including knitting them together and running another thread through them, but all such techniques produced fabric with selvages (i.e. edges) that were constructed differently. An alternative line of development first appears in an English patent of 1864 in which two weft-insertion mechanisms are used on opposite sides of the fabric, producing material with identical selvages.[8]

The preamble to an American patent of 1919 describes the situation as it seemed to E. Waite, the inventor, in the following terms:[9]

> Narrow woven fabrics are now manufactured almost universally in looms of the general character employed to manufacture wider goods. The output is limited by the speed at which the shuttle can be driven. . . . It has been attempted heretofore to overcome this difficulty by inserting the weft with a reciprocating needle instead of with a shuttle giving a much higher speed especially for short strokes. . . . Such looms, of course, lay a double length of weft in each shed of the warp thus forming a loop at the end of each of these double lengths . . . a shuttle is employed to run a selvage thread through each loop as it is formed. These machines also have encountered the difficulty of requiring very frequent shutdown to renew the bobbin in the shuttle . . . other difficulties . . . prevented machines of this type from going into commercial use to any substantial extent.

The patent then describes 'novel' features including a high-speed mechanism for locking the loops of the weft along one selvage with an independent selvage thread without a shuttle.

Negative support for Waite's claim that shuttleless looms were not successful is provided by the first book entirely devoted to narrow fabric weaving,[3] which was published in 1919. Thirty-three pages are devoted to 'Looms and Devices' but there is no reference to shuttleless looms.

Despite the lack of success, inventive activity, as demonstrated by patent applications, continued. Between 1897 and 1908, the English Patent Abridgement Indexes list 90 patents relating to 'Looms, kinds of, shuttleless'. After 1908, the Abridgement Index subdivides shuttleless looms into various subcategories. A simple count of patents indexed under all the subcategories will not give the total number of shuttleless loom patents as one patent can be indexed in two or three different sub-categories. However, taking just one sub-category, 'Looms, shuttleless, weft inserters, operating mechanism for', 25 patents were published between 1909 and 1915, 8 between 1916 and 1920 and 25 between 1921 and 1925.

Of course, not all these patents are applicable to narrow fabric weaving. Of the eight published between 1916 and 1920 and applied for approximately two years earlier, two are for circular looms used, for example, in the manufacture of incandescent gas mantles, one is a loom for weaving matting and one is a 'darning or weaving machine' apparently designed for use in the home. Of the remaining four patents, three are for shuttleless looms for no particular purpose stated in the abridgement and one is a 'loom for weaving a number of narrow elastic or other fabrics'. The latter patent[10] involves reciprocating needles and a loop-catching mechanism with two weft inserters on opposite sides of the loom. A similar concept was made use of by the German firm of Gabler & Co. who have been making a simple, not very fast, shuttleless loom since the 1920s.

Patent activity continued to increase into the 1930s. There was, for example, an average of $7\frac{1}{2}$ patents per year for weft-inserting mechanisms published between 1931 and 1934 compared with 5 per year for 1921–5 and $3\frac{1}{2}$ per year for 1909–15. The total number of patents relating to shuttleless looms averaged 16 per year for 1931 to 1934 compared with $6\frac{1}{2}$ per year for 1897 to 1900. It was not until 1938 that the first British patent[12] appeared that was to be exploited commercially in the field of narrow fabrics. This was the first of the Clutsom shuttleless loom patents.

(c) Invention without innovation

The amount of inventive activity over a hundred years with no commercial application requires some explanation. The main advantage of getting rid of the shuttle is an increase in speed of operation but this can only be achieved by improving the efficiency of all parts of the loom which then becomes more expensive. There is a need for such a loom only when labour is either scarce or expensive. Prior to the Second World War, machinery manufacturers would probably not have been justified in carrying out the development work needed to produce a high-speed loom for a somewhat limited market which in any case had a plentiful supply of relatively cheap female labour.

Most of the inventive activity was probably carried out by inventive individuals who were concerned with the production of narrow fabrics rather than the production of looms. Certainly Clutsom and Bonas were primarily fabric producers who were prepared to develop their own looms. From early days, fabric producers have employed people to carry out maintenance work on their looms and many of these people experimented with methods of improving the looms that they looked after. However, a really high-speed loom required the complete redesigning of the total loom, a task beyond the resources of the lone inventor.

In the early history of weaving, as discussed above, innovation in narrow fabric weaving preceded similar innovations in broad cloth weaving and it might have been expected that the same would be true for shuttleless looms, as it is a simpler problem to speed up a small narrow fabric loom. However, the pre-war Clutsom loom was only a partial success, being limited to weaving elastics, and it was not really high-speed. The conditions required before a machinery manufacturer would consider the cost of developing a really high-speed loom occurred first in the post-Second World War years in broad cloth weaving. The Sulzer shuttleless loom first manufactured commercially in 1950 was the product of development work extending back to the inventive activity of Rudolph Rossmann in the 1920s.[11]

The American Draper Corporation decided in 1945 that the market represented by mass-produced single-shuttle fabrics was so large that it justified the development of a high-speed shuttleless loom (for broad cloth). The research and development work took fifteen years and cost more than $6 million before the Draper shuttleless loom went into production in 1957 at the rate of 100 per month,[13] this rate being eventually increased. Before the war there was no corresponding market for a shuttleless loom and it can be concluded that one reason why none of the early patents was exploited is that there was not sufficient need to justify the commercial development of the patented inventions.

Another reason for the lack of exploitation of the early patents in the narrow fabric field is the difference between the two edges of a tape made on shuttleless looms other than those which use two weft-inserting mechanisms on opposite sides. It was the custom for commercial buyers of ribbon and tape to inspect the selvages (edges), which were regarded as indicators of the quality of the product. Even for applications in which the tape would not be seen in the final product, there was prejudice against a knitted selvage.[1]

(d) The Clutsom loom

The firm of Clutsom and Kemp was founded in Coalville near Leicester by Charles Clutsom in 1915 with a small team working in one room and weaving narrow elastic webs for foundation garments, which were about to develop into a major industry. The advent of covered rubber thread and circular knitting machines led to the production of roll-ons, and woven two-way stretch elastic garments produced further growth. Clutsom and Kemp became a public company in 1934 and through expansion and take-overs became the major English elastic fabric company. In 1966 the Clutsom and Kemp Group acquired the international Penn Elastic Holdings Ltd and the new Clutsom-Penn Group claimed to be the largest single group in the world in the field of elastic fabrics.[14]

Charles Clutsom was an inventive person as well as a good manager. His development of continuous pad dyeing for both narrow and broad elastics is perhaps a greater achievement than his shuttleless loom. The first Clutsom loom patent[12] was applied for in 1936 and involved a scythe-shaped weft inserter and a latch needle for the selvage. The patent claimed that the loom could be used 'for weaving elastic fabrics for corsets, belts, girdles ... and for ribbons, braids ... or for weaving ordinary fabrics'.

The Clutsom loom operated at about three times the speed of the then existing shuttle looms and in 1938 an initial batch of looms proved successful in the Clutsom and Kemp Coalville factory, producing narrow elastic webs. The machines were produced by the K.C. Engineering Co. Ltd, a subsidiary of Clutsom and Kemp. By the outbreak of the war some looms had been exported to France and the United States and after the war exports continued. A building licence was granted to the Italian firm TEXNOVO.[15]

In 1952 the Clutsom loom was discussed by Thompson,[2] the author of the second English book ever to be published on narrow fabric weaving. He lists the following advantages of the Clutsom loom:

(i) Higher speed because of removal of the limiting restriction of the speed of the shuttledrive.

(ii) One weaver can mind more machines because of the elimination of quill changing (quills are the reels of weft that fit into the shuttle). One weft package will keep the weft-insertion mechanism supplied for several days.

(iii) Weft waste is reduced. Looms with shuttles have to be stopped to change the quills and there is a tendency for the weaver to replace quills that are not completely exhausted, causing wastage.

(iv) Reduced floor space, lighting and power saves overheads.

(v) The weft-inserting needle is thin compared with a shuttle; therefore a small shed (the gap between the warp threads) can be used which is always conducive to good weaving.

Thompson also mentions certain drawbacks:

(i) Owing to the method of weft insertion, a double pick has to be used in each shed. This compels the manufacturer to use finer yarns (i.e. thinner) which is more expensive.

(ii) It is not easily practicable to produce a web with a large number of picks owing to the heavy beat-up tending to put the inserting needle out of register (i.e. the mechanism for packing the weft threads tightly together interferes with the needle). With elastic webs there is only a light beat-up and it is only after the rubber has been allowed to relax that the web becomes thick and heavy. (Weaving is carried out with the warp threads under tension. The elasticity of rubber produces contraction when the tension is released and hence the weft is pulled together.)

(iii) One edge is normal, i.e. woven, but the other is knitted, which does not matter for elastic webs and lighter fabrics but can be important for high-quality fabrics.

According to G. E. Whiteley, who worked with Clutsom, the principle which made Mr Clutsom's system work and others fail was simply that the fastenings for the weft laying-in mechanisms were made to a stationary part of the loom framework, whereas all the other methods continued to attach the weft laying-in mechanisms to the reciprocating batten. Shuttleless weaving relies on the consistent accuracy of the weft laying-in arm placing the weft yarn on to the needle at every pick, and this necessitates a far greater degree of engineering precision than previously found in narrow fabric weaving machinery. Other but less important reasons for the success of the Clutsom loom were due to the fact that it was designed primarily for weaving Men's Trunk Top Elastic Webs with a frilled edge, which meant that the weft packing factor would be less than for rigid tapes and webbing, and the appearance of the knitted selvedge was not critical because during garment manufacture the sewing stiches would cover this selvedge.[15]

The Clutsom loom had some success but was limited primarily to elastic web manufacture. In the 1960s, Clutsom Kemp engineers improved the Clutsom loom and in 1965 a more flexible version was announced;[16] speeds of up to 800 picks per minute were claimed. A Bonas Bros advertisement in 1966 was, however, claiming speeds of up to 1,250 picks per minute for their loom.[17]

(e) Bonas Bros Weavematic Looms (England) Ltd

The development of Bonas Bros Ltd is a story similar to that of Clutsom and Kemp's but on a smaller scale. The weaving of narrow fabrics resulted in looms being manufactured, first for the company's own use and then for sale. Bonas Bros Ltd first moved into machinery manufacture through the incorporation of the Gresley Machine Co. Ltd, a separate company set up to manufacture the shuttleless 'Weavematic' looms. The Queen's Award was received by a sales and development subsidiary, Bonas Bros Weavematic Looms (England) Ltd.

In 1954 Harry Bonas, the son of one of the original brothers, went to Canada to set up a mill weaving zip-fastener tape. There he met Silberman, who was primarily a user of zip tapes but moved into loom manufacture via weaving his own tapes and then making his own looms.[1] Silberman had been using Saurer looms which were small looms designed in Switzerland in the 1920s. An ex-Saurer technician, Ulrich, replaced the shuttle in the Saurer loom by a reciprocating needle and in 1952 two American patent applications were filed in the names of Silberman, Dellaquila and Ulrich.[18] Silberman used forty of these shuttleless looms for his own production and attempted to market the looms through Disco Industries, Inc. which he owned. Harry Bonas bought the patent rights in the shuttleless Saurer loom (the Disco loom) and in 1955 manufacture of the Disco loom was started by Bonas at Castle Gresley near Burton-on-Trent. However, the loom did not manufacture good zip tapes and Bonas could not continue to use them for his own zip-tape manufacture without modification.

In 1955 Bonas advertised for an engineer and W. C. Arnold joined the company and proceeded to redesign the Disco loom. By 1956, a loom was available which produced good zip tapes at high speed and was used in Bonas's zip manufacture. Harry Bonas is part of a world-wide community of zip-tape manufacturers among whom there are extensive personal contacts and, although he was sure the new loom would sell to some of these manufacturers, he also knew that zip-tape weaving was not a very profitable operation and that diversification into elastic web manufacture looked a more suitable area for investment. He therefore encouraged Arnold to build better and faster looms and extend their use into elastic web weaving. Arnold succeeded in doing this and he became Technical Director of Bonas Bros Weavematic Looms (England) Ltd with H. Bonas as Chairman. An American subsidiary, Weavematic Inc., marketed the looms in the United States.

Mr A. Thompson of the Derby College of Technology became interested in the Weavematic loom and he published a description[19] of the first two models to be offered for sale. This description can be compared with a publicity pamphlet[20] issued by Disco Industries to assess the advances in design made by Arnold. The Disco loom claimed a speed of 1,500 picks per minute, but this is more correctly described as 750 double picks per minute as the reed motion beats up two threads at one time instead of one thread as on conventional shuttle weaving. The Weavematic loom described by Thompson operated at a speed of 1,000 double picks per minute. Thompson[19] states:

> Bonas Bros Ltd have taken full advantage of the new materials, better metals and improved bearings which are available today. Almost frictionless bearings, lighter moving parts and improved lubrication have all played their part in achieving the high speeds of which the Weavematic is capable. The loom is a model of precision engineering.

A range of looms was gradually developed and the company expanded in 1959 by setting up a factory in the Pallion Industrial Estate, Sunderland. This move from the Midlands received the assistance of the Board of Trade.[21] The export market was attacked vigorously and eventually overseas sales reached 90 per cent of the total. Sales to Germany, an acknowledged home of precision textile machinery, were particularly encouraging[21] and a German subsidiary, Bonas Weavematic Looms G.m.b.H., was set up. Technical literature in German was first produced in 1959 as part of the expansion involving the new Sunderland factory.

A small development team in a converted hall at Netherseal, near Burton-on-Trent, is continuing to produce improved looms under the guidance of Arnold, who claims that new projects will keep them in the forefront of the needle loom industry and also enlarge their field of activities. Arnold attributes his successful design of a commercially acceptable shuttleless loom to 'the utilisation of good engineering principles' and 'the logical approach of an engineer' being more successful than 'the snap decisions usually met with in the small weaving mill from which most narrow fabric looms have sprung'.[22] Prior to joining Bonas Bros, Arnold had worked for a firm of warp-knitting machinery manufacturers and his experience in a field where machinery design was much more sophisticated was of great help in redesigning the shuttleless loom. The speeds used in the warp-knitting field, and the attention to detail which had to be paid in setting the knitting elements and warping of yarns etc., were invaluable assets when coming into the narrow loom industry.[22]

The success of the Weavematic loom encouraged other manufacturers to enter this field and the majority of narrow fabric weaving is now carried out on shuttleless looms.

(f) Comments

The early history of shuttleless looms has been described in some detail to illustrate the point that inventive activity on its own is not necessarily the prime mover of innovation. The number and variety of patents from 1844 onwards suggests that displacing the shuttle by some other form of weft insertion was a fairly obvious step which occurred to numerous people. These inventive people could not make a commercially viable shuttleless loom, however, until two other conditions obtained. First, to be able to make use of a faster weft-insertion mechanism, the techniques of precision engineering had to develop to the stage where the whole loom could be speeded up. Second, there had to be a market for a high-speed loom requiring higher capital investment than conventional looms. Both these conditions were fully realised only after the Second World War.

This case study also illustrates the phenomenon of technology transfer 'on the hoof', i.e. by movement of people rather than through the literature or by other means. Bonas became interested in shuttleless looms through travelling to America and meeting Silberman. The techniques of precision engineering were transferred from the field of warp-knitting machinery to that of narrow fabric weaving through the movement of Arnold, who was able to apply his knowledge of low-friction bearings, lightweight moving parts, improved lubrication, etc., in a new area.

The problems associated with attempts to define 'origins' of innovations are also illustrated. From a historical point of view, several strands of technical, commercial and social development meet together in the Weavematic loom. At the micro-level, from the point of view of the Award-winning firm, it is possible to answer questions of the type exemplified by 'Where did the technical ideas made use of come from ?' Such questions, however, cannot be answered at a macro-level. What was the 'origin' of low-friction bearings, making use of modern plastics ? There are many possible answers to this question, ranging from 'the military needs of the Second World War' to 'the rise of curiosity-oriented organic chemistry in nineteenth-century Germany'. To give an adequate account of all the factors contributing to the emergence of commercially useful shuttleless looms would be a lifetime's work involving a study of the histories of weaving and engineering, social changes and their effects on consumer demand and the nature of available labour, the psychology of key individuals, the role of science, the diffusion of new ideas, the role of education and so on. Focusing the study on Award-winning firms removes some of the problems associated with a study of 'innovations'. By concentrating on the role of one particular firm, some of the complex factors involved in the process of innovation can be described.

References

1. W. C. Arnold, interview, 1967.
2. A. Thompson, *Narrow Fabric Weaving* (Harlequin Press, Manchester, 1952).
3. Posselt, *Narrow Woven Fabrics* (Textile Publishing Company, Philadelphia, 1919).
4. Pigot and Slater, *Manchester Trade Directory* (1841).
5. Quoted in C. H. Richmond, *The History and Romance of Elastic Webbing* (Published privately by United Elastic Corporation, Easthampton, Mass., 1946).
6. Smith, British Patent No. 10,347 of 1844.
7. Index to British Patent Abridgements.
8. Lawson and Lawson, British Patent No. 719 of 1864.
9. E. Waite and Standard Woven Fabric Co., U.S. Patent 1,296,025 (1919).
10. S. Kendrick, British Patent 116,395 (1917).
11. G. F. Ray, *National Institute Economic Review*, May 1969, p. 60.
12. C. Clutsom, British Patent 478,529 (1938).
13. *Modern Developments in Weaving Machinery* (Columbine Press, Manchester, 1962). Chap. 6 describes the Draper shuttleless loom.
14. 'Clutsom-Penn Stretch Everywhere', publicity booklet (1967).
15. G. E. Whiteley, private communication, 1968.
16. Hanson, *Textile Recorder*, July 1965, p. 46.
17. Advertisement in *Textile Recorder*, Apr 1966, p. 36.
18. Silberman, Dellaquila and Ulrich, U.S. Patents 2,758,614 and 2,800,927.
19. A. Thompson, *Textile Weekly*, 18 Oct 1957.
20. 'The New Disco Narrow Fabric Loom', publicity leaflet (1954).
21. Arnold, copy of speech made on occasion of Queen's Award presentation.
22. Arnold, private communication, 6 Mar 1968.

4 BRITISH BAKING INDUSTRIES RESEARCH ASSOCIATION: CHORLEYWOOD BREAD PROCESS

The British Baking Industries Research Association (B.B.I.R.A. – now amalgamated with the Research Association of British Flour Millers to form the Flour Milling and Baking Research Association) gained the Queen's Award in 1966 for 'innovation in breadmaking processes'. The new method of breadmaking is the Chorleywood Bread Process (C.B.P.).

This study is an illustration of how old ideas can be made the basis of an important new method of manufacture. It also demonstrates the complex relationships, within a research association, between the member firms, the research organisation and machinery manufacturers. One of the difficulties facing research associations is how to get their discoveries adopted by industry or, alternatively, the difficulty is how to make discoveries that industry will want to adopt. The diffusion of an innovation through the industry is therefore of greater importance in the case of a discovery by a research association than it is in the case of research within a firm which can adopt the innovation itself; this later stage of the innovation process is therefore described in more detail in this study than in others.

Dough consists of a mixture of flour and water, together with smaller amounts of yeast, fat and other additives. Before dough can be baked to give satisfactory bread, certain physical and chemical changes have to take place. In technical terms, the dough has to be 'developed'. The traditional method of obtaining dough development has been to leave the dough for some hours. During this period, fermentation takes place, some 2 per cent of the flour is converted to alcohol and carbon dioxide, and the dough changes from a dense mass to an elastic material which is expanded by carbon dioxide.

The Chorleywood Bread Process developed by B.B.I.R.A. replaces this lengthy period of dough development by a combination of mechanical and chemical treatments which produce the desired effect in the dough. The process was announced in 1961 and by 1969 it was estimated that 65 to 70 per cent of the total bread production in the United Kingdom was by C.B.P. A substantial export market for British machinery has developed as many overseas countries have adopted C.B.P.

The mechanical and chemical aspects of C.B.P. are discussed in the following sections.

(a) Mechanical dough development

The origin of the use of mechanical work as an aid to dough development is lost in the unrecorded practices of a craft-based industry. In an emergency, many bakers used more yeast and more mixing to give a 'short-time dough'. Early scientific investigations were carried out by chemists and the mechanical effects of applying work to the gluten (the protein constituent of flour) were not realised. The effect of greater mixing was thought to be due to the trapping of more air in the dough. To practical bakers, the concept of applying intense mechanical work to dough would have seemed ludicrous. Jago, the author of a standard work on baking technology, wrote in 1911: 'Dough is not a material which may be ill-treated with impunity; it is, or should be, a living mass which may suffer irretrievable damage if handled with a trifling excess over the permissible severity.'[1] This belief had to be overcome before new methods of dough development could be developed.

In 1926 Swanson and Working,[2] at the Kansas State Agricultural College, published the results of a study of the effect of stirrer speed which showed that, on a laboratory scale, the lengthy period of fermentation could be replaced by seven minutes of mixing with an experimental stirrer operating at 120 r.p.m. At 60 r.p.m. fifteen minutes' mixing was required. From these data it can be suggested that the amount of work applied to the dough is critical.

Swanson and Working did not explore the scientific implications of their work but they were aware of commercial possibilities. They stated that mechanical dough development could save time, save material used up in fermentation and save space by the elimination of the 'dough room' in which the dough was left to ferment. Commercial application was attempted by some machinery manufacturers and in this country a patent[3] for the mechanical development of dough was applied for in 1927. No satisfactory commercial process was arrived at and in 1928 Working withdrew the suggestion that the discovery could be commercialised.

One of the reasons why this early work met with no success was the fact that mechanical work alone is insufficient to replace the effect of fermentation. Chemical oxidising agents or 'improvers' are also required.

(b) Chemical oxidising agents

Certain wheat varieties produce flours which, in the past, gave unsatisfactory baking performances unless they were matured by means of long periods of storage. It is now realised that this maturing process involves oxidation and that the baking performance of flour can be modified by the use of chemical oxidising agents.

In 1920, Kohman, Irvin and Cross patented[4] a salt additive containing a small amount of potassium bromate as an 'artificial maturing agent' which enabled freshly milled winter-wheat flours to be used without ageing. The oxidising nature of the maturing process began to be recognised in 1921 following the accidental discovery that another oxidising agent, nitrogen trichloride, when used as a flour bleach[5] also produced a maturing effect. Oxidising agents became increasingly used either as a gas applied to the flour in the milling process or as a solid added in the baking process and it was found that the time required for dough development by fermentation could be reduced. Several scientific investigations of the effect of oxidising agents on flour were carried out but in 1945, twenty-five years after the first published use of potassium bromate, a review article[6] admitted that 'efforts to obtain direct and positive evidence as to the mechanism by which these effects are produced have thus far failed to produce convincing results'. Following the Second World War, the Medical Research Council arranged for tests on the toxicity of oxidising agents to be carried out by the B.B.I.R.A. and the use of chlorine dioxide for the treatment of flour and potassium bromate as an additional oxidising agent or 'improver' became standard practice.

The replacement of the long period of fermentation by a short period of intense mechanical work only became commercially possible when mechanical development was combined with the use of chemical oxidising agents. J. C. Baker, who had patented the use of nitrogen trichloride as a flour bleach in 1921, was the first person to carry out an extensive study of the combined effects of work and oxidising agents on dough. During the 1940s and 1950s he carried out empirical studies involving a greater rate of doing work than had been used before as well as a systematic investigation of oxidising agents.[7] His research was aimed at a continuous process for bread manufacture. Obviously, the lengthy period of dough development had to be eliminated if the process was to be continuous and he succeeded in doing this. Baker's results were incorporated in the Wallace and Tiernan 'Do-Maker', a continuous bread-making process[8] which was adopted in the United States. The 'Do-Maker' was tried in this country but not used to any great extent as it produced bread with a texture unacceptable to the British housewife.

The fact that the 'Do-Maker' process was continuous tended to obscure its reliance on the application of work to the dough until in 1958 B.B.I.R.A. workers at Chorleywood began an investigation of mechanical development in its own right using a small batch-mixer. This investigation led to the Chorleywood Bread Process.

(c) The British Baking Industries Research Association

The B.B.I.R.A. was set up in 1946 and in 1947 the conversion of a country house into a research station was started at Chorleywood in Hertfordshire. The first Director was Dr J. B. M. Coppock, who was faced with the problem of convincing a craft-based industry that a research association could be

of value to 'down-to-earth' bakers.[9] In the early years of the
R.A., the emphasis had to be on solving the immediate problems
of industry and providing Liaison Officers who could visit
members of the Association. The R.A. was also responsible to
the Government as represented by the Department of Scientific
and Industrial Research (D.S.I.R.) who provided about a third
of the Association's income.

In the years 1948–50, workers at Chorleywood published
seventeen papers. Four were in the *Journal of the Science of Food
and Agriculture*, three were in the *British Medical Journal* and the
rest were scattered among seven other journals.[10] Nine of the
seventeen papers were concerned with the purity of food or
analytical tests associated with the determination of purity and
few, if any, were of direct relevance to the practical baker who
was helped not through the scientific and technological literature but through the *Bulletin* issued to members and the visits
of the Liaison Officers.

Basic research, when carried out, was a part-time activity for
research workers who were also concerned with the day-to-day
problems of industry as represented by the following examples
of titles of articles published in the *Bulletin*: 'The heat required
to bake a loaf of bread', 'The case of the bread with purple
spots', 'American-type cookies', 'Deep-freeze storage of baked
goods', 'Cellulose films and waxed papers for biscuit packaging',
'Ants in the bakery', 'The food hygiene regulations', 'Baking
practice for meat pies'.

Coppock wanted to be able to carry out more long-term basic
research but he did not have enough staff to be able to do this
as well as coping with the other activities that were required.
In September 1957, just before leaving the R.A. to become
Research Director of Spillers, Coppock wrote in his annual
research report:[11]

Nevertheless, despite these difficulties, we continue to carry
out a great deal of basic research which is the insurance for
the future. For example, our first experiments on the deep
freezing of bread were reported in the February Bulletin of
1952; now the topic is in many bakers' thoughts. Similarly,
in our 1955 Annual Report we mentioned . . . continuous
dough mixing. These are two clear illustrations of what

basic research means in contributing to the future and even highly scientific investigations into the rheology of doughs, particularly their behaviour under compression and relaxation, will quite possibly yield data of considerable application to our mixing studies.

In the year 1956–7, the R.A. employed a technical staff of thirty-two, of whom four had research degrees, nine had first degrees only and four had technological qualifications (e.g. the National Bakery Diploma). According to Coppock's annual report, forty-two research projects were being carried out in that year in addition to the other activities of the R.A.

In May 1958 a successor to Dr Coppock was appointed. The successor was Dr G. A. H. Elton, who had formerly been Reader in Applied Physical Chemistry at the Battersea College of Technology and had also at one time worked for the Printing Research Association. Coppock told Elton that some selection of research projects was required and over a drink he suggested that there were certain areas of research that should be concentrated on. This would require some reorganisation at Chorleywood but a new Director should be able to achieve this. One of these areas of research was the mechanical development of dough.[9]

In September 1958, in his annual report, Elton wrote:[12]

I feel that it is incumbent upon me as the new Director to attempt to produce new and original thinking. . . . I now have in hand a careful survey of the existing situation at Chorleywood . . . I feel there is scope for a good deal more fundamental research into projects likely to be of long-term benefit to the Industry. This point has in fact been emphasised by Dr Coppock in previous Reports.

There was a substantial increase in membership of the Research Association and hence an increase in income in the year 1957–8. The increase was continued in the following year. (Membership of R.A.s is voluntary and the amount of financial assistance from government sources depends on the amount contributed by industry.) The total income for 1958–9 was £67,000 against £51,000 for 1956–7, which enabled an increase in qualified staff to take place. There were thirty-three members of the technical staff at Chorleywood in 1958–9, of whom nine had research degrees, six had first degrees only and six had technological qualifications. ('Technical staff' excludes library and administrative staff.)[13]

Towards the end of 1958, Elton reorganised the research being carried out at Chorleywood. Six basic research projects were selected, all containing certain aspects which had previously been looked at, usually as a part-time activity. Under the reorganised research plan, greater resources were allocated to these projects than had been available before. One of the basic research projects was described in the 1958–9 report as 'The Continuous Mixing and Mechanical Development of Bread Doughs'. Previous work on continuous mixing had taken place but this was the first time that an investigation of mechanical development in its own right had been carried out and it was to lead to the development of the Chorleywood Bread Process.

(d) The Chorleywood Bread Process

At a symposium held in 1949, Dr D. W. Kent-Jones, the author of *Modern Cereal Chemistry*, suggested that dough ripeness could be brought about by physical means.[14] He asked if the loss of flour solids which took place during fermentation could be saved by the use of some other means of achieving dough ripeness and he suggested that this would be a good research topic for the 'newly formed Research Association'. (Dough is said to be 'ripe' after it has been 'developed'.)

As mentioned above, it was not until 1958 that the Research Association was in a position to carry out what seemed to be a long-term basic project on the mechanical development of dough. The previous work on continuous mixing was an empirical investigation of the effect of using various types of flour, various recipes and other factors on continuous mixing and was carried out by A. W. James, an assistant in the bakery, under the direction of J. J. Devlin, the Principal Bakery Officer. T. H. Collins had also carried out some work, but as he was both an Experimental and a Liaison Officer responsible for contacts with some of the R.A.'s members, the amount of time available for experimental work was limited. Under Elton's reorganisation in 1958, Collins was put in charge of the mechanical development work and allowed to concentrate on this one job.

The previous work had used a laboratory-scale Buss Ko-Kneader continuous mixer which, like other continuous mixers, resembles a plastics extrusion machine. (A continuous mixer has a continuous stream of dough-making materials entering at one end with dough leaving at the other end. A 'batch' mixer is a scaled-up version of the housewife's bowl and mixer which mixes one batch at a time.) In order to gain a better knowledge of the effects of various factors, Collins was set to work on batch-wise experiments using a laboratory-scale Morton Duplex mixer. The reasons for carrying out work with a batch mixer instead of a continuous mixer were given as follows:[15]

> As the ... Morton Duplex ... gave satisfactory results with flour 1 (which had also given good results on the continuous mixer) and could be operated without difficulty on a batch process, most of the work was carried out on this machine. Formula and mixing time could be altered easily and quickly, and a larger number of experiments carried out than when using the Buss Ko-Kneader continuous mixer.

One of the factors that was investigated was 'the energy requirements of mechanical dough development' and a recording wattmeter was attached to the Morton mixer with the help of S. J. Cornford, a physics graduate who was a Senior Scientific Officer in charge of a small physics section. Thus the energy in horsepower minutes per lb of dough required to give the best bread could be measured. The energy required varied slightly according to the type of flour and the formula used. The addition of oxidising agents slightly reduced the required energy but the general conclusion was that 'to mechanically develop a dough, approximately 0·4 h.p. min/lb of mechanical work is required'.[15]

Another factor investigated was the effect of oxidising agents and it was found that a fast-acting improver was required in addition to the standard use of potassium bromate. Various fast-acting improvers were tested including potassium iodate, which was used in other tests, and ascorbic acid (vitamin C) which was not quite as effective.

In December 1959, as part of the increase in qualified staff,

Dr N. Chamberlain joined the R.A. (he had previously worked for Boots as a section leader in charge of screening for new antibiotics). Chamberlain was given the title of 'Senior Scientific Liaison Officer' but he asked to take a particular interest in T. H. Collins's work on mechanical development. The two approaches of using laboratory-scale continuous and batch mixers continued and it was found that about 0·4 h.p. min/lb gave the best results in continuous as well as batch mixing.[16] Continuous mixers in use in industry were found to be operating empirically at similar figures.

Further work on the effect of various factors was continued and it was realised gradually that the batch mixing process originally used as a method of investigating factors of importance to continuous mixing could, in fact, be a process in its own right, offering advantages associated with the replacement of a long period of fermentation by a short period of intensive work. Before a practical baking process could be developed, a suitable mixer was required of larger scale than the laboratory Morton Duplex mixer. At this stage, the bread machinery manufacturers who were associate members of the R.A. were not interested. Mixers used in the rubber and cement industries were tried but found to be unsuitable.

Chamberlain then suggested that a 'Tweedy' mixer be examined. This mixer was produced by a small engineering company, George Tweedy and Co., which was run by K. Pickles. It was a dry-materials mixer used for breaking up cake crumbs and similar applications. In April 1961 a Tweedy mixer of 56-lb capacity was lent by Pickles to the R.A. and was used to make satisfactory bread.

In July 1961, the results of the experiments on batch mixing were summarised and presented to the R.A. as the Chorleywood Bread Process.[17] In addition to the need for the correct amount of work and the presence of a fast-acting improver, the inclusion of fat and extra water were described as essential features of the new process. The fast-acting improver recommended was ascorbic acid (vitamin C) as the Food Standards Committee considered it to be preferable to potassium iodate.

The process still had to be scaled up for large-scale bakery use and Pickles worked extremely hard with great enthusiasm to produce a mixer of 450-lb dough capacity. The scaling-up

process was largely empirical, with Collins giving opinions on the
bread produced by a variety of designs of mixing action.

English Electric Ltd, through their Meter Sales Department,
was able to supply an automatic electrical control system that
could be fitted to mixers in order to control the amount of work
done. The established machinery manufacturers and members
of the R.A. began to take an interest and in May 1962 a full
report was issued[18] containing practical details of how to use
C.B.P. commercially.

(e) The adoption of C.B.P.[19]

Between the first published report of the C.B.P. in July 1961
and the report in May 1962 considerable activity took place,
aimed at the member firms of the association. At Chorleywood
a newly created Member Services Division under the direction
of Dr Chamberlain made use of a special D.S.I.R. scheme
instituted in 1959 for grants to improve the liaison of research
associations with industry. A van was obtained which toured
the country with a Tweedy pilot-scale mixer. Demonstrations
to local bakers were given in various places including Belfast,
Dublin and Edinburgh. By February 1962 about a thousand
people had seen C.B.P. demonstrated and most of them had
been able to see, feel, taste and smell bread made by the new
process. For the small baker, this had far more impact than
any number of B.B.I.R.A. reports.

A very important factor in the adoption of the C.B.P. by the
baking industry was the early involvement of British Bakeries.
In 1961, parts of British Bakeries, who produced 22 per cent
of British bread, were considering re-equipping. Experiments
with continuous mixers were being carried out but were not
entirely successful. British Bakeries are a subsidiary of Rank,
Hovis and McDougall whose Chairman, Lord Rank, was
persuaded to visit Chorleywood where he saw C.B.P. in operation on a Tweedy pilot-scale mixer. Lord Rank was enthusiastic
and, through his involvement, Pickles was able to collaborate
with British Bakeries in the design and production of full-scale
mixers. Pickles realised that a scaled-up version of the Tweedy
mixer required a partial vacuum and eventually a successful
mixer was obtained. Tweedy and Co. eventually sold a large

number of mixers for C.B.P. before they were acquired by the American group, Ward Bakeries.

By 1966, 35 per cent of the total bread production of the United Kingdom was being made by C.B.P. This means that, within five years of its introduction in the form of a report, over 1,000 million loaves per year with a retail value of over £60 million were produced by a new method.[20] By 1969, the estimated use of C.B.P. had further increased to 65–70 per cent of the total United Kingdom bread production.[21]

The advantages predicted by Swanson and Working in 1926 were obtained and in addition it was found that C.B.P. could use a higher proportion of weaker flour (i.e. flour of lower protein content), which reduced the need for Canadian imports. A further benefit to the country was the development of an export market for mixing machinery as C.B.P. spread abroad. Dr Elton helped this process by carrying out a tour of Australia and New Zealand.

(f) *Comments*

Project selection. The Chorleywood Bread Process is an example of innovation stemming out of a planned programme of research. The allocation of resources to research that is likely to produce a 'winner' is the central problem of research management, and effort is being expended on ways of improving this allocation, in particular on the development of techniques for project selection. It is interesting to speculate what would have happened when Dr Elton carried out his review of research if he had employed someone to carry out estimates of likely benefits, probabilities of technical success and so on. Such estimates applied to the research which led to the C.B.P. would probably have given the wrong answer because they would have been concerned with a continuous process whereas, in fact, a new batch process was the result. The reasons for carrying out research with a batch mixer, as described in section (d) above, make it clear that the original intention in studying the mechanical development of dough was to gain more control over the continuous process rather than éto develop a new batch process.

Although, from the point of view of project selection, it seems that the successful outcome of the research was partly fortuitous, it remains true that mechanical development was selected as an area of research likely to produce useful results. This selection of an area of research which is thought likely to produce useful results of a sort that cannot be precisely predicted is the method used by research management in what is loosely described as 'basic' research. The selection of such an area is usually a matter of individual intuition or the growing up of a collective climate of opinion in favour of that area.

Timing. Dr Kent-Jones had suggested as far back as 1949 that the Research Association should look at alternative methods of dough development, but it was not until 1958 that the research on mechanical development started. This raises the question whether the C.B.P. could have been developed earlier by adopting Kent-Jones's suggestion more quickly. The answer is probably no, because the process requires treated flour and the use of standard improvers such as potassium bromate. The use of chlorine dioxide for the treatment of flour and the addition of improvers was not established in this country until the R.A. had carried out tests for the Medical Research Council. The type of non-treated flour used in 1949 could not have been used for the C.B.P. There is a good illustration here of the general point that it is not sufficient to select an area likely to produce useful results: the timing must also be correct. This is a delicate matter, for it involves not being too late as well as not being too early; it is quite possible that the C.B.P. would not have been as successful as it has been if it had been developed later when improvements in continuous mixing by the industry could have taken the industry in another direction.

The importance of correct timing in the selection of basic research areas has perhaps been underestimated in the past. Could it be, for example, that some of the areas selected by grant-giving bodies as being worthy of special support are in fact areas which have already produced useful results and are now 'too late' ? It is, of course, more difficult to select potentially useful areas in the sort of research typically financed by the Research Councils than it is in the case of a research association set up to serve a particular industry.

Input from curiosity-oriented research. As mentioned on p. 36,

F

the C.P.B. was studied to see what imputs it contains from 'curiosity-oriented' research. This was done in connection with a suggested method for quantifying the economic benefits from such research and we were looking in particular for research results delay in which would have delayed the innovation.

Some indirect links with the pure science of the last century were found. For example, tests were carried out on unbaked dough using a Simon Extensiometer, which measures some physical properties of dough. These tests were devised technologically in the 1930s on a theoretical basis which stems from the last century. Similarly, the use of a recording wattmeter can be seen as an input from electrical research in the nineteenth century. (It could even be seen as stemming back to James Watt's measurements in the eighteenth century of the rate of doing work of a steam engine – but then Watt's work can hardly be described as curiosity-oriented.) Such inputs were, however, probably not essential to the success of the C.B.P.

We considered the possibility that the use of ascorbic acid (vitamin C), found to be a dough improver in 1935, might represent an input from curiosity-oriented research to an essential feature of the process. The use of fresh fruits to prevent scurvy dates back to the 1750s or even earlier and the research in the earlier part of the twentieth century on the identification, isolation and synthesis of the specific dietary factor was clearly motivated by more than curiosity. It is true that the synthesis of ascorbic acid and its subsequent manufacture in 1937 could not have been achieved without the knowledge accumulated by academic chemistry since the nineteenth century, in particular the work of Emil Fischer and others on carbohydrates. However, by the time the C.B.P. was being developed, ascorbic acid was but one of several fast-acting improvers that could have been used, so that its availability was not crucial.

We were therefore forced to the conclusion[22] that there is no identifiable curiosity-oriented research of the present century which, had it been delayed, would have occasioned a corresponding delay in the development and adoption of the C.B.P.

References

1. W. Jago, *Technology of Bread-Making*, 3rd ed. (Simpkin, London, 1911) p. 647.
2. C. O. Swanson and E. B. Working, *Cereal Chemistry*, **3** 65 (1926).

3. H. A. Kohman, Mellon Institute of Industrial Research, U.K. Patent 244,489.
4. Kohman, Irvin and Cross, U.S. Patent 1,325,327.
5. J. C. Baker, U.S. Patent 1,367,530.
6. M. J. Blish, *Advances in Protein Chemistry*, **2** 351 (1945).
7. J. C. Baker, *Bakers Weekly*, **161**(11) 60 (1954).
8. Wallace and Tiernan Process Co., U.K. Patents 735,184 and 741,526 (1955).
9. J. B. M. Coppock, interview, 1967.
10. B.B.I.R.A. Annual Report 1955–6, list of publications.
11. B.B.I.R.A. Annual Report 1956–7, p. 5.
12. B.B.I.R.A. Annual Report 1957–8, p. 5.
13. List of staff and accounts of B.B.I.R.A. in Annual Report, 1958–9.
14. D. W. Kent-Jones, 'Symposium on Recent Advances in the Fermentation Industries', *R.I.C. Special Report* (1950) p. 47.
15. T. H. Collins, D. J. Devlin and A. W. James, B.B.I.R.A. Report No. 47 (1960).
16. N. Chamberlain and T. H. Collins, B.B.I.R.A. Report No. 48 (1960).
17. N. Chamberlain, T. H. Collins and G. A. H. Elton, B.B.I.R.A. Report No. 59 (1961).
18. N. Chamberlain, T. H. Collins, G. A. H. Elton and S. J. Cornford, B.B.I.R.A. Report No. 62 (1962).
19. G. A. H. Elton and N. Chamberlain, interviews, 1967.
20. B.B.I.R.A. Press handout, 'The Queen's Award to Industry 1966'.
21. G. A. H. Elton, private communication, 1969.
22. M. Gibbons, J. Greer, F. R. Jevons, J. Langrish and D. S. Watkins, 'The value of curiosity-oriented Research', *Nature*, **225** 1005 (1970).

5 BRITISH INSULATED CALLENDER'S CABLES: CONTINUOUSLY TRANSPOSED CABLES

This case study shows the extent to which a firm may depend, before making any large-scale investments, on pressure from their customers. The original survey showed that there was not a significant market for continuously transposed cables. A few years later, however, there were signs of an incipient market and a decision was made by B.I.C.C. to invest in this type of cable. The technical development was very smooth and no major organisational changes were involved.

(a) Cable for transformer windings

Transposed cable has been increasingly used in large transformers during the last few years, its basic advantage being that it reduces stray current losses. It consists of a number of

rectangular copper strands, assembled together in two rectangular stacks and transposed continuously. The current losses
are made up of two components: eddy current loss within each
insulated strand and circulating current loss between the
individual strands. The former loss can be reduced by using
many thin strands while the latter can be reduced by the
transposition of the insulated strands. Each strand is covered
with a thin film of polyvinyl acetate enamel, and paper insulation is applied to the assembled group. A transposed winding
occupies less space than one formed from strips individually
wrapped with paper, while the elimination of hand transposition reduces labour costs.

The heart of the transposing plant is the transposing head.
There are differences between the firms engaged in manufacturing transposed cable, but basically the head is a device
incorporating a system of cam-operated fingers which continuously interchange the relative positions of the conductors
as they pass through, an action very similar to braiding. It is
usually equipped to accommodate up to 31 conductors.

(b) The B.I.C.C. innovation[1]

In the late 1940s, the General Electric Company of America
(G.E.C.) began to manufacture continuously transposed strip
for high-voltage transformers. The Americans developed all
their own machinery, including the all-important transposing
heads, and acquired extensive patent protection. At this time
the European market was small and only the Brown Boveri
Company attempted to design their own equipment for making
transposed strip. On the other hand, the British Thompson
Houston Co. (B.T.H.) licensed the American equipment from
G.E.C. and began manufacturing transposed cable for use only
in their own transformers. There was no wide commercial need
at this time.

In March 1953 B.I.C.C. received an enquiry from its
Australian subsidiary about the possibility of supplying transposed strip for transformer windings. The enquiry engendered
a market survey into its commercial (as distinct from technical)
feasibility. The existence of the B.T.H. continuously transposed
cable was well known to B.I.C.C. Although the processes of

transposed cable manufacture were well understood, B.I.C.C. could find no appreciable market for the cable and as a result decided not to attempt to manufacture it.

The issue seems to have been dropped after this initial enquiry until December 1957, when a letter from the Ferranti Co. requested some continuously transposed strip, suggesting that, if B.I.C.C. did not undertake to supply this type of cable, they might do it themselves.

There was also pressure from elsewhere. Early in 1958 D. McDonald, a former employee of B.T.H. and at this time with Bruce Peebles Co., was in Australia trying to get contracts for some transformers from the Electricity Commission of South Australia. In order to get these contracts, he was forced to stretch transformer technology to the limit. McDonald knew that the only way in which his specifications could be met would be to use continuously transposed strip:

> In 1958, when I was in Australia, I began to feel the hard impact of competition, particularly in loss levels and began to feel the need to use continuously transposed cable in disc windings – in almost any kind of winding – to minimise loss and create gain in capitalised performance. We took orders for the largest transformers then to go into Australia on very tight losses and I could see no [other] way of accomplishing the performance we had set ourselves.[2]

McDonald approached 'his friends' at B.T.H., the London Electric Wire Co., and B.I.C.C. to persuade them to manufacture the transposed cable for him.

On the strength of contracts from Peebles, Ferranti and A. C. Parsons, B.I.C.C. decided in March 1958 to undertake the manufacture of continuously transposed strip. In November 1958 the first strips were being run off at the B.I.C.C. plant at Kirkby, near Liverpool, and by Christmas 1958 the first lengths were ready to be delivered for the Peebles Australian contract. The total time from the decision to embark on the cable manufacture to the first delivery was about nine months. According to McDonald, 'I had required of them [B.I.C.C.] a very high performance and there was no doubt that they managed to produce a cable which was for many years unequalled anywhere in the world.'[2]

(c) Comments

The speed with which the production stage was reached is due partly to the fact that much of the necessary technology had been well developed for some time and B.I.C.C. had extensive experience with it. The major design effort concerned the transposing head, which (at least in its final form) has an elegant simplicity.[3] The design team numbered about ten, of whom four were university-trained; a major role seems to have been played by Mr R. Hinds. the chief mechanical development engineer

Essentially, the innovation represents modification of existing processes and no major organisational changes or recruiting and training were required. Except for the enamel coating equipment, no great outlay of capital was involved.

Pressure from Ferranti and Bruce Peebles seems to have come at a high (managing director) level. There is a feeling within B.I.C.C. that when a project comes 'from the top downwards', as it did in this case, progress is smoother because of support in human terms like enthusiasm and encouragement as well as in money and manpower. The Managing Director, Mr (now Sir Ronald) Fairfield, appears to have taken it as his personal responsibility to see that nothing stood in the way of meeting the deadlines.

The transposed strip has become firmly established as a B.I.C.C. product. 'During the first five to six years after the plant was completed, sales increased by 15 per cent per year. In the following three years, the increase rose sharply to 45 per cent per year and it has now levelled out.'[4]

References

1. Interview with V. F. Warr, Works Manager, Winding Wires Division, B.I.C.C., 1968.
2. D. McDonald, Bruce Peeble Co. Ltd, private communication, 1968.
3. British Patent 963,833 (1963).
4. V. F. Warr, private communication, 1968.

6 DAVID BROWN CORPORATION (VOSPER THORNYCROFT): GAS-TURBINE-POWERED FAST PATROL BOATS

Traditionally, the shipbuilding industry has lagged behind most engineering industries in the application of new concepts. Its craft-based structure has sometimes hindered its transition into the present age of technological innovation. The Portsmouth-based shipbuilding firm of Vosper Ltd is, however, a case where an awareness of the importance of seeking and applying technological developments to traditional products is evident. This firm was presented with the Queen's Award to Industry for technological innovation in 1966 in recognition of its development of gas-turbine-powered fast patrol boats (F.P.B.s). The work was carried out on behalf of, and in conjunction with, the Directorate of Naval Construction. Several of the Admiralty's testing and development establishments were involved in the evolution of the gas-turbine-powered F.P.B.'s hull, power plant and transmission systems. In addition, the Admiralty participated in armament development.

(a) The company

Vosper Ltd was founded by an engineer, Edward Vosper, in 1868 at Portsmouth. Among the specialised craft which he built was a petrol-engined launch in which King Edward VII reviewed the fleet in 1904. The company had the further distinction of producing the Royal barge for King George VI which accompanied him on his 1938 tour of South Africa.[1] In the late 1920s a naval officer named Kidston resigned from the Royal Navy and decided to have an 'ideal' fast launch boat designed for naval service. He financed its design and employed Lieutenant Peter Du Cane as a consultant naval architect. Du Cane had been, in succession, a midshipman, sub-lieutenant and lieutenant in the Royal Navy and had taken advanced engineering courses at the Royal Naval College at Keyham and Greenwich. After a time at sea as lieutenant he resigned and it was during the following six months that he became associated with Kidston's project. The boat was constructed by Vospers but the Admiralty refused to buy it.

In 1930, a coal firm in the Camber, Portsmouth – Fraser and White – decided to acquire some additional sea-frontage in Portsmouth. To do so, its Chairman, Major Gilbert, bought out Vospers and installed Du Cane as its Managing Director. Gilbert had little interest in the shipbuilding industry and so he sold the company to Du Cane (who held two-thirds of the shares) and a non-active partner, who held one-third. The firm was subsequently formed into a publicly owned enterprise with Du Cane retaining control through his shareholding. He sold his shares in 1958 to Mineral Separation Ltd, a holding company, and they in turn sold the firm to the David Brown Corporation in 1963. In 1969 the Chairman of Vospers was Sir David Brown and Du Cane was Deputy Chairman. In March 1966 Vospers, employing 1,000 people and worth £1 million, took over the Southampton-based firm of John I. Thornycroft which employed 2,300 people and was worth £3 million. The two firms have retained their identities although many aspects of their operations have been made complementary. In general, Vospers build craft up to 200 ft in length and Thornycroft concentrate on steel-hulled vessels of up to 5,000 tons.

(b) Developments in patrol boat design

The modern F.P.B. has a genealogy extending back to H.M.S. *Lightning*, a motor torpedo boat built in 1876 by Thornycroft.[2] This craft was small and fast, qualities which were to be characteristic of all future F.P.B.s. The operating role for which this type of craft was evolved required the ability to carry out offensive and defensive operations in coastal waters where speed and manœuvrability were essential requisites for survival. A milestone in the development of F.P.B.s came when, in 1936, the Portsmouth-based firm of Vosper Ltd under the influence of its Managing Director, Commander P. Du Cane, produced a 70-ft patrol boat. This prototype was developed for use by the

Royal Navy. Over 360 were built by Vospers and by other firms to Vosper's designs between 1936 and 1946 and they fulfilled a number of operational roles, including those of coastal patrol, air/sea rescue and coastal offence.

The engines for these were initially supplied by British firms such as Thornycroft, and by Packard in the United States. Limited supplies of engines had been obtained from Italy but these were stopped when hostilities broke out between Britain and Italy. The dependence on external supplies of engines for its F.P.B.s, together with the demonstrated fire hazard of equipping lightly-armed patrol boats with petrol engines, prompted Britain's war-time Government to set up, in 1941, a committee under the chairmanship of Sir Roy Fedden to examine the question of providing engines for F.P.B.s and recommend what action should be taken with regard to the development of a British engine. As a result of its deliberations, the Admiralty in 1946 awarded a contract to D. Napier Ltd to develop a triangular configuration diesel engine. This concept, aimed at increasing the power output from an engine whilst reducing its weight and volume, led to the development of Deltic engines. The first Deltic began test running in April 1950 and in September 1953 the first of the production engines was run.[3]

Writing on the development of F.P.B. propulsion, the Managing Director of Vospers said:

Development immediately before and during the war brought the top speed of naval planing craft capable of carrying a useful weapon load to about 40 knots. Reciprocating petrol engines were still the main type of propelling power units, although many of the German war-time fast naval small craft had diesel engines. In larger naval ships, still largely relying on steam turbines, a speed of about 40 knots was again the upper practicable limit, and most warships having any pretension to speed had top speeds of 32–36 knots. After the war continued development, notably improved propeller design, made possible speeds of some 42 knots with petrol engines, on much the same displacement as the war-time craft had.[4]

F 2

Gas-turbine developments. Even with the development of highly rated reciprocating engines such as those by Packard, Napier and Mercedes-Benz, the possibility of achieving operating speeds of 50 knots was extremely remote and so the gas-turbine engine became the focal point of an intense programme of development by the Admiralty from 1942 onwards. It offered the promise of greater power for less weight and less bulk than any other potential marine engine. On 14 July 1947 the world's first marine gas-turbine-powered motor gun-boat (M.G.B.) was demonstrated at Portsmouth.[5,6,7] This was designated M.G.B. 2009 and was powered by two 1,250-b.h.p. Packard reciprocating petrol engines and a 2,500-b.h.p. Metropolitan Vickers Gatric-G1 gas-turbine engine. The appearance of the craft was compared in significance to the first demonstration of the S.S. *Turbina* – the first ship to be powered by steam turbines – when it appeared at the Spithead Naval Review of 1897.

The reasoning behind the Admiralty's interest in gas-turbine installations was outlined in an Admiralty press announcement[8] in January 1947. It was expected that

a gas turbine plant especially developed for naval use is likely to have the following advantages over steam turbine machinery:

(i) Probable reduction in weight and space for a given horsepower, with eventually a gain in overall efficiency, allowing greater radius of action or more weight for weapons or armour.

(ii) Less time required for starting machinery from cold.

(iii) When satisfactorily developed, gas turbine machinery is likely to be less complicated and less vulnerable than steam machinery.

The announcement also mentioned that 'the information gained in the development of jet propulsion engines has been made available by the Ministry of Supply to the Admiralty which has taken steps to interest firms other than those engaged in the aircraft industry in gas turbine work for warships. Some aircraft firms by agreement with the Ministry of Supply are forming marine wings.' The National Gas Turbine Research Establishment was to co-operate in the development of the marine units. In view of subsequent developments it is of interest to note that the Admiralty 'emphasised that marine gas turbines present problems which are not encountered in aircraft practice and that generally speaking such units are not suitable and cannot be used for marine purposes'.

Following the successful demonstration of M.G.B. 2009, the Admiralty ordered a series of extended trials and by 1949 over 400 hours of testing had been completed. As a result of these tests, contracts were awarded to four firms for the development of naval gas turbines in the 5,000-b.h.p. class. These were Metropolitan Vickers, English Electric, Rolls-Royce, and W. H. Allen, Sons & Co. Ltd.

Metropolitan Vickers developed a successor to the Gatric engine of M.G.B. 2009 which was fitted to two F.P.B.s in conjunction with Deltic diesel engines.[9] The English-Electric project was terminated when it became apparent that the Rolls-Royce gas turbines, which were lighter in weight, showed greater potentialities.[10,11] W. H. Allen undertook the development of a gas turbine and base-load generator.[12] Although the turbine was not specifically designed as a propulsion unit, it is of interest here for two reasons. Firstly, it could have been used as a prime mover to drive a generator/motor system.[13] Secondly, it marked the beginning of a collaborative venture between W. H. Allen and the Bristol Aeroplane Co. Ltd which served to introduce the latter to the problems and requirements of marine gas turbines. A consultancy agreement was made which entitled W. H. Allen to design data, advise on production methods and the use of certain testing facilities.[14]

(c) The Vosper innovation

Stimulated by the severe economic difficulties which Vospers went through in the years 1950–4, Commander Du Cane set about to obtain further work for the company. He had available a staff of approximately five hundred people skilled in the production of F.P.B.s and there were few opportunities to use these skills for other work. He approached the Admiralty at a high level with the following suggestions:

(i) It should re-examine its present policies which were relying on the successful development of a controllable-pitch propeller to permit high-speed diesel engines such as the Deltic to be used efficiently in F.P.B.s. In these craft, large variations in load carried and operational speed are the norm and so a fixed-pitch propeller mechanically coupled to the engine, as is the case with the diesel engine, is less efficient than a similar propeller indirectly coupled to the engine as in the case of the gas-turbine where a separate power turbine is provided to drive the propeller.

(ii) It was an opportune moment to embark on the construction of an advanced F.P.B. that would incorporate the lessons learned since 1944 in the construction of such craft.[15]

Following these suggestions, a meeting was held in March 1954 between representatievs of the Admiralty headed by W. G. John, Assistant Director of Naval Construction, and of Vospers, headed by Commander Du Cane. The general requirements which were to be met by an advanced F.P.B. were outlined. A contract was awarded to Vospers by the Admiralty on 3 June 1954 for a design study listing possible configurations of the craft and stating what the expected performance of each type would be. The most important of the Admiralty's requirements for the F.P.B. were given as:

(i) The boat to be of hard chine type and as small as possible commensurate with sea-keeping and speed requirements.

(ii) The top speed to be the maximum obtainable and not less than 44 knots in the gunboat version with half fuel consumed; this speed to be maintained for a quarter of an hour in waves 3 ft high, the ultimate aim being to produce a boat capable of 50 knots under these conditions.

(iii) Slow speed to be not more than 12 knots (8 knots preferred) and capable of being maintained for four hours. Harbour manœuvring at 4–6 knots required.

(iv) Armament to be suitable for quick conversion to the role of torpedo boat, gunboat, minelayer or raiding craft.

(v) Complement to be twenty-two persons.

To meet these and the other requirements, three engine configurations were proposed by the Admiralty and one by Vospers. They were:

   (i) two 5,000-b.h.p. Metropolitan Vickers G4 gas turbines;
  (ii) three 2,500-b.h.p. Napier Deltics;
 (iii) two Napier diesels, each compounded with a gas turbine;
 (iv) Vospers' proposal: three Bristol Proteus-type 1250 gas turbines.

The Proteus engine had been developed from the Bristol Theseus[16,17,18] of 1946 and was intended to power the Bristol Brabazon Mark II aircraft and the Saunders-Roe S.R. 45 Princess flying-boat with ten engines in six installations. Both of these projects were cancelled shortly after the prototypes had flown (the Princess completed 100 hours of test flying on derated engines after its first flight in August 1952 before being scrapped[10]). Subsequently, the engine was used to power the Britannia aircraft.

After careful study, all concerned came to the conclusion that the optimum arrangement was represented by an installation consisting of three Proteus marine gas turbines in conjunction with Allen's reverse reduction gear drives. Basically the case for the gas turbine as a prime mover rests upon the very high power to weight ratio available combined with the torque output characteristics inherent in the use of the free power turbine. The latter feature is of outstanding importance for the case of the high-speed planing type of craft where it is necessary that the machinery output in the form of available torque must always exceed the requirements of the hull at any particular load, trim, or speed. Piston engines possess a characteristic torque/speed curve which is rather sensitive to overloading. For an increase in hull displacement of 20 per cent, the reduction in power available from the diesel is nearly 30 per cent whereas in the case of the turbine it is negligible. For this reason one can design the propeller for the turbine to absorb full power at the lightest condition, while for the diesel one must design the propeller for the heaviest and most resistful condition likely to arise due to fouling, etc. This feature in itself can account for a difference in speed available from a given horsepower of anything up to 15 to 20 per cent.[20]

Some important problems had to be resolved as a result of this decision to use gas-turbine engines for propulsion, notably that of reducing the rotational speed of the engine (11,000 r.p.m.) to that suitable for the propeller (1,700 r.p.m.) by means of some gearing system and that of ensuring that the engines were protected from the effects of salt-water spray in the intake air.

Gearbox development. The gearbox adopted for the gas-turbine F.P.B. project (the *Brave* class) was an epicyclic reverse reduction unit designed by W. H. Allen, Sons & Co. Ltd of Bedford according to the principles outlined by Dip. Ing. W. G. Stoeckicht of Munich in his patent[21] of June 1931. Conventional gearboxes have to be made both rigid and accurate in order to function properly. Inaccuracies in manufacture show themselves in gear failures and with high-power transmissions the required accuracy increases although the size has correspondingly increased and made this a more difficult requirement. Stoeckicht's concept was simple. If there was a danger of gear failure due to irregularities in the gearset, then the solution was to make the set flexible enough to deflect away from the irregularity without failing.

> Stoeckicht was undoubtedly one of the world's outstanding engineers. He had the rare attribute of being able to combine exceptional technical ability with sound business acumen. He was also something of a diplomat. In engineering history he will take his place as the father of high-power epicyclic gears for industrial and marine use.[22]

In 1946, W. H. Allen undertook the development of epicyclic gear trains on behalf of the Admiralty.[23] It was obvious that one of the applications for which this type of gear should be developed was as the reduction gear for a gas-turbine engine where the high-speed reduction ratio obtainable would be most advantageous in saving space and weight. Allens inspected Stoeckicht's designs, had tests carried out on them by Mr R. J. B. Keig of the Royal Naval Scientific Service and, on the basis of the encouraging report received, concluded a design consultancy agreement with Stoeckicht.[24] By 3 June 1949 Stoeckicht had further enhanced the high-power capabilities of his gearsets by introducing the concept of double helical gears.[25] In this, two helical gears were permitted to flex relative to each other to ensure that they carried equal loads while at the same time doubling the load capacity of the gearset. This type of unit was used in the 1,000-kW gas-turbine generator built in 1950 for the Admiralty[14] and it formed the basis on which the company received a contract in 1951 from the Admiralty to develop an epicyclic primary gear (11,400 r.p.m. to 5,000 r.p.m.) and a reverse reduction gear (5,000 r.p.m. to 1,720 r.p.m.) for use in the *Brave*-class boats.

After some initial difficulties had been overcome with regard to gearbox bearings, the gears were fitted to the *Braves* in 1958 and tested under service conditions. They had, however, two main disadvantages: they were expensive and heavy, the latter due in part to the fact that 'the gearboxes were originally designed for an output shaft speed of 850 r.p.m. at full speed. Subsequently this output speed was changed to 1,720 r.p.m. which involved the gearbox designers in a reduction to 850 r.p.m. and then a further step up to 1,720 r.p.m.'[20]

Air intakes. A gas turbine such as the Proteus consumes vast quantities of air when in operation. At full power the combined air consumption of the three engines fitted to the *Braves* is 200 tons per hour. It was therefore necessary to ensure that the engines were adequately protected from the ship's spray, etc. The solution adopted for this was provided by Mr J. T. Revans, of the Director of Naval Construction's office:

One of the major troubles experienced had been the entry of sea water into the engine room and engine intakes. The work which was undertaken to solve this problem served as a good example of the teamwork which went into the whole *Brave*-class project. We made some scale experiments with water separation and sketched various intake arrangements, but these looked like hen-coops erected on the deck. Then one day, Mr Revans came into my office, followed by two men carrying something which at first I thought was a coffin but which turned out to be a rough wooden model of the *Brave* on which he had built in plasticine an air intake arrangement which looked right and which was the type that was adopted. His arrangement not only provided a good intake but also a first-rate crew shelter. So far from being noisy, as was feared by some, it has proved to be the place where people will go for a quiet smoke or to eat their sand-wiches during trials.[31]

Propeller design. The propeller is a vital part of any ship design and is a particularly difficult component to design for F.P.B.s, where its performance must be optimised for high speeds:

> When the gas turbine became available about 1950, it was clear that to take full advantage of the high powers available from this source the propulsion arrangements would be of paramount importance. After consultation with Dr Gawn and others experienced in the field, we decided to install a cavitation tunnel in our Portchester shipyard with the object of exploring the high-speed case. Much encouraged by the Director of Naval Construction of the day, who instructed us to carry out a methodical series of tests in this range, we have now gained invaluable experience which has done much towards enabling the latest fast patrol boats of the *Brave* class to exceed 50 knots on trials.[26]

It is of interest to note that this privately operated facility duplicated the cavitation tunnel facilities at the Haslar Experimental Works of the Admiralty which had been under the direction of Dr R. W. L. Gawn since 1938. The Admiralty had been testing propellers since 1873, the year in which William Froude carried out his first experiments on behalf of the Admiralty.[27]

To staff the new experimental facility, Du Cane recruited a German hydrodynamicist from the Haslar works, Mr H. P. Rader. He had been brought to England to assist in the develop-ment of hydrofoil craft but at Vospers he turned his attention to the design of supercavitating propellers. Much of this experimental work was reported[28] by Rader and R. N. Newton, Superintendent at Haslar, in October 1960. Rader returned to Germany and was replaced by Dr Kruppa who came to

Vospers from Berlin University. He too returned to Germany to take a university chair in hydrodynamics and was in turn replaced by Dr Klaus Suhrbier from Hamburg University.[15]

Drag reduction. Allied with the problem of optimising propeller form for maximum efficiency was the problem of reducing hull drag. A significant development in this was made by applying a technique of mounting the propeller on its shaft without the use of keys (requiring slots in the propeller and shaft). This method had been developed by the S.K.F. Bearing Company Ltd, Luton. It consists of expanding the propeller by means of oil pressure in a pair of wedge-shaped sleeves. When the pressure is released, the whole assembly (shaft, sleeves and propeller) is locked solid due to the contraction of the propeller.

Hull form and construction. In Commander Du Cane's words: A year or two prior to the 1939 war, we were authorised by the Director of Naval Construction of the day, Sir Stanley Goodall, to undertake what really amounted to a methodical series [of tests] to arrive at an optimum form for a motor torpedo boat, having regard to the stiff requirement for speed to be combined with sea-keeping ability. A number of model hull forms were first tested outside our shipyard in Portsmouth Harbour, towed by a speed boat at Froude Scale speed from an outrigger. The best of these from the point of view of steady running and performance in waves were then tested in the Haslar tank by Dr Gawn. Finally . . . we constructed and ran in the Solent or at sea a manned scale model about 25 ft long, incorporating machinery capable of producing 'scale' speed in waves.[26]

As a result of war-time experiences a prototype hull was built by Vospers in 1944 for the motor torpedo boat 538. Only one of this type of hull was made because of the war-time requirement that F.P.B. hulls and equipment should be interchangeable for maximum operational effectiveness, but it was subsequently adopted as the basis of the *Brave* hull. The higher speeds attainable with the *Braves* (over 50 knots) showed the necessity for increasing its performance with regard to slamming in head seas. An improved version of the hull was developed and used for later F.P.B.s, including the smaller

Ferocity class, powered by two Proteus engines combined with two diesel engines for long-range cruising. Instead of the composite aluminium hardwood construction used for the *Braves*, a cheaper and simpler laminated wood construction was used for the *Ferocity*. This avoided the necessity of using stainless steel structural bolts for corrosion resistance and the problem of the difference in elastic moduli of the aluminium and timber tending to overload the metal whilst underloading the wood. In addition, the all-wood construction meant that instead of having to shore-up the planking and decking while the structural adhesive set, barbed nails could be used to hold the planks in position. These nails were not load-bearing structurally but saved over £2,000 per boat in construction costs.[29, 30]

(d) Comments

The stimulus to undertake the development of the *Brave*-class F.P.B. was provided by the severe economic difficulties in which Vospers found itself after the 1939–45 war. It was oriented almost exclusively towards the production of launches, air/sea rescue craft and patrol boats for the Royal Navy and so was particularly vulnerable to cuts in military expenditure. For the three years prior to 1952 the company declared no dividend and it became obvious that the company would have to place more emphasis on the problem of diversifying its activities. With any specialised firm, the consequences of a lull in orders is generally more serious than would be the case with a broadly-based organisation because of the natural reluctance of the firm to lose the specialist skills it has paid to generate in its employees.

For Vospers, the lull in orders was particularly serious. There was no labour 'pool' to which they could dismiss their highly skilled employees and from which they could have subsequently recruited them. Anyone who left the firm was likely to find permanent employment only in an industry other than precision shipbuilding and so a surplus labour force was retained by the firm until large-scale work was begun on the *Braves* in 1954. At that time the firm had approximately 500 employees. Since then this number has grown to approximately 1,000, part of which is accounted for by the firm's activities in other fields.

Although the need for diversification in the company's products had been recognised for some time, its practical realisation was a process occupying twenty years. The company has developed a series of ship-stabilisation schemes of the non-retracting type and had sold over 250 of these by 1967. In addition to the one-, two-, and four-fin installations which it had been marketing for some time, the company developed a mini-fin installation suitable for yachts of up to 40 ft. These units were developed by the Control Systems Department of Vospers which was originally concerned with the development of rudder controls, etc. A more marked form of diversification has been the utilisation of the firm's considerable experience in the field of wood and aluminium fabrication to produce catering equipment, furniture, fuel tanks and water tanks both for ships (including the *Queen Elizabeth II* liner) and for land installations. The Vosper-Thornycroft Group obtained government sanction and 'official encouragement' at the end of 1967 to enter the field of hovercraft production. In March 1968 construction was started on a 45-ft scale model of the firm's proposal – the VT1 gas-turbine-powered water-propelled hovercraft ferry. The manned model was being tested by June 1968, and the full-scale VT1 craft began its trials in 1969. The VT1 project marked the first attempt to bring to hovercraft production the techniques of precision shipbuilding as opposed to the aircraft engineering standards adopted by other manufacturers to the detriment of their products' costs.[32]

In the past, many of the key positions in Vospers were occupied by ex-naval officers. This is not unconnected with Commander Du Cane's observation:

It is probably true that the best and most seaworthy craft are produced by those who are themselves practical seamen. The observant individual who has a superficial knowledge of the dynamics of control surfaces, etc., will, unfortunately, usually find it difficult to convey his meaning to the naval architect.[33]

It is interesting to note that, in all, six engines (five gas turbines and one diesel) were developed by the Admiralty for naval service in the years immediately preceding the design of the *Braves*. None of these was adopted for the 50-knot F.P.B.

that the Admiralty required and instead an aero engine, the Proteus, which had been proposed to the Admiralty by Du Cane, was accepted to power the F.P.B. Subsequently two further engines built for the aero-engine market (the Olympus by Bristol Aero Engine Co. and the Tyne by Rolls-Royce) were adopted as power plants for the Navy's warships. The personal commitment which Du Cane had to the concept of gas-turbine-powered F.P.B.s, together with the authority he could wield over his firm, were vital factors in the successful development of the *Braves* and later of the smaller *Ferocity*-class F.P.B. This work was done at all times in association with the Admiralty (through the Department of the Director of Naval Construction) who paid for the work done by Vospers, W. H. Allen, and the Bristol Aero Engine Company and provided the use of test facilities such as those at the Admiralty Experimental Works at Haslar, the Naval Construction Research Establishment, Rosyth, and the Admiralty Engineering Laboratory at West Drayton (London). F.P.B.s similar in general design to the *Brave* or *Ferocity* classes have been sold by Vospers to Libya, West Germany, Sweden, Denmark, Greece, Malaysia and Brunei in addition to those sold to the Royal Navy.

Commenting on the use of patents by Vosper Ltd, Commander du Cane wrote:

> We have patented very few items and then mostly of a minor mechanical or electrical aspect of F.P.B. design and construction. This however would probably not apply to manufacturers of turbines, gearboxes, or installation items such as radar and armament.[30]

References

1. *The Engineer*, **183** 213 (1947).
2. I. Cooper, 'Company Profile – Vosper Thorneycroft', *The Times Review of Industry and Technology*, Sep 1967.
3. A. R. M. Oliver of D. Napier and Son Ltd, private communications, 10 Oct 1968 and 13 Dec 1968.
4. J. Rix High Speed Craft for Naval and Commercial Duties', p. 53, British Shipbuilding Today Supplement to *Motor Ship*, Nov. 1967.
5. 'Gas Turbine Propelled M.G.B. 2009', *The Engineer*, **184** 218 (1947).
6. 'Gas Turbine Propelled M.G.B. 2009', *The Engineer*, **185** 622 (1948).
7. 'Naval Development of the Gas Turbine', *The Engineer*, **185** 144 (1948).
8. 'Admiralty Press Announcement', *The Engineer*, **183** 126 (1947).
9. 'Review of Naval Developments 1953', *The Engineer*, **197** 32 (1954).

10. G. F. A. Trewby, 'British Naval Gas Turbines', *Trans. Inst. Marine Engineers*, **66** 126 (1954).
11. 'Gas Turbines for the Royal Navy', *The Engineer*, **194** 873 (1952).
12. 'Allen 100 kW Generating Set', *The Engineer*, **188** 562 (1949).
13. 'Gas Turbine Alternator Propulsive Machinery for Tanker "Auris"', *Engineering*, **171** 209 (1951).
14. '1000 kW Marine Gas Turbine Set', *The Engineer*, **192** 614 (1951).
15. Interviews with Commander P. Du Cane, Deputy Chairman, Vosper-Thornycroft Ltd, 12 Sep 1968 and 18 Dec 1968.
16. 'Bristol "Theseus" Propeller Turbine', *The Engineer*, **182** 258 (1946).
17. 'Lincoln II Bomber with Propeller Turbine Engines', *The Engineer*, **183** 517 (1947).
18. 'Bristol "Proteus" Propeller-Turbine Aero-Engine', *The Engineer*, **188** 260 (1949).
19. H. Knowler, 'Flying Boats', *Journal of the Royal Aeronautical Society*, **70** 137 (1966).
20. P. Du Cane, 'The Development and Running of the "Brave" Class Fast Patrol Boat', Symposium Swedish Inst. of Naval Architects (Stockholm, 1960).
21. W. G. Stoeckicht, German Patent No. 556,683, 24 June 1931.
22. T. P. Jones, 'Fifteen Years Development of High Power Epicyclic Gears', *Trans. Inst. Marine Engineers*, **78** 273 (1966).
23. H. N. G. Allen and T. P. Jones, 'The Application of High Powered Epicyclic Gearing for Industrial and Marine Use', Paper to Engineering Institute of Canada, May 1960.
24. 'Epicyclic Gearing', *Allen Engineering Review*, no. 23 (Jan 1950) 3.
25. W. G. Stoeckicht, British Patent 660,497. Applied for 3 June 1949, published 7 Nov 1951.
26. P. Du Cane, 'Research in a Small Shipyard', *International Design and Equipment* (1961).
27. R. W. L. Gawn, 'The Admiralty Experimental Works, Haslar', Paper No. 1, Autumn Meeting of Inst. of Naval Architects (Torquay, 1954).
28. R. N. Newton and H. P. Rader, 'Performance Data of Propellers for High Speed Craft', *Trans. Royal Inst. Naval Architects*, **103** 93 (1961).
29. 'Fast Patrol Boats for Malaysia', *Engineering*, **204** 11 (1967).
30. P. Du Cane, private communication, 4 Nov 1968.
31. A. A. C. Gentry and J. T. Revans, 'The Brave Class Fast Patrol Boats', *Trans. Royal Inst. Naval Architects*, **102** 367 (1960), a contribution to the discussion by B. G. Markham of Bristol Aero Engine Co. Ltd.
32. Interview with Mr D. Blower, Shipyard Manager, Vosper-Thornycroft Ltd, 26 July 1968.
33. P. Du Cane, *High Speed Small Craft* (Temple Press, London, 1st ed., 1951; 3rd ed., 1964).

7 CAMBRIDGE SCIENTIFIC INSTRUMENTS: STEREOSCAN ELECTRON MICROSCOPE

The role of university research in stimulating technological innovation has received much attention from both government and industry. The former are

concerned that some of the results of university research be turned to the common good, while the latter
seek, in scientific and technological activities, new
opportunities for investment.

This case study is an illustration of an innovation
whose technological component derived from contact
with a university laboratory. It also shows that opportunities for innovation occur in unwonted places and
how extensive may be the company reorganisation
required to deal with new product lines.

(a) Scanning electron microscopy[1,2]

Scanning electron microscopy is based on the principle that
electrons are emitted by a solid when a beam of high-energy
electrons impinge on it. These secondary electrons contain
information about the nature of the emitting surface and can be
collected and used to form an image of the surface. One of the
advantages of this method over the more conventional transmission electron microscopy is that specimen thickness is not a
limiting factor, and there is no electron optical system located
behind the specimen, leaving a large space for the specimen and
specimen manipulation. In terms of resolution, the best obtainable using a scanning electron microscope (S.E.M.) is of the
order of 50 Å to 500 Å, somewhat less than with a transmission
microscope, although instruments have been developed with
resolutions of the order of 3 Å using field emission guns
($1 Å = 10^{-8}$ cm).

In a S.E.M., electrons from an electron gun are accelerated
under vacuum along a column. This beam is focused by means
of electromagnetic lenses and directed on to the specimen.
Interaction of the electrons with the surface of the specimen
releases a spray of secondary electrons which are collected in a
detector. The resulting current is amplified and used to modulate
the brightness of a spot on the face of a cathode ray tube. The
electron beam is made to scan the surface of the specimen in
synchronism with the spot on the cathode ray tube. As the beam
moves over the specimen, varying numbers of electrons are
emitted depending on the nature of the specimen surface.
Consequently, the brightness of the spot on the cathode ray

tube varies according to the features of the specimen, while the location of the spot on the screen is correlated with the position of the electron beam at any time. As a result, an image of the surface is built up on the face of the cathode ray tube, in a manner very similar to that in a television set.

The S.E.M. has many uses but should not be regarded as a direct competitor with the higher-resolution transmission microscopes. Its main advantages are seen in situations where preparation of thin specimens is inappropriate or difficult – for example, in fibre technology or metallurgy. In addition, it has been found that the S.E.M. can be used to detect potential variations across surfaces, or variations in charge collection inside the specimen, and this has had important application in determining the integrity of various semiconductor devices.

(b) The Cambridge Scientific Instruments innovation[3]

The Cambridge Scientific Instruments company (C.S.I.) innovation involves the interaction of two distinct but complementary aspects. On the one hand, C.S.I. itself has undertaken considerable development since the Second World War; on the other, the activities of the Electrical Engineering Department of Cambridge University in electron optics contributed greatly to the technical aspects of the 'Stereoscan' which C.S.I. subsequently marketed.

To explain the origins of Stereoscan, it is necessary also to describe the earlier developments of a scanning instrument of a different type. In the early 1950s, C.S.I. began development of a new product line – Microscan, a scanning electron probe X-ray microanalyser. This instrument was based on work done in the Cavendish Laboratory by Dr P. Duncumb under Professor V. E. Cosslett, and on subsequent work carried out at Tube Investments Laboratory by Dr D. A. Melford. The Microscancombined the scanning techniques developed in the Cavendish Laboratory at Cambridge with the static electron probe X-ray microanalyser which had just been developed by Professor Castaing and others in France.

The company. The C.S.I. company was founded in 1881 by Sir Horace Darwin. The firm began as a one-man enterprise making scientific measuring instruments for the Cambridge University laboratories. At this time, the university laboratories had only small workshop facilities, and as these grew, C.S.I. provided a source of manpower. From very early days, then, there have been close ties between C.S.I. and Cambridge University. For example, in 1887 C.S.I. offered for sale a servo-pen recorder designed by Professor Callendar of Cambridge University, which sold well for many years and was the first potentiometric pen recorder in the world.

Between 1886 and 1955, the company developed considerable skill in electrical and precision mechanical engineering, and had established a reputation as makers of precision scientific instruments as well as the more robust instruments needed for industry. Over this period C.S.I. acquired a diverse product range and, although they did not have a dominant share of any one market, there was scarcely a type of instrument with which they had no experience. The main skills in the company were related to precision mechanical engineering and precise low-frequency electrical measurements.

But the technology on which C.S.I. was based involved highly skilled use of passive elements – resistors, capacitors, induction coils, etc. – and it was widely felt that this technology was rapidly being superseded by electronics, an area with which they had only a slight acquaintance. In general terms, then, the company was interested in acquiring an electronics-based product, but the development was unlikely to come entirely from within since the firm had little expertise in the new technology. During the late 1950s such a product appeared in the field of electron optics. This instrument was the X-ray scanning microanalyser, Microscan.

The Microscan innovation. The development of Microscan dates from about 1947 when Hillier in the United States stated that impurities in metals could be detected by the wave-length of emitted X-rays. In the early fifties, Castaing in France made a more sensitive electron probe but did not employ any scanning techniques.

At the Cavendish Laboratory of Cambridge University one of Professor Cosslett's Ph.D. students, P. Duncumb, had in 1953 acquired the use of an old Radio Corporation of America

(R.C.A.) transmission microscope. He adapted it for scanning and the measurement of the emitted X-rays. This instrument may be regarded as a prototype microanalyser, although it was by no means in a suitable form for the commercial market. It was intended to be used as part of Duncumb's experimental apparatus to examine the properties of matter and was based on the principle that each material gives off a characteristic spectrum of X-rays when bombarded with energetic electrons.

While this work was under way, Duncumb received a query from Dr D. A. Melford of Tube Investments (T.I.). At T.I., work was under way using a transmission microscope to study various problems in metallurgy. In particular, one metal consistently seemed to exhibit an array of very fine cracks and the group at T.I. were anxious to find the reason. The transmission microscope itself could not give the topographic information required. Melford at T.I. had heard of Duncumb's work and requested permission to use his instrument. When the sample was put into the instrument, it was found to contain a high concentration of non-ferrous metals along the fissures. It seems that the concentration was high in the region of the fissures because the iron had oxidised and that this had caused the splitting. It was subsequently discovered that these steels were made using a quantity of scrap containing copper and chromium impurities.

Duncumb's X-ray scanning microanalyser thus provided T.I. with a valuable instrument. Subsequently Cosslett built a similar machine for their laboratory. As soon as T.I. began to publish papers using results obtained with the instrument, other firms began to enquire if they could obtain one.

T.I. were not really interested in becoming instrument makers and, as we have noted above, C.S.I. were looking for a new, preferably electronics-based, product. T.I.'s Research Director, Mr T. Hughes, and C.S.I.'s Managing Director, Mr H. C. Pritchard (both of whom were Welsh), were friends, and Hughes offered Pritchard the manufacturing option on the Cosslett instrument, which, after some discussion, C.S.I. accepted.

Little remained to be done other than development for production and within a year they sold their first machine to the Atomic Weapons Research Establishment at Aldermaston in

May 1960. By December 1965 about 100 Microscans had been sold at approximately £16,000 each. For C.S.I., Microscan was a new departure in that it was an electronic instrument requiring different types of skills from those employed previously. In addition, the selling price of the instrument was about sixteen times the firm's average product price.

Microscan was very successful and sold well for four or five years. During that time the company had, through Microscan, become familiar with the potentialities, both technical and commercial, of a scanning-type instrument. Production was saturated to meet demands for Microscan and more than 50 per cent were exported, largely to the American and German steel industries.

The Stereoscan innovation. The original idea of using secondary electrons to produce an image is not new: Knoll in 1935 took out a patent on a scanning electron beam instrument. However, techniques in electronics were not sufficiently advanced and consequently development moved slowly. By 1942 von Ardenne and a group at R.C.A. Laboratories under Zworykin had both produced scanning microscopes, but these were limited by the techniques available at that time and the subject was allowed to drop.[4]

In 1952, however, Professor Oatley and D. McMullan, a Ph.D. student, both of the Department of Electrical Engineering at Cambridge, were able to produce a working instrument using some radar techniques developed during the war. Coming to electron optics from radar technology, Oatley was perhaps less influenced than some of his colleagues by the general trend towards high-resolution electron microscopes.[5] He seems to have looked at the problem of microscopy in an essentially different way. Instead of focusing his attention on high resolution, Oatley argued from first principles in trying to build up a picture from information deriving from point sources. The point sources were the sprays of secondary electrons emitted from each area of the sample as a result of bombardment of the sample by energetic electrons. This was 'radar thinking' and led away from more conventional approaches to the problem of microscopy. Thus Oatley arrived at the principle of the S.E.M. at a time when others had little reason to envisage the possibility of reconstituting the secondary electrons from the sample to form a pattern resembling the sample.

The design of a second S.E.M. was initiated in the Department of Electrical Engineering at Cambridge University by another of Oatley's research students, K. C. A. Smith. Smith's instrument was subsequently sold to the Canadian Pulp and Paper Research Institute in Montreal. As the Engineering Department had no facilities for selling or transporting the instruments, Oatley chose to arrange these details with A.E.I. In addition, seeing the commercial promise of the instrument, he asked A.E.I. if they could undertake its manufacture. This they agreed to do in 1953–4 but they seem to have found difficulty in obtaining orders for the instrument. Possibly this was due in part to the fact that they were heavily committed to a healthy and expanding transmission microscope market and had little time or manpower to devote to opening the markets for the scanning electron microscope. Oatley, disappointed by the lack of success achieved by A.E.I., turned in 1961 to C.S.I. who because of their experience with Microscan agreed to manufacture the S.E.M. as soon as the existing agreements with A.E.I. could be terminated.

In 1962 Du Pont in America, who were about to embark on a large programme in artificial fibres, sent out a specification for an instrument closely resembling Oatley's S.E.M., but there was no one manufacturing such an instrument at this time. Du Pont's interest was an additional stimulus to C.S.I.'s efforts to extend their Microscan expertise to the related but different field of scanning electron microscopy. C.S.I. were experiencing difficulties because the scanning microscope is an instrument involving higher precision than a microanalyser. The original Microscan had been based on a design worked out elsewhere, whereas with the scanning microscope there was no suitable prototype design to copy for commercial production. The company consulted with the Department of Engineering on technical matters and, in addition, recruited new staff, including A. D. G. Stewart who had been working on scanning microscopes with Professor Oatley. He developed a new design of instrument suitable for commercial sale which was eventually called the Stereoscan.

(c) Comments

Operational requirements. While it seems fair to say that the invention of the Stereoscan had its origins in university research, the commercial success of the product depended on several additional factors. From the sales point of view, Oatley's instrument was in need of considerable development. The development team headed by Stewart attended to many details, small and large, which were necessary to make the equipment acceptable to eventual users. 'In particular, great emphasis was placed on simplicity of operation and use. By comparison with transmission electron microscopes, which require expertise both in preparing specimens and in operating the instrument, Stereoscan is simple to use and in the majority of cases requires no special specimen preparation techniques. The two-year development period seems to have been spent as much on building in customer appeal as on ironing out technological bottlenecks. The initial size of the development team numbered three or four, all of whom had university training in science or engineering. In addition there was a back-up staff of about ten.

Relationship with the university. The relationship between C.S.I. and Cambridge University has changed over the years since the foundation of the company. In the early days of the company C.S.I. was set up partly to supply the university with precision measuring equipment. At this time most universities did not have extensive workshop facilities and so C.S.I. filled a real need. In parallel with the growth of university science, technical services were also augmented with the result that C.S.I. became a source of specialised manpower to fill vacancies in the university's technical departments. More recently, however, C.S.I. have drawn many of their highly skilled personnel from the university.

As an instrument company C.S.I. is especially dependent on new inventions and, since it is likely in the field of advanced instrumentation that research laboratories will provide an important sales outlet for the new equipment, the company has kept a close liaison with the university at all levels. It is interesting to note that, although C.S.I. employees generally regard the university as their 'hidden research laboratory', there are no formal arrangements for exchange of ideas or information between the two institutions.

New management. The movement away from more conventional electrical engineering products towards electronics-based ones seems to have been initiated by some major organisational changes. In 1954 Dr Percy Dunsheath was appointed, at a relatively young age, as a member of the board. This in itself was somewhat of an innovation since prior to this it seems that such appointments were made largely on the basis of seniority. Dunsheath in turn hired Pritchard, an Oxford mathematician, and others, and he seems to have given them the task of finding new products, based on modern technology, for the company. Although there appear to have been certain organisational difficulties with the new regime, it seems clear that they created an environment conducive to innovation. In the opinion of some, the atmosphere created by Pritchard allowed the company to take risks on Microscan, and to a lesser extent on Stereoscan, that would not have been allowed under a more rigid management structure.

In July 1968 C.S.I. was taken into the George Kent Group who appear to have adopted a more market-oriented approach than previously. Sales of Stereoscan as well as Microscan have continued to increase as novel uses for these instruments have appeared. After its Queen's Award for technological innovation in 1967, C.S.I. won an Award in 1968 for exports, a large percentage of which comprised sales of Microscan and Stereoscan.

References

1. C. W. Oatley *et al.*, 'Scanning Electron Microscopy', *Advances in Electronics and Electron Physics*, **21** 181 (1965).
2. The author would like to thank Mr A. Thornley for his help in preparing this section.
3. Interviews with Mr J. Morgan, Mr J. G. Hammond and Mr L. A. C. Dopping-Hepenstal, Cambridge Scientific Instruments Ltd, 1968.
4. G. Möllenstedt and F. Lenz, 'Electron Emission Microscopy', *Advances in Electronics and Electron Physics*, **18** 251 (1963).
5. D. McMullan, 'An Improved Scanning Electron Microscope for Opaque Specimens', *Proceedings of the Institute of Electrical Engineers*, **100** (2) 245 (1953).
6. T. E. Everhart and R. F. M. Thornley, 'Wide-band Detector for Micro-microampere Low-energy Electron Currents', *Journal of Scientific Instruments*, **37** 246 (1960).

8 COLCHESTER LATHE CO.:
METHODS OF LATHE MANUFACTURE

The Colchester Lathe Company won the Queen's Award in 1967 for technological innovation and export achievement and for the latter only in 1968. The award for technological innovation was concerned in general with the company's methods of lathe manufacture and assembly and specifically for successful development of its hydrostatic 'Flowline' method of assembly of lathes. The firm's success in obtaining very large orders for its products, such as the sale of over 200 lathes for training purposes to the Government of Thailand, and its sustained growth of exports to a level of over 50 per cent of its production earned for it the export achievement Awards.

(a) Background[1,2]

The Colchester Lathe Company was founded as a private firm in 1907 to produce a range of machine tools including centre lathes. By as early as 1908 the decisions had been made to rationalise production and concentrate efforts on the production of a range of centre lathes. This is of interest, for in departing from the then normal practice of the machine-tool industry, which was for firms to produce a full range of types and sizes, it foreshadowed the rationalisation efforts of the industry so evident in the 1960s.

In 1908 the firm was producing lathes at the rate of approximately 100 per year and its output grew steadily until it stood at 20 lathes per week in the late 1940s. The founder of the firm died in 1945 but the company remained small and privately owned until 1948 when Mr A. J. Hayward was appointed as Works Manager. He had occupied a similar position in the Wellworthy Company which specialised in the mass production of pistons and piston rings for the motor industry. Shortly after this, Mr P. A. Long joined the firm as Assistant Works Manager. In 1954 the firm was taken over by The George Cohen 600 Group Ltd, and the production rate of the firm was gradually expanded until in the early 1960s it stood at a level of approximately 50 lathes per week.

(b) The Problem

The assembly method then used in the production of the company's lathes was one that has been traditional to the machine-tool industry – sequenced batch production. A batch of lathe beds were placed in their assembly bays at the beginning of each production cycle, and the various sub-assemblies added to each machine in unison. At the end of the production cycle, the machines were tested and only then were they removed from their assembly position in the factory. In 1960 the Colchester Lathe Company was operating on the basis of a weekly production cycle and at the start of this cycle a batch of fifty lathe beds would be moved into position and assembly started by a team working on each lathe. The firm had effectively reached the ceiling of its production capability using the batch method and existing work schedule. It was clear that some alternative production means would have to be used if the company was to increase its share of the centre-lathe market. This problem was the particular preoccupation of Mr Long, who had become Works Director to the company, and Mr Hayward, who had become its Managing Director. To both these men with early associations with the mass-production automobile industry it was clear that the solution to the problem of expansion lay in the adoption of an assembly line where the product would move continuously past a series of assembly stations to have its sub-assemblies fitted. In effect, they wished to make the same transition in assembly methods as had been made by Henry Ford with his Model T car when he changed from the traditional batch assembly methods to an assembly line scheme.

To accomplish this appeared to be extremely difficult, however, because of the precision called for in the assembly of the machines. It is vital that a machine tool be accurately assembled and that it be free from distortion, otherwise it will reproduce these errors on the work-pieces that are machined on it. It was considered of fundamental importance that any assembly line for the production of lathes should have a maximum deviation from the horizontal of 0·001 in. in 10 ft if the lathes were to have an acceptable accuracy when assembled. Board approval was sought and obtained for the installation of an assembly line when such a system would be available and it was left to Long as Works Director to recommend the type of system to be adopted.

Several possibilities existed. Firstly, there was the possibility of assembling the lathes on wheeled rafts that could be moved from workpoint to workpoint but this was considered unsuitable for the degree of accuracy that was required. In Russia a system for the large-scale production of lathes had been in operation for some time. In this system the lathes were assembled on rafts mounted in an accurately levelled trench, and at the end of each production stage the combined unit was lifted and advanced to the next work station. But it was not an assembly line in the sense that was desired by Long and Hayward. They then considered the possibility of building an accurate version of a conventional assembly line with slats attached to ropes or chains and drawn between guide rails. This scheme was already extensively used in the engineering industry and appeared to be more promising. Long spent nine months checking the various manufacturers of conveyor assembly lines and concluded that it would be impossible using these conventional techniques to obtain accuracies of the required standard. Yet another scheme to receive serious consideration was the three-point support system. With this, the lathe would be supported either directly or indirectly (standing on a raft) on three support points. These three points would always define a plane in space and so ensure that no distortion would be imposed on the lathe by the assembly line. The support points would form part of a carriage that would move along guide rails past each of the work stations with the lathe. But this system too has a major disadvantage for it meant that the checking of the alignment of each of the units assembled on to the lathe bed would be dependent on the use of clock-gauges and would in practice be both difficult to carry out and time-consuming. Thus there was no apparent solution to the clearly identified need of making an assembly line to the required degree of accuracy capable of carrying loads of over 5,000 lb.

(c) Origin of the 'Flowline' concept

It was about this time that several events took place of significance to the ultimate development of the 'Flowline' assembly
system. Mr Eric Moss was appointed as Chief Engineer to the
company, having graduated in engineering from Birmingham
University and joining the Colchester Lathe Company after
relinquishing a position with Alfred Herbert Ltd. It was
intended that he would be largely responsible for the further
development of the company's machine-tool range. Shortly
before this a young engineering graduate, Mr D. E. Shaw-
Stewart, returned to the Colchester Lathe Company after a
year's seconded study at the Cranfield Unit for Precision
Engineering at what is now the Cranfield Institute of Technology. The (then) Director of the Unit, Professor J. Loxham,
later commented:

> I recommended to Mr Long that he should sponsor a student
> to a one-year course at Cranfield with a specialisation in
> Precision Engineering. I was of the opinion that hydrostatic
> lubrication could be applied to the new lathe which I knew
> the Colchester Lathe Company were developing, and I
> recommended to Mr Long that Shaw-Stewart should under-
> take a thesis study in this subject. Mr Long agreed and
> because we had received a grant from the Science Research
> Council for research into the application of hydrostatic
> lubrication to a machine tool, I arranged for Shaw-Stewart
> to undertake the design and testing of a hydrostatic slide as
> his thesis study. The work proceeded satisfactorily and at the
> end of the academic year the slide had been made and tested.
> The slide worked very satisfactorily and the thesis written by
> Shaw-Stewart, entitled 'The Development of Hydrostatic
> Bearings for Slides, Spindles, and Leadscrews on High
> Precision Machine Tools', expounded the benefits of hydro-
> static lubrication. I was very satisfied with this and I thought
> that when Mr Long read the thesis he would decide to apply
> hydrostatic lubrication to certain models of the new lathe
> he was developing. A few months passed and Mr Long
> asked if he could come and see me concerning Shaw-Stewart's
> thesis. He came and to my surprise said that if what was

G

written in the thesis was true he could build an assembly line for lathes which was a development that he had for many years hoped would become a possibility. The hydrostatic slide in this case would be about 300 ft in length. We undertook to examine this project for him and it was abundantly clear that the large area that was available would support the load. We also found that if arrangements were made to adjust the elements of the slide so that they produced a long straight surface, the whole project appeared to have a high probability of success. I must give full credit to Mr Long, who was then Works Director, and Mr Hayward, who was Managing Director of the company, for having the courage to go ahead with this project which involved the digging of two wide trenches the full length of the new erection shop that was under construction. The hydrostatically lubricated ways were designed and in this detail design Mr Shaw-Stewart and Mr Moss, the present Chief Engineer of Colchester Lathe Company, took a very active part. When the work was completed the equipment operated very satisfactorily without any serious 'teething' troubles. It was named the 'Flowline' and the new erection shop complete with 'Flowline' was opened by the then Minister of the Board of Trade and what I believe to be a very positive step forward was made in the erection of lathes produced in large quantities. It may be that Mr Long has mentioned to you that the fairly high cost of installing this facility was more than saved by the reduction in the value of stock and work in progress and the further increase in their output must have increased this saving still further.[3]

It is worth noting that two large engineering projects were under way in Britain during the development of the 'Flowline' that may be related to it. The first[4,5] was the programme of work started in 1953 by the National Engineering Laboratory (N.E.L.) on the subject of air lubrication. This work was under the direction of Dr C. Timms, the (then) Superintendent of the Machinery Group, and led to the production (in 1954) of air bearings for an Admiralty high-speed ball-bearing testing machine, and in 1961 to the production of an experimental air bearing wheelhead for one of N.E.L.'s Churchill grinding machines. This latter work led to the development in 1962 by Churchill of a commercial prototype machine with air-lubricated bearings. Long was aware of this work, and it was on a modified Churchill machine that Shaw-Stewart carried out his tests in Cranfield. (Churchill subsequently won a Queen's Award for both technological innovation and export achievement for the design and production of their grinding machines which incorporate air bearings.)

The second development programme of note was that of the hovercraft. In December 1955 C. S. Cockerell filed his basic patent[6] on the design of a vehicle which was to be supported on a cushion of air. His objective was to reduce the friction between the vehicle and the surface over which it was to travel – an objective that was similar to that of the *Flowline*, as was stated on the patent granted to the lathe company: 'in brief the invention is concerned with the smooth and precise movement of objects, with immunity from disturbance, unlevelling, or change of stress'.[7] The first considerations of the possibility of designing a 'Flowline' centred on the possibilities of using air as the pressurising medium. It would have been possible – the hovercraft and the Churchill grinder had already demonstrated it – to build a 'Flowline' with compressed air as the fluid. But it was not considered desirable for several reasons, including the fact that such a suspension would have a much lower stiffness than the equivalent oil-supported raft. It would also have been more expensive to run because of the greater costs of compression for the lubricating air.

(d) Realisation of the concept

To think of the 'Flowline' concept was the matter of a moment. The idea is reported to have come to Long at three o'clock in the morning, which is in keeping with the best romantic traditions of inventing. The development and realisation of the concept, however, was a much more protracted stage, the engineering of which was the full responsibility of Moss. Advice was needed on a wide range of topics. Professor Loxham was consulted and he reported that

I discussed with Mr Long, Mr Moss and Shaw-Stewart the design of the elements that should be used for constructing the long hydrostatic slide. I also discussed with them the design of the sliding platen and made recommendations on the tests that should be made to ensure the construction and maintenance of a highly accurate slide which, as you will appreciate, being 300 ft long, must be made up of a large number of separate elements.[3]

Advice was also sought by the company from the British Cast Iron Research Association on the design of the rafts that were to form the assembly floor. These were obviously a vital element of the system and it was clear that unless they could be built so that they would not distort in service they would be unsuitable for the task they had to perform.

A ribbed box-like structure is used for the rafts, giving a high resistance to bending when carrying heavy loads but still retaining a desirable degree of torsional flexibility. The overall dimensions for each raft are 96 in. long by 42 in. wide, with a depth of section of 15 in. The total raft weight, including ballast, is just under $2\frac{1}{2}$ tons.

At each of the four bottom corners of the raft is a flat pad some 30 in. long by 8 in. wide. The four pads are precision ground flat in one fixed plane, and float freely on the pressurised film of oil that is interposed between the raft pads and the corresponding fixed pads formed in a pair of precisely positioned cast-iron base rails.

Short guide strips on the bottom of the raft work in conjunction with guide faces on the line of rails to provide a side location.

Flexible sealing strips on the sides of the rafts prevent the ingress of dust or small stray components. A water- and dustproof seal is also provided across the end faces of the rafts to ensure that abrasive particles do not drop through to damage the hydrostatic pads.[8]

On the advice of the Machine Tool Industry Research Association (M.T.I.R.A.) the firm sought the assistance of Professor W. Eastwood, Head of the Department of Civil and Structural Engineering, University of Sheffield, on another vital element of the 'Flowline' system – the concrete foundations on which each of the two assembly lines are supported.

Professor Eastwood wrote in reply to an enquiry on the part he played in the design of the foundations:

I think that the 'Flowline' foundation is novel in so far as the limitations on movement due to settlement, shrinkage and creep of the concrete and deflection under load were more stringent than any other design specification of which I am aware.

Concrete normally exhibits considerable shrinkage and creep movement during the early months in its life and, in addition, foundations inevitably settle by small amounts under load. In this particular instance the movements had to be considerably less than the inaccuracies which could be tolerated in the setting up of a lathe.

No specific testing was required in connection with this foundation design. The expertise involved was the outcome of some 10 to 15 years' experience as a designer of precise foundations, and in carrying out research in the field of soil mechanics, and in shrinkage and creep movements in concrete.[9]

The care required to ensure that the 'Flowline' would be technically successful depended largely on the precision with which relatively simple operations were carried out. This is well reflected in the construction of the foundations:

Excavations were made to a suitably stable sub-stratum. Weak-mix concrete was poured in to provide a basis for the main foundations, which were cast in sections of 25 ft nominal length. The first sections cast were allowed several weeks to cure before intervening sections were also filled in to give two final continuous runs, each over 300 ft in length.

A rigidly controlled specification of sand, cement and stone was combined with heavy steel reinforcement to give maximum ultimate stability with minimum shrinkage.

Possible movement between the cast-iron base rails and the foundations – through changes in temperature – was also minimised by ensuring that the concrete mix had a temperature coefficient of expansion closely corresponding to that for the rails.

Finally, the foundations were completely sealed with a

waterproof paint to give a controlled slow curing of the mix without cracking and to retain the high stability of a full saturated construction. Without such a seal, movement of the foundations would be likely through uneven drying.[8]

Further research information and design advice was obtained from the Production Engineering Research Association (P.E.R.A.), and from M.T.I.R.A. – assistance being provided as part of the service normally provided to members of the associations. P.E.R.A. were by that time 'particularly expert' in the design of hydrostatic bearings and could therefore advise on the design of the basic element of the 'Flowline', the hydrostatic pad. There are twenty-eight of these cast in each 10-ft length of base rail which in turn is mounted in a steel channel trough rigidly grouted and bolted to the main foundation. One rail and trough is accommodated in a unit length of 10 ft and is repeated to give a continuous system for the entire length of the assembly line.

The pitch of the pads is $4\frac{1}{2}$ in. so that six or seven pads give support to each of the four 30-in-long faces on the underside of the raft – the effective number of pads alternating from six to seven according to the position of the raft on the rails. Each pad is rectangular in shape and measures 6 in. by 3 in. A $\frac{3}{4}$-in. land is provided to give a central pocket $4\frac{1}{2}$ in. long by $1\frac{1}{2}$ in. wide.

The central pocket in each of the pads provides the means for exerting the required pressure on the ground underfaces of the raft to obtain the initial lift off the slideway pads. When the raft sits firmly on the pads, with no gap, then the relatively high pressure in the oil gallery is transmitted virtually without loss to provide a maximum lifting force. Once the raft lifts, an oil film is established across the pad, achieving an equilibrium condition in which the gap between the raft and pads restricts the film of oil sufficiently to maintain a load-supporting pressure in the pocket. The nominal design gap for the system under consideration is 0·005 in. when supporting a raft weighting $2\frac{1}{2}$ tons. The related pocket pressure is about 21 p.s.i., with a main oil gallery pressure of approximately 105 p.s.i.

The 5:1 pressure drop from gallery to pocket not only imparts a high load-bearing capacity, but also limits the oil flow from an uncovered pad to no more than a 25 per cent increase on that for a covered pad.[8]

The absolute simplicity sought in the design of the 'Flowline' is again well illustrated by the means adopted to control the oil flow to each of the pockets of the hydrostatic support pads. Instead of valves and sensors to control the flow, measured lengths of capillary tubing were used such that the amount of oil that could flow to the open hydrostatic pads would be restricted by the pressure loss in the tubing but would not impair the build-up of a high pressure in each pad once the oil flow was sealed off by the advancing raft.

When a receptacle is unloaded or generally speaking uncovered, the oil overflow should be controlled so as not to be exorbitant, yet the pressure rise in a covered receptacle should be adequate for the requisite hydrostatic support of a craft. The use of lengths of cheap and easily cut pipe (plastic tube) is consequently a matter of practical utility. We have found that in practice ... about 20 in. of nylon piping having a bore of 0·080 in. is successful.[7]

A more elegant and simple solution to the problem could hardly be desired.

Further vital but conventional elements of each 'Flowline' are the oil collecting, filtering and pumping system, the reciprocating ram for advancing the rafts at their normal speed of 3 in. per minute with a quick return stroke, the friction roller and cam units at the start and finish of the line for separately lifting off and replacing the rafts evenly on to the hydrostatic pads, and an electrical pick-up system so that the assembled lathes may be operated and tested whilst still on the 'Flowline'. The instruments used to obtain the correct alignment of the system were of conventional design (spirit levels, alignment telescopes, straight edges, slip gauges etc.) but they were used to obtain accuracies in a civil mechanical structure that were more in keeping with aeronautical engineering standards. Without these accuracies, the 'Flowline' could not have been successful.

(e) Comments

Success of the system. Many aspects of the system are of interest but perhaps the most important is the degree to which it satisfied its objective. Before installing the 'Flowline', the company was nearing its maximum production capacity of 60 lathes per week and further expansion would have required major changes in working hours and organisation for the labour force of approximately 1,000. By 1968 production had grown to 150 lathes per week with a labour force of 1,147 and in 1969 a peak production rate of over 200 lathes per week was reached. These increases in throughput, with their disproportionately small increases in the labour force, brought with them an immediate reduction of over £100,000 worth of lathe components that had to be held in stock, and so the new hydrostatic 'Flowlines' were effectively obtained at zero capital expenditure by the firm. With the new system the assembly cost per lathe was reduced by 18 per cent and the average time from start of lathe assembly to completion reduced from between three and four weeks to two days. The large volume of throughput has enabled the firm to strengthen its competitive position further by reducing its average delivery time for lathes to between four and six weeks. The increased throughput is further reflected in the performance of The George Cohen 600 Group because the castings for the lathe bed and other fittings are produced for Colchester Lathe Ltd by The 600 Group Meehanite Foundry at Morriston, Glam.[10] A measure of the impact of machine tools on The 600 Group is given by the fact that in 1965 machine tools earned 23 per cent of the Group's profits whilst in 1968 they accounted for 50 per cent of the Group's profits on 27 per cent of the Group's turnover of £26·5 million.[11] It must, however, be remembered that apart from Colchester there are some ten other machine-tool companies within The 600 Group.

Markets. Factors that must be associated with the success of the Colchester Lathe Company are the developments which have taken place in its markets. At home, a major stimulus to purchases of standardised machine tools of proven reliability and minimum cost was provided by the Industrial Training Act which accounted for a large proportion of domestic sales in 1966, 1967 and 1968. This Act placed levies on a wide range of industries, including the engineering industries, to pay for the training of new entrants to these industries. It also made provision for grants to be paid to firms who built and equipped training schools and workshops to an officially recognised standard. Colchester Lathe was well equipped to deal with this relatively sudden expansion of the market because of the flexibility of its production system, the standardisation of components which it had built into its products, and the design philosophy which had dictated that the firm would concentrate on building inexpensive machines suitable for being used at maximum capacity for five years and then economically discarded. It is a philosophy which differs radically from the traditional practice of the machine-tool industry and one which is in line with the developing views of machine tools' purchasers generally.

A second and important aspect of the company's success, and one which generated the continuing need for the 'Flowline' system, was the redesign of its products which was carried out on the basis of the market requirements of three major sales areas – America, Continental Europe and Britain. The resulting redesign exercise led to the production of the company's *Mascot 1600* lathe – a product which won for the company one of the seven awards made by the Council of Industrial Design in 1967 for products displaying 'aesthetic and ergonomic advantages'.[12] It is interesting to note, too, that the redesign of the *Mascot 1600* machine was carried out in collaboration with a firm of consultant industrial designers (F. C. Ashford and Associates), a consultant ergonomist (Ronald Easterby) and the Ergonomics Research Unit of M.T.I.R.A.

A further extension of its range of centre lathes took place when the Company announced the '*Triumph 2000*' machine, which was again based on a reappraisal of its previous products in the light of market data from its three principal sales areas.

Government aid. A topic that attracts frequent comment is the question of finding the optimum mechanism for government agencies to support research. Rarely can the granting of direct economic aid to firms be justified because of the uncertainties of research and the difficulty of selecting suitable recipients without being accused of unfairly subsidising some firms at the expense of the taxpayers. Granting aid to bodies such as universities and research institutes can help to ensure that research is done in the interests of the nation rather than sections of it. But here the objection is frequently raised that the time-lag between granting the aid and its incorporation in the production function is too long and that in many cases the potential benefits of the work carried out are lost. In two instances referred to in this study, the development of the hydrostatic air bearings for grinding machines in N.E.L. and the development of hydrostatic slideways for machine tools at Cranfield, the transfer to industry of technological developments originally funded by government agencies was particularly effective. In both cases, industry took an active part in the project, by supplying equipment, data and trials assistance, and by supplying some of the personnel engaged on the studies. Shaw-Stewart's work at Cranfield was transferred to the Colchester Lathe 'Flowline' partly through his thesis, but largely through his personal involvement with both the centre of learning and the centre of application. Without the provision of Science Research Council funds for the study of hydrostatically lubricated slideways, it is highly unlikely that the development of the 'Flowline' would have taken place when it did, if indeed at all.

The machine-tool industry. The past decade has seen the implementation of consistent policies of rationalisation of both firms and their products within the machine-tool industry. A measure of the degree of fragmentation which existed is given by the fact that even in 1968, after several years of amalgamation and restructuring of the industry, the Ministry of Technology reported that there were 250 firms operating with machine tools as a major but not dominant element of their production and that 50 per cent of the national production of machine tools was accounted for by nine groups, one of them The George Cohen 600 Group. The Ministry's investigations showed that

in many cases the industries turned out to be fragmented to the point where the advantages of volume, both in manufacture and in sales outlets, were not being enjoyed and where the turnover was unable to support the appropriate level of research, development and particularly production technology. There is also the problem that the multiplicity of small firms, with their limited resources, find it difficult to attract top-quality technologists on to their staff. The ability to even out the order cycle variations and to make the large capital investments (such as the purchase of numerically controlled machine tools) are also related to the size of the company.[13]

It is in the context of the moves to rationalise the machine-tool industry and to increase its international competitiveness that the policies of the Colchester Lathe Company, and the success of its products, must be judged.

References

1. P. A. Long, Managing Director, Colchester Lathe Co. Ltd, interviewed by W. G. Evans on 19 May 1969.
2. P. H. Baines, Assistant Sales Manager, Colchester Lathe Co. Ltd, interviewed by W. G. Evans on 19 May 1969.
3. J. Loxham, Cranfield Unit for Precision Engineering, Cranfield Institute of Technology, letter to W. G. Evans dated 21 Aug 1969.
4. G. Bishop, Technical Director, B.S.A.-Churchill Machine Tools Ltd, in *New Technology* (Ministry of Technology), no. 7 (1967).
5. 'Innovation: A Case History', *Mintech Review*, no. 16, Aug–Oct 1969, p. 10.
6. C. S. Cockerell, 'Improvements in or related to Vehicles for Travelling over Land and/or Water'. British Patent No. 854,211, applied for 11 Dec 1956, published 16 Nov 1960.
7. E. F. Moss, P. A. Long and D. E. Shaw-Stewart, 'Improvements relating to Slideways for Transport'. British Patent No. 1,093,051, applied for 10 Jan 1964, published 29 Nov 1967.
8. Colchester Lathe Co. Ltd, Submission to the Office of the Queen's Award to Industry, 27 Oct 1966.
9. W. Eastwood, Head of Dept. of Civil and Structural Engineering, University of Sheffield, letter to W. G. Evans, 15 July 1969.
10. 'Building Centre Lathes on a Flow Line Basis', *Machinery*, 8 July 1964, p. 1.
11. R. Sanders, *Sunday Times Business News*, 16 Nov 1969.
12. *Seven Attitudes to Design in Engineering*, Joint Publication of Council of Industrial Design and Ministry of Technology (H.M.S.O., London, 1967).
13. I. Maddock, 'Stimulating Technological Innovation in Industry', *Proc. Inst. of Mech. Eng.*, **182** (1) no. 32 (1967–68).

9 CONCRETE LTD:
INDUSTRIALISED BUILDING

Both the use of concrete in buildings and the concept of manufacturing components in a factory prior to assembly on the building site originated in the last century but were only used when special conditions created an advantage over other forms of building. This case provides an example of innovation in response to a social need, the need for more homes. A complex mixture of technical ideas, some new and some old, was utilised to produce a system capable of satisfying the needs of local authorities and other customers.

(a) Industrialised building

The first buildings to be assembled out of large components made elsewhere were in iron or wood but were usually 'one off' jobs like the Crystal Palace. The Californian Gold Rush provided an early example of an urgent demand for homes quickly erected and many countries exported easily erected wooden buildings to California. The Gold Rush also demonstrated that special conditions of demand can evaporate very quickly; after only two years, the Californian building market was so flooded that prices fell to the level where many building companies made losses.[1]

Industrialised building is not just a synonym for prefabrication; it implies a highly organised system with the continuous production of components, the transport of these components to a site and their final assembly all under one control. Such a system involves a high level of planning, and industrialised building can be said to have originated in the American Army with the special conditions involved in the rapid construction of wooden huts to form an Army camp. During the First World War, American Army engineers developed systems of control that enabled wooden huts to be constructed with great efficiency. Timber is, of course, quite different from concrete but the principles of modern techniques of organisation can be seen in the publications of the Army engineers. The Army engineers also developed factory methods for the production of concrete piles.[2]

The First World War produced a shortage of steel and timber and focused attention on the use of concrete in some applications which had been technically possible since the previous century. The housing shortage which followed the war also provided an incentive to the use of prefabricated concrete components. Concrete Ltd was founded in 1919 mainly for the purposes of manufacturing concrete floor units. Other components were manufactured in the inter-war years – prefabricated balcony units for cinemas, for example – but such operations cannot properly be described as industrialised building.

A major obstacle to the development of completely industrialised systems was the fact that traditional methods of on-site construction, although slower, continued to remain cheaper. In economic terms, industrialised building involves the substitution of temporary labour by capital invested in factory production units with no guarantee of a continuing demand for the products of the factory, and there was no incentive for factory-based home-building without government intervention.

A form of industrialised building system involving prefabricated wooden houses was used in Stockholm in the 1920s. This was assisted by the municipal government, who were attempting to provide houses for poorer families.[1]

The need for government intervention was recognised by Wilson Wyatt, who was put in charge of the American housing programme in 1946. He provided loans to enable factories to be set up and also gave market guarantees.[1] However, with the return of the Republican Party this housing programme was stopped.

In Britain, various enthusiasts for industrialised building worked at convincing government bodies of the need for government support. Prejudice against the type of 'prefab' used in the late 1940s had to be overcome and it was gradually realised that the only type of housing construction for which industrialised building offered an economic advantage over traditional methods was the tower block of flats. Flats outnumber houses in several European countries and techniques developed abroad were capable of being used to attack the British housing problem. The Conservative Minister of Housing and Local

Government, Sir Keith Joseph, became enthusiastic about the possibilities of industrialised building and by 1963 several systems were available in this country. In 1964, following the return of a Labour Government, the new Minister of Housing and Local Government, Richard Crossman, warned local authorities that 'the amount of encouragement they get from the Ministry will depend very largely on their readiness to adopt modern techniques of system building'.[3] A National Building Agency was set up to advise local authorities on various systems and the National Plan aimed at 500,000 new houses a year by 1970. The 'white-hot heat of technology' was to be unleashed on the housing problem. The years 1964 and 1965 saw the start of a boom in industrialised building systems aimed at local authority housing projects. With an apparently assured market in this area, other applications could be considered and by 1967 there were over 200 different systems available. In 1967 the enthusiasm for high blocks of flats began to decline and the Ronan Point tower block disaster marked the end of an era. By 1969 some factories established for housing systems were reduced to producing precast kerbs and lamp-posts.[3]

Concrete Ltd were aware of the pressures developing against tower blocks of flats and developed the 'Bison' Wall Frame system for 'low-rise' housing, a system which is claimed to be economically competitive with traditional housing methods. Ronan Point seems to have hit Concrete Ltd, whose profits for 1968–9 showed a drop for the first time in ten years, but according to the firm there was a drop in contracts even before the disaster.

(b) *Concrete technology*[4,5]

Industrialised building systems in concrete such as Concrete Ltd's 'Bison' system were made possible by technical advances in the use of concrete. High-strength concrete was obtained in the last century by reinforcing the concrete with metal bars. The first patent for reinforced concrete was taken out in England in 1854 by a Frenchman and commercial techniques were developed in the 1880s and 1890s in France, Germany and Denmark. Pre-stressed concrete in which the steel reinforcement is stretched until the concrete has set was patented in 1868, made practicable in the 1920s and developed in Germany during the 1930s. Pre-stressed concrete was first used in Britain in an Army ammunition depot and then for the production of railway sleepers in 1940 when there was a shortage of timber. The development of pre-stressed concrete involved new equipment for stretching and anchoring wires and moulds for casting.

An important development in the 1930s was a new design concept. Previously, high buildings were built round a frame, usually of steel, which supported the building and to which walls were added to keep out the weather. In the 1930s, high reinforced concrete buildings were designed in which walls and floors were treated as extensions of columns and beams. This meant that the buildings were supported by the walls instead of by a frame and such designs could not be accepted in this country until building specifications were modified. These buildings used concrete cast on the site, but the design concepts form the basis of industrialised systems like Concrete Ltd's 'Bison' Wall Frame system.

The shortage of steel after the Second World War and its higher price led to an increased interest in reinforced concrete. Improved cranes and other apparatus made the use of pre-cast concrete components more acceptable in high building construction and improved methods of joining together the pre-cast units were developed. The industrialised building of multi-storey concrete buildings was technically possible.

(c) Concrete Ltd and the Bison Wall Frame system[6, 7, 8]

Concrete Ltd was founded in 1919 by Ambrose and Mathews, who had met on the Somme, where they were engaged in the construction of pre-cast concrete pill-boxes. From 1919 onwards, Concrete Ltd supplied prefabricated components such as floor beams which were used by builders together with on-site methods of construction. Concrete columns were usually cast on site as this was the easiest way of forming a strong joint. After the Second World War, Concrete Ltd developed the use of pre-stressed concrete for flooring units, but by 1954 orders were decreasing and it was decided to attempt the production of pre-cast units for frames in order to gain more contracts. A three-storey high pre-cast column was designed to be competitive with on-site construction. This system was used in 1955 for a warehouse which had originally been designed in steel. Delays in steel delivery gave Concrete Ltd the chance to use their system with pre-cast columns. The cost of the system and the construction time was better than had been expected with steel, and Concrete Ltd then gained contracts for various buildings including a sixteen-storey block of flats. Each building was designed separately for a specific contract and the pre-cast concrete units were limited to the components necessary to construct a frame, the walls being cast on site or built of brick. With the experience of designing several buildings, Concrete Ltd's designers were able to produce a standardised system, the Bison Preferred Dimension Frame, which was developed in 1960 and launched in 1962. The development of this system was assisted by a contract for twelve identical blocks of flats in Birmingham. At the same time, the possibility of replacing the *in situ* walls by pre-cast units was seriously considered. Systems had been available on the European continent since the 1930s in which load-bearing walls and floors took the place of a frame structure, but calculations by Concrete Ltd showed that the Continental systems were only economical for schemes of 2,000 or more homes. Increasing emphasis in government and other circles on the need for improving the productivity of home construction pointed to a serious consideration of such systems.

Early in 1961 the Board of Concrete Ltd, under the Chairmanship of K. Wood, decided to go ahead with the Preferred Dimension Frame system for schools, factories and similar buildings and also to develop a new system for the building of blocks of flats. The possibility of licensing one of the Continental systems was considered. British requirements were somewhat different but there was many years' experience behind the Continental systems. Concrete Ltd managed to use some Continental know-how without paying royalties by an arrangement with Malmstrøm, a Danish firm of consulting engineers who were consultants for both the Jespersen and the Neilsen systems in Denmark. This arrangement allowed one of Concrete Ltd's design engineers to spend six weeks with Malmstrøm studying panel jointing problems in return for the payment of a consultancy fee. (Industrialised building had developed in Denmark with the aid of government subsidies. Danish systems had more variety than the French and German systems which were aimed at larger markets and were therefore more likely to be used in Britain. However, Concrete Ltd claim that their system is superior to Danish ones.)

In 1963, after two years of development, Concrete Ltd were able to launch the Bison Wall Frame system for 'high rise' housing, i.e. blocks of flats of eight storeys and upwards. In November 1963 a block of flats was completed at Kidderminster and officially proclaimed by Sir Keith Joseph, the Minister of Housing, as 'the first fully industrialised block to be completed in Great Britain'.

From a European point of view, there was little that was technically new in the Bison Wall Frame system but it was well organised and well presented for British local authorities. The economics of industrialised building demand standardisation, but architects acting for local authorities prefer flexible systems that avoid regimentation and allow for some degree of individuality. The Bison system has some flexibility in it through the optional inclusion of balcony units and a variety of external finishes. To the layman, one tower block of flats looks pretty much the same as any other, but the flexibility of the Bison system has been used as a selling point to local authorities who like to feel that their tower blocks are better than someone else's. Many systems were introduced into this country, the majority being based on Continental systems. Concrete Ltd made the most of being the first into the field, the 'British' nature of their system and its flexibility, with the result that, up to 1967, 20 per cent of the high blocks of flats for local authorities had been constructed by Concrete Ltd.

From 1961, Concrete Ltd had as a declared aim the industrialisation of 'low rise' building, an area in which it was much harder to compete with traditional on-site methods of construction. Murmurs in some quarters against the use of tower blocks encouraged Concrete Ltd to extend its Wall Frame system into the low-rise area and in 1965 the Bison Wall Frame system for low-rise housing was announced. Bison publicity claims that the low-rise system contains two components which are unique in the world. One of these is a long span floor unit in pre-stressed concrete; its advantages over other methods of construction are open to question.

By 1967, the year of the Award, Concrete Ltd was operating

nine factories in different parts of the country including Northern Ireland, and was employing some 4,000 people. It could claim to be 'the world's largest organisation specialising in structural pre-cast concrete'. Publicity material stated that the 'Cathy Come Home' building system had supplied homes for 30,000 people in four years and homes for another 40,000 were on the way. Since 1967, the fortunes of Concrete Ltd have fluctuated. In 1967 5,600 dwellings were included in local authority approved tenders; this figure fell to little more than 3,000 dwellings in 1969. The fall would have been much greater if Concrete Ltd had not developed its low-rise system.

(d) Comments

The Concrete Ltd case can be seen as an illustration of the fact that invention is not always the prime mover in innovation. It was not the invention of reinforced concrete that led to the formation of Concrete Ltd but the First World War which, besides bringing together the two young engineers who founded the firm, produced a housing shortage which encouraged new methods of construction. The developments in concrete technology and the design of concrete structures which took place in the 1930s were not used in this country until there was a need for them. It might even be argued that inventions not already available would have been made when the need for them arose. This must, of course, be purely speculative. It is worth noting in this connection, however, that the housing shortage produced by the First World War led to building research being financed by public funds; some of the nineteenth-century inventions might well have emerged from this research if they had not already been available.

Concrete Ltd produced the first fully industrialised tower block in this country, not because they invented anything but because they took steps to satisfy a need. An important factor in the firm's success was the good management provided by K. Wood as Managing Director and Chairman. K. Wood studied mathematics at Cambridge and wanted to be an engineer. This was not possible so he worked for I.C.I. as an accountant instead. He married the daughter of John Ambrose, one of the two founders of Concrete Ltd, and was then able to combine his interest in engineering with his accountancy knowledge so that his enthusiasm for new ideas had a self-imposed financial brake.

Other factors responsible for Concrete's ability to compete in a market with too many suppliers were the competence of their design team and an excellent sales force.

References

1. B. Kelly, *The Prefabrication of Houses* (Chapman & Hall, London, 1951).
2. *Engineering News Record*, **82** 397 and 1139 (1919).
3. B. Moynahan, 'Building: Systems in a Snarl-up', *Sunday Times Business News*, 23 Mar 1969.
4. M. Bowley, *The British Building Industry* (Cambridge University Press, 1966).
5. K. Billig, *Structural Concrete* (Macmillan, London, 1960).
6. Press release, 'Queen's Award for Industry to Bison Wall Frame Building System', 21 Apr 1967.
7. K. M. Wood, 'The Bison Wall Frame System', in *Housing from the Factory*, Proceedings of a Conference organised by the Cement and Concrete Association Oct 1962.
8. P. J. Schryver, Director of Concrete Ltd, interview, 1969.

10 DAVY-ASHMORE: AUTOMATIC CONTROL OF STEEL STRIP THICKNESS

The Award to Davy-Ashmore Ltd in 1967 was for innovation by its wholly owned subsidiary, Davy and United Instruments Ltd, in developing the 'Gauge-meter' system of automatic gauge control (A.G.C.) for steel strip production. This system is designed to ensure that the thickness of steel strip produced on a Gauge-meter-equipped rolling mill shall be automatically held to close tolerances throughout its length of several thousand feet. A notable feature of the innovation is the part played by the British Iron and Steel Research Association (BISRA) (which later became the Inter-Group Laboratories of the British Steel Corporation). Not only was most of the crucial early work done within BISRA but the Research Association also provided, from the members of its staff who worked on the project, people who joined the firm to help to develop the system of gauge measurement and control into a viable engineering design capable of withstanding the difficulties of commercial service.

(a) Rolling mill design

Although one could say that work on rolling mill design and practice has taken place for centuries, the foundation of the research which is of significance in this case study can be considered to be the setting up in 1930 of a Rolling Mill Committee of the Iron and Steel Industrial Research Council.

> Its early work was concerned essentially with factors determining efficiency of rolling mill operation, including studies of cost data, power consumption, production studies and roll design. In 1936, following a discussion by the Metallurgy Research Board of the Department of Scientific and Industrial Research, a communication was addressed to the [Iron and Steel] Research Council emphasising the desirability of research in this country on rolling mill design. Great Britain had been the pioneer in the development of the rolling mill... but initiative in the design of new plant, except, perhaps, in some aspects of cold-rolling technique, was largely in the hands of the Continent and the United States. Rolling mill plant design was still very much a matter of experience, and even in countries where considerable research into the problems involved had been made, the fundamentals on which the efficiency of rolling mill plant were based were still only imperfectly understood.[1]

Following surveys of rolling mills on the Continent and in Britain by a panel set up to investigate the possibilities and requirements for rolling mill research in Britain, it was concluded that no suitable mill was available, and that pending the provision of such a mill, surveys of the available research experience should be made, correlated and disseminated, that the instruments available for rolling mill research should be examined and a report submitted, and that a limited programme of research should be instituted at selected mills in Britain where facilities might be obtained for working on ferrous and non-ferrous metals. To direct this work, a Research Subcommittee was formed representing the ferrous, non-ferrous and constructional interests of the rolling mill industry.

Besides directing these programmes, the Research Subcommittee appointed Dr E. Orowan as a consultant and, through the assistance of Sir Lawrence Bragg, facilities were made available to Orowan at the Cavendish Laboratory, Cambridge. Orowan, a Hungarian, had been concerned for several years with the study of the fundamentals of the plastic deformation of metals, and his research objectives at the Cavendish were:

(i) A study of the plastic properties of the materials rolled and the friction between the rolled stock and the rolls.
(ii) A study of the behaviour of the material between the rolls, with the ultimate aim of calculating roll pressure, torque, and other quantities from fundamental physical data.[1]

It was an ambitious programme, and one which was potentially of great value to the rolling mill industry. 'If it is desired to reduce the thickness of a strip by rolling, what force and torque need be applied to the rolls? This is the problem which all the theories of rolling aim to solve. The data required are the initial and final thicknesses of the strip, the yield stress at each point in the roll gap, the radius and elasticity of the rolls, and a value of the coefficient of friction'.[2] To solve this theoretical problem, experimentally derived data were needed, and so Orowan, with the team of researchers he had gathered, designed a range of instruments as part of their programme. Among these was a load meter to measure the separating force applied to the rolls of the mill as the material passed between them. On 2 December 1944, Orowan and his co-workers, E. A. W. Hoff, an Austrian, and J. Los, a Pole, applied for a patent[3] to protect their new design of load meter in conjunction with BISRA, which had in 1944 assumed responsibility for the research efforts of the iron and steel industry. This load meter design formed the foundation on which the 'Gaugemeter' control system was later to be built. It consisted of a machined column of steel with strain gauges bonded to its surface and arranged in a Wheatstone bridge configuration such that the application of a load to the cell produced a change in its electrical characteristics which could be calibrated to indicate the value of the load. Orowan and his co-workers continued to investigate the theoretical basis of rolling mill behaviour until September 1950 when Orowan left the Cavendish Laboratory and the research contract for work on the theory of rolling terminated.[4] Orowan's earlier work,[5,6] together with the work of H. Ford,[7,2] provided the theoretical background against which the findings of the 'Gaugemeter' rolling mill development team were analysed and judged.

Orowan's work was one line of development that was initiated by the Rolling Mill Research Subcommittee, but they were also concerned with a second investigation which was being carried out in Sheffield University. Following the circulation of a review[1] of the state of the art and science of roll mill design by Dr L. R. Underwood in 1939 (revised and published in 1946), it was concluded by the Subcommittee that an investigation of rolling mill parameters should be undertaken as soon as possible. Owing to the outbreak of the war this work did not begin until 1942, when research facilities and the use of a 1928 rolling mill were made available to Dr Underwood. This mill was redesigned and re-equipped to extend the range of experiments which could be performed upon it. In particular, it was equipped with load meters to measure the rolling load, and tension reels to vary the tension in the steel strip as it was drawn through the mill.

The installation and direction of the new facilities passed to Ford when Underwood left the British Iron and Steel Federation to join the firm of W. H. A. Robertson and Co. of Bedford. Ford was appointed Head of the Mechanical Working Division of BISRA after its formation in June 1944 and directed the research programmes of the Division until January 1947. He was then succeeded by Mr W. C. F. Hessenberg, who had represented the British Non-Ferrous Metals Research Association on the Rolling Mill Research Subcommittee.

Thus, two teams, one in the Cavendish Laboratory at Cambridge University and another at Sheffield University, both supported by BISRA, were attacking the problem of improving the theory and practice of rolling mill design in the 1940s. Such improvements were seen to be necessary in the context of the increasingly sophisticated requirements of the sheet metals market. After the experience gained in the Second World War there was a rapid increase in the demand for sheet metals, particularly in the automobile, food-processing, electrical transformer and aircraft industries. With the increasing sizes of the automobile panels being produced, it had become urgent to ensure the uniformity of the sheet material to be pressed into shape. Variation in thickness of the raw material would cause irregularities in the shape of the finished product and a high rate of rejection of finished panels. The increasing degree of automation being applied to the assembly of food containers and similar products made it imperative that the machines be fed with sheet metals of consistent characteristics. Transformers of greater size and higher performance were being sought by the electricity authorities to keep pace with the expansion of the national electrical energy demand, and this too reflected on the sheet metal industry's need for a higher quality of control of the rolling process. A more sophisticated requirement for sheet metal was being proposed by the aircraft industry, which had always had a keen interest in reducing the structural weight of aircraft. It was suggested that for the more advanced aircraft of the future a double benefit could be reaped if the wing skins and other parts could be produced from sheet materials of tapering thickness. This would not only reduce the quantity of expensive material required per aircraft (titanium was then being widely advocated for future high-performance craft), but the structural weight of the aircraft would also be reduced, making it possible to carry additional payload.

Commenting on the performance of a conventional cold rolling mill used to produce a coil of steel strip 0·0625 in. thick and 3·5 in. wide, some of the BISRA workers stated:

The British Standard Specification for the gauge of this class of material is to within ± 0·0015 in. of the nominal thickness. Although these tolerances are wide, almost half the length of the coil is close to or just outside these limits, and this is not considered to be exceptional. Mill operators experience great difficulty in controlling gauge closely at high strip speeds, for it is well known that if the screws are adjusted quickly a series of under- and over-corrections occur which may give rise to gauge errors greater than those that were to be corrected. The thickness profile of this coil illustrates one of the major problems of the cold-rolling industry; the speed at which it is economical to run the mill is too great for the operator to correct the variations occurring in the rolled-strip thickness due to imperfections both in the ingoing strip and in the cold-reduction process, but at the same time new markets for strip are demanding closer tolerances in gauge. The need for automatic gauge control on rolling mills is, therefore, both obvious and urgent.[8]

Apart from the speed limitations on the ability of the mill operator to control a mill, a variety of other factors had to be considered. The incoming coil of strip would probably exhibit different hardness properties along its length, depending on the quality of the thermal treatment it had received; welds between coils would upset the mill setting; the acceleration and deceleration of the mill could adversely affect the gauge of the material – not only because of the properties of the sheet, but also because of the effect of the speed of the mill on the roll gap; any eccentricity in the rolls would manifest itself in a waviness imparted to the strip; and the coefficient of friction and the degree by which it could be modified by lubricants and coolant was also known to have an important effect in the attainment of uniform gauge material. The stiffness of the mill and the design of its control mechanism were also of obvious importance if the standards of sheet production were to be raised.

On the Continent attention had been given to the problem of improving the quality of metal strip production. In May 1935, E. Meyer, a German, applied for patent protection for his roll-control mechanism in which the gap between the rolls would be held constant by the application of pressure to a hydraulic cylinder mounted between the roll bearing block and the frame of the mill. 'The regulation is such that as the rolling pressure increased (so causing the rolls to spread) the rolls tend to be moved together, whilst in the case of a decreasing roll pressure they are caused to be moved apart.'[9] The Swiss firm, Aluminiumwerke A.G., Rorschach, devised a method for improving strip quality based on controlling the tension applied to the strip as it was being drawn into the mill. A gauge measured the output thickness of the strip, and when the thickness departed outside the acceptable limits, a brake was applied to the unreeling coil of strip via a hydraulic control system and brake.[10] In France, Blain[11] had developed a method of control based on the direct measurement of the gap between the rolls and control of this parameter by means of a hydraulic signal applied to a cylinder between the bearing chocks and the frame of the mill. But each of these methods suffered from disadvantages that limited their application to production units. Only Blain's method had been effectively used in practice, and it had a limited sensitivity.

That, then, was the situation in 1948. Efforts were being made on a wide variety of fronts to improve the quality of rolling mill practice, specific proposals had been made to assist this but they suffered from a variety of practical limitations, and in Britain a research association, BISRA, had taken on the task of investigating the theoretical and practical aspects of rolling mill technology in an effort to develop mills capable of meeting the needs of the market.

(b) The evolution of the Gaugemeter [12,13,14]

In January 1947, W. C. F. Hessenberg became Head of the Mechanical Working Division of BISRA. The Sheffield Laboratory of the Division contained sections concerned with rolling mills, forging, wire drawing, solid mechanics, metallurgy and electronics, and in 1950 its research staff consisted of 28 persons.[4] The South Wales Laboratory based at Swansea contained sections dealing with chemistry, electrochemistry, metallurgy, physics and engineering; it had a staff of 21 persons. Between them the two laboratories had 29 staff members with a B.Sc. qualification or its equivalent, and 8 of these had M.Sc. or Ph.D. degrees in addition. Ten of the staff members at Sheffield were directly concerned with rolling mill design and performance, making up the largest section in the Division.

Hessenberg began to recruit members for his various sections and in November 1948 he recruited R. B. Sims as Head of the Rolling Mill Section. Sims had previously worked with

Baker-Perkins on the design of biscuit ovens; he has been described by Hessenberg as a person with 'extraordinary energy and extraordinary technical courage'.[12]

In June 1949, Hessenberg wrote to Sims and asked 'why could I not design an improved form of interstand tensiometer for five-stand tinplate mills? His view was that if interstand tension could be held constant then gauge would be improved. This set me thinking about the role of tension in the gauge variations and I replied to his memorandum in July 1949, putting forward the suggestion that if tension was varied and the rolling load held constant then gauge could be completely controlled'.[13] In a multi-stand mill, the steel strip passes through successive sets of rollers, each succeeding set running at a higher speed to draw the elongated material away from the preceding rolls. It was known that, by varying the forward tension on the strip as it emerged from the mill or the back tension as it was drawn into the mill, the quality of the strip could be controlled in terms of uniformity of thickness, absence of bulging in the centre of the strip and absence of ruffled edges. The experienced rolling mill operator would adjust these parameters as well as allowing for the speed of the mill, the effects of acceleration or deceleration or the effects of worn rolls, to ensure that a uniform product would emerge from the mill. But the rolling mill operator's ability to allow for these factors was limited by his ability to detect and respond to errors in the product. Describing the problem, Hessenberg and Sims wrote in 1951:

> At present, the usual method of controlling gauge is to mea-sure the thickness of the strip continuously as it leaves the mill and, by means of a manually controlled, power-assisted roll-adjusting mechanism, to try to adjust the rolls whenever a persistent change of gauge is observed. The thickness measurement is, of necessity, made some distance away from the roll gap so that an indication of a change is not manifest until some time after it has occurred. There is a further delay whilst the operator becomes aware of the change and attempts an adjustment of the rolls; the response of the operator is neither rapid nor sensitive enough to prevent several under- or over-corrections, so that a substantial length of strip may

have passed through the mill before the correct thickness is restored.[15]

The manual methods of roll control were considered inadequate, as were the proposals of Meyer, Blain and the Aluminiumwerke A.G. It was in this context that Hessenberg made his proposal to Sims. On the basis of this and of Sims's later work, two methods were developed 'based on certain elementary principles of rolling which do not appear to have been described before'.[15] Hessenberg's suggestion as developed by Sims 'was put to the test within days and the proposition proved to be correct. The tension method of gauge control was established'.[13] Initially the proposition was tested using a manually controlled rolling mill with the tension applied to the strip as it emerged from the mill being varied to keep the rolling load constant. Shortly after this, however, an automatic link between the roll force and strip tension was built and installed with the assistance of L. N. Bramley, Head of the Electrical Engineering Section of the BISRA Plant Engineering Division, and of P. R. A. Briggs of the Electronics Section of the BISRA Mechanical Working Division.[15]

Within a few months it was clear that the tension method of gauge control was capable of improving the quality of rolled material, and in October 1949 Hessenberg and Sims filed a joint application for patent protection on the method in association with BISRA.[16] By 'monitoring the separating force between the rolls or dies and adjusting the tension applied to the sheet or strip in accordance with any variations in the separating force from a predetermined value ... the thickness of the outgoing material may be substantially uniform along its length'.[16] But this method of gauge control, known as the 'tension method', was of limited capability. Only very low tensions could be applied to hot strip, and it was not a convenient method to use when rolling sheet products. So, though the tension method looked promising for the cold strip rolling industry, it was clear to Hessenberg and Sims that an alternative method of gauge control would have to be adopted for hot strip and sheet metal rolling.

The conventional method of gauge control, where little or no tension could be applied to the strip or sheet, was to vary the gap between the rolls of the mill. With constant material quality and mill conditions, a uniform output could be expected. But if, for instance, a portion of material passing through the rolls was harder than the rest a greater force would be applied to the rolls by the material and, in consequence, the mill would stretch so that the roll gap increased. If, however, the setting of the rolls could be adjusted by the mill operator to compensate exactly for the stretching of the mill, then the roll gap would remain constant and the gauge of the output material would also be constant. The essence of the 'setting method' of control was, therefore, the measurement of the force applied to the mill by the material as it passed through the mill, and the adjustment of the roll position control to ensure a constant-thickness product. 'The separating force [acting on the rolls] is proportional to the difference between the outgoing strip thickness and the roll setting. The factor of proportionality is the elastic constant of the mill, commonly known as the mill spring.'[17]

In July 1950, Hessenberg and Sims filed another patent application to protect their invention of the setting method of gauge control. They proposed that 'load responsive elements which will respond to variations in the separating force are incorporated in the mill and linked by mechanical, electrical, and/or hydraulic means with the roll adjusting mechanisms in such a way that the desired relationship between the changes in separating force and roll setting is continuously maintained'.[17] They proposed that the roll position adjusting mechanism should consist of hydraulic rams rather than the conventional screw and nut arrangement. In the BISRA annual report for 1950, it was stated that 'slow-speed experiments on the experimental mill have shown that this method, in which the screwdown gear is adjusted, has possibilities. Hydraulic roll adjustment is required, however, for the proper application of this method, and discussions are being held with suppliers of hydraulic equipment.'[4]

Two experimental mills were used by the BISRA team. In their investigations of the 'tension' method they used a two-high 10 in. × 10 in. mill at Sheffield University that had been modified by Ford in the early 1940s for his investigations into the theory of rolling. For the 'setting' experiments the Association's two-high 6 in. × 5 in. mill was used. Both mills were under the direction of Sims as Head of the Rolling Mill Section of BISRA. Commenting on these early investigations into the 'tension' and 'setting' methods of gauge control, Sims wrote:

> I do not doubt now that modern exponents of control theory would be vastly amused at our halting steps towards realising the mathematical analysis in terms of suitable control systems and hardware, but in 1950/1 these things were in their infancy and we had a lot of troubles. Also about this time it was realised that a fast-response roll adjustment system would be desirable; it was clear enough that the best way of doing this would be with hydraulics and a small experimental mill was equipped with hydraulic control. The control mechanism was the hydraulic relay developed by Ford of the Admiralty for gun turret control, allied to a valve for power amplification. The method was not successful for two reasons. The experiment demonstrated that a hydraulic controller was feasible but it was not practicable on production mills because synchronisation between the two control units, one on either side of the top roll bearings, could not be achieved, neither could we get one controller to do both jobs reliably. Twenty years later the Moog valve, developed primarily for aeronautics application, has solved this problem but it must be confessed that we were somewhat in advance of time to attempt hydraulic control in 1951.[13]

In 1951, then, the BISRA team had proposed and to a limited extent proved two novel methods of gauge control based on the use of load meters to measure roll load and the use of an automatic control system to modify either the tension applied to the strip, or the position of the roll-adjusting screws. Of these, the 'tension' method appeared the most promising system in terms of its demands on the control technology available to the BISRA team. But

> important doubts in the minds of production personnel arose over the possibility of the mill crew handling this complex mechanism without assistance from trained scientists, and without loss in production of strip. Others doubted the stability of the controller when on a mill rolling at high speeds, as the prototype laboratory experiment was limited to 300 ft/min. The only way to resolve these doubts was to

carry out a long-term investigation on an industrial mill rolling strip of sufficient width and at high speed with a large throughput of material. Such an experiment would also demonstrate the reliability of the equipment and instruments, all of which were designed in the BISRA Rolling Mill Laboratory and, at the commencement of the experiment, had not been tried over long periods under works conditions. The Directors of John Summers and Sons Ltd made available the four-stand tandem mill at their works at Shotton for this project and gave every possible assistance and encouragement. The experimental work was commenced in October 1951.[18]

The tests showed that the tension method was stable at rolling speeds of over 1,400 ft/min. and that the quality of the strip was better than could be produced by manual control alone. But it also highlighted deficiencies in the system particularly with respect to the problems of bringing the gauge of the material quickly to the required value, and

it was abundantly clear that the control in its present form was not suitable for use by the mill operator, for there are too few independent safeguards for the mill drive in the event of a maladjustment of the controls. On the electrical side, also, there must be further research to reduce the effect of acceleration and retardation on gauge due to the inertia of the machines and sluggishness in control at low speeds when controlling through the motor field.[18]

The problem of bringing the material on to gauge quickly was of fundamental importance to the development and acceptance of automatic gauge control (A.G.C.), for it would determine the amount of reject material that would be produced by the mill at the start of the rolling operation.

None of the instruments for measuring the thickness of rolled strip which are available at present is entirely satisfactory. The chief disadvantage of existing instruments is that they work close to the strip and are therefore easily damaged by what is known to those acquainted with the art as a cobble. Another disadvantage is that the instruments cannot be placed close to the roll gap so that there is a delay between the rolling of the strip and the indication of its thickness.[19]

But Sims, in a patent application, explained how these problems could be overcome by converting the rolling mill itself into a flying micrometer: 'the measurement of strip thickness is indirect, being derived from a measurement of the initial roll setting and of the force tending to separate the rolls'.[19] This was the 'Gaugemeter' method of thickness measurement. It was derived from an analysis of data already known but now understood with particular clarity. The derivation of the rolled material's thickness depended on only three factors, the force acting on the rolls, the mill-spring, and the setting of the rolls. The mill spring could be derived experimentally, and was known to be linear except at relatively low loads. The roll setting was dependent on the mill operator, and the roll force could be obtained directly as a monitoring and controlling signal from the roll force meters. 'Trials of the "Gaugemeter" strip thickness measurement unit took place during 1953 in the hot strip mill at the Ebbw Vale works of Messrs Richard Thomas and Baldwins Ltd. The accuracy of the "Gaugemeter" was affected to some extent by roll wear but the Ebbw Vale trials showed that normally the "Gaugemeter" worked well.'[20]

With the invention of the 'Gaugemeter' and its demonstration under operational conditions, the way was clear to developing the 'screwdown' method of gauge control. This method was closely dependent on the 'Gaugemeter' principle in that it was the error in thickness of the strip that was to be used as the control signal and it was similar to the 'setting' method of control in that it was the position of the rolls which was to be adjusted to eliminate the error. But whereas the 'setting' method of control adjusted the rolls hydraulically, the 'screwdown' method of control was based on electric motors being used to adjust the screw positions. It had the advantage that it could be fitted to practically all existing modern mills (whereas the setting method would involve considerable modifications) but it suffered from the limitations inherent in using an electric motor for roll adjustment – the most serious of which was the speed response available from the system.

Late in 1953, Sims therefore proposed that a combined system of automatic gauge control should be adopted. He suggested that the combined system would have the best features of the 'screwdown' and 'tension' methods and so give a system of sufficient flexibility to be useful as a production unit. A 'Gaugemeter' unit would be used to measure the output strip thickness and compare it with the desired value. The tension applied to the strip would be varied in an attempt to bring the gauge of the material to its correct value. If the tension that could be applied was insufficient to overcome the error, then the coarser 'screwdown' method would be employed to bring the gauge back to the desired value and reset the tension control to zero.

(c) Development in Davy and United

This system looked promising. 'At the end of 1952 it was decided that the time had come to launch this invention on to the commercial market, and with Sir Charles Goodeve's blessing and active participation I [Sims] joined Davy and United Engineering Co. Ltd [as Head of the Research Department] and the team of men who worked on gauge control joined the company with me shortly afterwards.'[13] The members of the BISRA team who transferred to Davy and United Engineering (D. & U. E.) with Sims were P. R. A. Briggs, W. H. Bailey, K. H. Slack and W. Bagshaw. J. A. Place, who had been responsible for directing most of the experimental trials of the 'tension', 'setting' and 'screwdown' systems, also left BISRA in 1953, but did not join the team at D. & U. E. Bailey had been responsible for most of the mechanical designs, including the precision equipment used in the controllers, Slack and Bagshaw had been concerned with the carrying out of tests on the experimental mills, and Briggs with the development of the electronics instrumentation. Davy and United Instruments Ltd was later formed from the instruments research activities of Davy and United Engineering Ltd with Briggs initially as Deputy General Manager and later Managing Director.

It was on the successful development of the instrumentation that the A.G.C. project was critically dependent. The metal rolling industry was largely unfamiliar with electronic instrumentation and it had to be convinced that the sensing and control units could be built sufficiently robust to withstand the heavy usage of the industry whilst at the same time being accurate enough to provide the mill performance being demanded and simple enough for easy repair and adjustment. The instrumentation had to be stable in operation although it would be used for rolling a wide variety of materials with different hardnesses, at different speeds and with widely varying cross-sections.

The work of developing the instrumentation fell largely to P. R. A. Briggs. After graduating in physics in 1944, Briggs was directed to join the Royal Aircraft Establishment where he spent eighteen months associated with transducer development. After the war he joined the newly-formed BISRA, in the Electronics Section of the Mechanical Working Division under A. L. M. Douglas, and working mainly on the problems of developing the electronics packages for transducers for load, torque and distance measurement.[22] There were many problems to be overcome in developing the electromechanical equipment for the A.G.C. systems:

> We were demanding accuracies of at least an order of magnitude better than anything hitherto achieved. There was another area where a lot of pioneering work had to be done. We found early on that we could not make a D.C. network solve the basic equations of the control mechanism and efforts were made to use a 400 cycles per second A.C. network. The team took advice from the Royal Radar Establishment, Malvern, and the National Physical Laboratory and received most discouraging comment. Nevertheless we proceeded, and largely due to Briggs and Mr S. S. Carlisle [now (1970) Director of the Scientific Instruments Research Association] the desired very high orders of accuracy were achieved.[13]

Carlisle was Head of the Instruments Section of the Physics Department of BISRA until 1953 when he became Head of the South Wales Laboratories at Swansea; in 1955 he became Assistant Head of the Mechanical Working Division of BISRA.[23]

Looking back on their earlier efforts, Briggs commented on the fact that they had been using inadequate instruments and an inadequate understanding of control theory (because little of it existed) to solve a sophisticated control problem. Their inability to obtain hydraulic relays capable of giving the required mill control performance forced them to adopt an electromechanical system and there they encountered limitations

H

on components such as magnetic and rotary amplifiers, relay reliability and the speed response of the control motors. Though the development of the various units of the control system was not perhaps as exciting in terms of the novelty of the designs as the earlier patents on the 'Gaugemeter' system, the importance of this development work cannot be overlooked.

The transfer of the BISRA team to D. & U.E. in 1952–3 was accompanied by the hope that 'gauge control methods will become available towards the end of 1954'.[20] By January 1954, Sims had arranged for the first A.G.C. unit with 'screwdown' control to be installed at the works of the Lancashire Steel Manufacturing Company at Corby. This was installed in July 1955 and the annual report for that year stated that 'the results so far obtained indicate that this type of automatic control will keep the coil very closely to the required constant gauge along its whole length. The resulting higher yield of on-gauge product will give a direct saving and the work of the mill operators will be simplified.'[24] But there were problems, and the most important of these was the fact that the A.G.C. system was found to be

> susceptible to temperature variations in the mill, which manifested itself as a slow drift in the datum gauge setting. As a crash programme we built and commissioned the supervisory loop operated from a contact thickness gauge and, thanks be, it worked superbly. This system of automatic gauge control with the supervisory loop operated from the end of 1955 and established the basic system of gauge control which is in action today.[13]

Thermal drift was a problem which had been recognised from the earliest days of A.G.C. In 1952 Hessenberg and Sims had written:

> Certain difficulties in gauge control . . . have been given greater prominence by these methods [of A.G.C.]. The thermal expansion of the rolls has the same effect as an alteration in roll setting. The control methods described . . . are not capable in their present form of correcting for either roll eccentricity or thermal expansion, but it is not impossible in principle to design additional compensating mechanisms to take care of them.[15]

By 1954 the problem appeared somewhat less important. 'It has been found by experiment that changes in gauge due to roll temperature and wear occur only slowly and the controllers may be reset to correct for such variations without difficulty.'[21] But on the production mill, the problem was not easily dismissed, and so yet another invention was necessary. The mill could be used to control the thickness of the material issuing from it only as long as the control signals were related in a known way to the material thickness. When the rolls expanded more rapidly than the mill housing, it was equivalent to reducing the roll gap and so would alter the relationship between the position of the screws and the thickness of the product.

What was needed was some control that would constantly compare the thickness signal from the 'Gaugemeter' with the actual thickness of the strip, and adjust the 'Gaugemeter' signal accordingly. Sims proposed to do this by placing two thickness gauges of complementary characteristics on the strip line.

> The first measuring means is located close to the mill gap and has a relatively short-term stability; because of its short-term stability it can be a relatively inexpensive device. The second measuring means is located at a safe distance from the mill and has a relatively long-term stability; the latter ... may therefore be relatively expensive, but, being disposed remotely from the mill, is not liable to damage. The thickness of the strip is controlled mainly by the first measuring means, so that there is little time-lag between the occurrence and the detection of an error in the thickness of the material but its operation is monitored by the second measuring means so that errors, resulting from the short-term stability of the first measuring means, may be compensated.[25]

The monitor was an elegant solution to the problem; it worked well, and may be considered as a further reason why Hessenberg considered that Sims was 'an extremely good developer';[12] the essential characteristic of the development phase for any project is the repeated detection and elimination of problems as the project is brought to fruition.

Yet another problem became apparent as the D. & U.E. team began to increase the capabilities of the 'Gaugemeter'/

A.G.C. system. The team had shown that for loads above a certain value, the mill stretched with a linear relationship between applied load and mill elongation.[21] Below that critical value the relationship was predictable though non-linear. Sims had made use of this linear relationship in his 'Gaugemeter' patent,[19] but 'recent developments in the design of rolling mill housings have produced instances where the normal rolling load is lower than the critical load . . . and in this case the signal will differ from the true gauge error'.[26] By designing a servo-controlled balancing circuit, Briggs modified the output signal from the 'Gaugemeter' so that the mill could be used to roll materials accurately even into the non-linear portion of the mill extension roll load curve.

With the gradual development of the A.G.C. system and with the elimination of the problems of thermal drift and non-linear extension of the rolling mill, only one major problem was left to solve. This was the problem of roll eccentricity. Although considerable efforts had been devoted to improving the methods used for grinding the rolls, difficulties were still being experienced through the rolls imparting a waviness to the strip corresponding to their eccentricity. To overcome this problem, Smith of D. & U.E. invented[27] a summation device to detect and then subtract the errors due to roll eccentricity from the 'Gauge-meter' signal prior to it being fed to the controller. It was yet another way of improving the system, building on the foundation of the three primary patents [16,17,19] of 1949–51.

Sims has commented:

> I think it would give better recognition to the fine work done by Briggs and his colleagues if it was mentioned that I left Davy and United Instruments in 1959 to become Director of Engineering of Davy and United, and subsequently joined the National Coal Board and Director-General, Research and Development, in 1964. I took up my present position (Chief Executive, Finance, British Oxygen Company) in November 1969. I think this would give a clearer measure of the responsibility carried by Briggs and his colleagues in meeting the commercial exploitation of the invention which could be said by now to have reached its plateau in marketing.[28]

(d) Comments

Many factors are of interest in this study, but perhaps the most important and most interesting is the way in which a research association fostered first within itself, and later in the market place, a development which was to be of major significance to the rolling mill industry. The charge is often made in Britain that there is too little expertise available in commercially exploiting the inventions which are said to be in relatively plentiful supply. In this instance, not only were the inventions of the gauge control methods made, but in order to exploit them, the team responsible for their initial proving moved to the commercial environment to complete the process. The invention and initial proving of the novel methods of gauge control were carried out relatively removed from the glare that they would have been subjected to in the works either of users or of manufacturers of rolling mills. The seedlings were transplanted when they were big enough to survive, but with the added advantage of having the gardener travel with them.

In the history of the 'Gaugemeter'/A.G.C. system one can see once more the influence of scientific investigation on the methods and equipment used in a traditional industry. When Sims started work on the design of improved rolling mills he found a relatively unresearched field of investigation waiting for him. Following his consideration of Hessenberg's proposal that strip gauge uniformity could be improved by holding strip tension constant, Sims went one step further and advocated using strip tension as a means of control, the load on the rolls being monitored and held constant by varying the roll tension. This method, 'the tension method', gave encouraging results and led directly to the 'setting' method where the load on the rolls was again measured and the position of the rolls was adjusted to maintain a constant roll load. From the setting method was developed the screwdown method of control where the position of the rolls was varied by an electromechanical drive system. The 'Gaugemeter' concept emerged from the realisation that the thickness of the product could be determined once the roll load, mill spring constant and roll position were known. This made it possible to develop a rolling mill system which controlled the thickness of the material being rolled whilst it was still in the roll gap. Further refinements were necessary to develop the instrumentation to the required accuracy and repeatability, but they were refinements based on the fundamental work of Hessenberg and Sims in particular.

Support for the A.G.C. team came from the top – one of the original inventors of the scheme, Hessenberg, was the Deputy Director of BISRA, as well as being the Head of the Mechanical Working Division. He enjoyed the whole-hearted support of the Director of the Association, Sir Charles Goodeve, and so, when the moment came to transfer the invention to industry, it was encouraged at a high level. Similarly, in 1953 the representative of D. & U.E. in BISRA was Mr (now Sir Maurice) Fiennes, the Managing Director of D. & U.E. (and later Chairman of the Davy-Ashmore Group) and it was with Fiennes that the agreement for the transfer of the team under Sims was concluded.

Initially D. & U.E. were granted a non-exclusive licence by BISRA for the sale of units incorporating elements of the A.G.C. and 'Gaugemeter' concepts, but this was later modified to an exclusive licence for sales to the rest of the world and a non-exclusive licence for United Kingdom sales. D. & U.E. subsequently sub-licenced four firms in the United States, two in Canada, two in Germany and one each in Japan and Switzerland to produce A.G.C. installations based on the BISRA/D. & U.E. patents. In the United States, the General Electric Co. was particularly keen to obtain a licence, for they were interested in obtaining orders for hot rolling plant. In the hot mills, many motors are used to ensure that the product will have a minimum cooling period prior to being processed. Since General Electric were actively engaged in the production of motors, taking out a licence on the A.G.C. was a good way of ensuring that they would retain their command of the motor market for this type of application.

As is true with nearly all inventions, the innovation process – that of introducing the invention into the production function – was not without its difficulties. The rolling mill industry is an old one and it was clear in the early days of A.G.C. development that

Many problems still remain to be solved in applying the controller to industrial mills; not least are the retraining of mill crews in the use of the device, the rearrangement of rolling schedules, and the abandonment of practices that have been hallowed by time and custom. A more important problem is that of maintenance; the controller is a complex mechanism, and includes several electronics units that will need checking and adjustment by properly trained personnel.[8]

Both these problems – the retraining of personnnel and the replacement of practices and skills that formed the basis of the hierarchy of workers – had to be overcome by the A.G.C. team in its efforts to promote the system.

Yet another source of resistance to the system was provided by the financial and operating sectors of the rolling industry. Once one rolling mill user adopted the A.G.C. system successfully there would be considerable pressure on all the other manufacturers to adopt similar plant with equally high standards of product quality. Few wanted to start a re-equipment race that would involve all of the mill owners in an expensive round of mill purchases unless it was absolutely necessary. There could, of course, be economic benefits from installing A.G.C. plant: it could be expected to give a higher yield of saleable product, it should stimulate sales, and it should lead to greater productivity in the industry. But, as Sims remarked, 'people remained to be convinced',[13] and so the first delivery of the A.G.C. system was made on a 'sale or return' basis. With the establishment of the system as a viable proposition, however, the sales of A.G.C. installations increased and few rolling mills are now installed without some form of A.G.C., at least in the finishing stages.

The patents on the A.G.C. and 'Gaugemeter' systems were drawn up by Mr J. R. Batchellor 'with such skill that they have never been broken'.[13] From the efforts to get around these basic patents have come the more recent hydraulic controllers which, by 1969, had been developed to give response times over twenty times faster than the original electromechanical systems with forces of over 1,000 tons being applied by each hydraulic capsule. Such response characteristics open the way for the elimination of the problem of roll eccentricity by modulating the force on the rolls to compensate fully for the roll shape. But just as there was resistance to the introduction of electronics into the mills because of the problems of retraining, etc., so there is evidence of similar resistance to the introduction of hydraulic controllers.

References

1. E. C. Evans *et al.*, *First Report of the Rolling Mill Research Sub-committee, of the Iron and Steel Industrial Research Council*, Iron and Steel Institute Special Report No. 34 (1946).
2. B. A. Bland and H. Ford, 'The Calculation of Roll Force and Torque in Cold Strip Rolling with Tensions', *Proc. Inst. Mech. Eng.*, **155** 144 (1948).
3. E. Orowan, E. A. W. Hoff, J. Los and BISRA, 'Improvements in or related to Stress Indicators'. British Patent No. 626,206, applied for 2 Dec 1944, published 12 July 1949.
4. BISRA, *Annual Report of Council for Year Ending 31 Dec 1950*, pp. 19–21.
5. E. Orowan, 'Graphical Calculation of the Roll Pressure with Assumptions of Homogeneous Compression and Slipping Friction', *Proc. Inst. Mech. Eng.*, **150** 141 (1943).
6. E. Orowan, 'Calculation of the Roll Pressure without the Assumptions of Homogeneous Compression and Slipping Friction', *Proc. Inst. Mech. Eng.*, **150** 146 (1943).
7. H. Ford, 'Researches into the Deformation of Metals by Cold Rolling', *Proc. Inst. Mech. Eng.*, **159** 115 (1948).
8. R. B. Sims, J. A. Place and P. R. A. Briggs, 'Control of Strip Thickness in Cold Rolling by Varying the Applied Tensions', *Journal of the Iron and Steel Inst.*, Apr 1953, pp. 343–54.
9. E. Meyer, 'Improvements in or relating to Means for Regulating the Relative Adjustment of the Rollers of Rolling Mechanisms'. British Patent No. 441,974, applied for 13 May 1935, published 30 Jan 1936.
10. Aluminiumwerke A.G. Rorschach, 'Installation pour laminer des bandes en métal dont l'épaisseur est réglée automatiquement'. French Patent No. 892,237, applied for 17 Dec 1942, published 31 Mar 1944.
11. P. Blain, *Revue de Métallurgie*, **45** 8 (1948).
12. W. C. F. Hessenberg and B. A. Jessop, Research and Development Dept of British Steel Corporation (previously BISRA), interviewed by W. G. Evans, 8 July 1970.
13. R. B. Sims, letter to W. G. Evans, 10 Aug 1970.
14. P. R. A. Briggs, Director and General Manager, Davy and United Instruments Ltd, interviewed by W. G. Evans, 20 Oct 1968.
15. W. C. F. Hessenberg and R. B. Sims, 'Principles of Continuous Gauge Control in Sheet and Strip Rolling', *Proc. Inst. Mech. Eng.*, **166** 75 (1952).
16. W. C. F. Hessenberg, R. B. Sims and BISRA, 'Improvements relating to the Production of Sheet and Strip Material'. British Patent No. 681,373, applied for 11 Oct 1949, published 22 Oct 1952.
17. W. C. F. Hessenberg, R. B. Sims and BISRA, 'Improvements in and relating to the Production of Metal and Other Sheet and Strip'. British Patent No. 692,267, applied for 5 July 1950, published 3 June 1953.
18. R. B. Sims, J. A. Place and P. R. A. Briggs, 'Works Trial of the "T" Method

of Automatic Gauge Control', *Journal of the Iron and Steel Inst.*, **173** 354 (1953).

19. R. B. Sims, 'Improvements in or relating to the Measurement of Thickness in the Production of Sheet and Strip Metal'. British Patent No. 713,105, applied for 27 Nov 1951, published 4 Aug 1954.
20. BISRA, *Annual Report of Council for Year Ending 31 Dec 1953*, pp. 33–40.
21. R. B. Sims and P. R. A. Briggs, 'Control of Strip Thickness in Hot and Cold Rolling – Application of Automatic Screwdown Technique', *Iron and Coal Trades Review* (1954) pp. 559–66.
22. R. B. Sims, J. A. Place and A. D. Morley, 'Loadmeter for Industrial Mills', *Engineering*, **173** 116 (1952).
23. BISRA, *Annual Reports of Council for years ending 1952, 1953, 1954.*
24. BISRA, *Annual Report of Council for Year Ending 31 Dec 1955.*
25. R. B. Sims and Davy and United Instruments Ltd, 'Improvements in or relating to the Control of Thickness of Material'. British Patent No. 945,058, applied for 24 Mar 1959, published 18 Dec 1963.
26. P. R. A. Briggs and Davy and United Instruments Ltd, 'Improvements in or relating to the Measurement and Control of Thickness in the Production of Sheet and Strip Material'. British Patent No. 946,820, applied for 24 Mar 1959, published 15 Jan 1964.
27. J. P. Smith and Davy and United Instruments Ltd, 'Improvements in or relating to Rolling Mills'. British Patent No. 947,525, applied for 8 May 1961, published 22 Jan 1964.
28. R. B. Sims, Chief Executive, Finance, British Oxygen Co. Ltd, letter to W. G. Evans, 20 Oct 1970.

11 ENGLISH ELECTRIC:
BRUSHLESS GENERATORS FOR AIRCRAFT

Aircraft electrical equipment is subject to design constraints considerably different from those found in normal industrial practice. The need for the lowest possible weight and bulk is always present and of overriding importance. The operating environment is subject to the extremes of temperature, pressure and humidity associated with variations in altitude and geographical location. The levels of vibration and acceleration are often severe. Ease of maintenance is desirable and the reliability requirement is as stringent as would be expected for equipment on which depends not only the operating economy of the aircraft but its very safety.

The Queen's Award won by English Electric's Aircraft Equipment Division (A.E.D.) at Bradford was for technological innovation in reducing the weight and increasing the reliability of generating systems for aircraft. These developments enable the electrical systems of aircraft to operate at very high temperatures and altitudes above 60,000 ft, which are within the normal range of conditions for supersonic civil and military flight.

(a) Development of electricity generating equipment for aircraft [1, 2]

The evolution of modern generators and their associated equipment has been influenced by the two major aircraft technology developments of recent years: jet propulsion and supersonic flight. The capabilities of the most recent generating equipment have been developed in direct response to the increasing demands made by progressively faster and more sophisticated aircraft. To get a clear appreciation of how generating equipment has evolved, it is necessary to examine the growth in this demand for electrical energy in some detail.

Up to 1935, aircraft electrical supplies had almost invariably been direct current (d.c.). The first major application was associated with early radio; the transmission which was made from the airship *Beta* in 1911 was among the earliest uses of d.c. Prior to this, the only use of electricity in aircraft had been in the engine ignition system.

Although aircraft generators and their associated equipment progressed steadily from this early stage to the relatively sophisticated engine-driven, blast-cooled unit of the large four-engined aircraft of the 1940s, they suffered from the inherent limitations common to all commutator-type d.c. generators. From the aircraft manufacturers' point of view, the most important effect of this was a limitation on the maximum power – 'the maximum installed capacity'. In any electrical system, power requirements determine the operating voltage. With increasing demands, this voltage had already been raised from 12 volts to 24 volts, then to 94 volts and in some aircraft even to 112 volts, in an effort to reduce cable weights and large current switching problems. In addition, the need to run at high speeds and to reduce generator and gearbox weight led to limitations in machine size. In practice, by the late 1940s, four-engined aircraft had a maximum installed capacity of about 50 kilowatts. By the mid-1950s, the demand was for about 100 kilowatts and by the mid-1960s a maximum installed capacity of some 300 kilowatts was required.

The alternative to d.c. generators was to change to alternating current (a.c.). This permitted the voltage to be increased to 200 volts. Various arrangements were tried during the 1940s but the main difficulty with a.c. generation was that the frequency of oscillation of the voltage is a function of the speed of rotation of the generator armature. Constant-frequency alternating current was, however, necessary to achieve minimum weight and to enable the generators to be operated in parallel. This difficulty was resolved with the successful development in the United States of precision lightweight hydraulic constant-speed gearboxes which overcame the problems of variation in the speed of the generators when driven direct from the main engines. This type of system operating at 400 cycles per second was introduced by English Electric in Europe in the early 1950s on the Vulcan Mark 1 bomber and has subsequently emerged as a standard choice for large aircraft in the 1960s.

However, a more specific problem emerged. It was already well known that commutator brushes wear more rapidly at low air pressure and humidity. One of the aspects of improved aircraft performance was sustained flight at increasingly higher altitudes at which these very conditions exist. Reduced brush life began to present problems in certain applications. The outcome was the development of the brushless generator in which the rectifiers are mounted on the generator shaft. This generator is now standard for the majority of applications.

The concept of brushless machines is not new and patents for them extend at least as far back as 1890.[3] In the conventional generator, a coil of wire is forced to rotate in a permanent magnetic field. A voltage is induced in the coil which in turn gives rise to a current. The current is removed from the coil and transmitted to a load by means of a commutator and carbon brushes. It is the friction between the commutator and carbon brushes which gives rise to the problem of brush wear.

By contrast, a brushless generator is a two-stage machine. The first stage is like a conventional generator except that it is inside out. Instead of rotating a coil in a magnetic field, the coil (called the stator coil) is fixed and the magnetic field is made to rotate. The second stage of the machine is concerned with setting up this magnetic field. A constant magnetic field may be established by passing a direct current through a coil of wire. The second stage, then, is a conventional generator in which another coil, mounted on the same shaft as the electromagnet coil in the first stage, is rotated in the magnetic field of a permanent magnet. The resulting alternating current is rectified by means of diodes mounted on the shaft and passed through the electromagnet coil in the first stage. This direct current sets up a constant magnetic field, which, because it is rotating, induces an alternating current in the stator coil. The need for the carbon brushes is thus eliminated.

The moving parts of the generator – the two coils and the semiconductors – are all mounted on the same shaft and are driven at high speed by the aeroplane engines. The rectifiers in particular are subjected to extremes of acceleration and consequently must be of sound construction.

Besides increased rates of brush wear, operation at high altitudes introduced the problem of generator cooling. Owing to the extreme rarification of the air, great penalties of size and weight began to be involved in retaining the conventional means of cooling by forcing air through the generator. The solution to this was to devise some means of rendering the machine free of its environment. In effect, this meant hermetically sealing the case of the generator, pressurising with gas and cooling by some secondary means not associated with the external atmosphere. For most applications, indirect liquid cooling with an oil coolant is used.

(b) The English Electric innovation[4]

During the 1940s, English Electric at Bradford were involved in manufacturing industrial motors and generators. Manufacture was undertaken of aircraft actuators for English Electric at Preston and a separate section was formed to deal with this equipment.

When the decision was taken in 1953 to extend the activities of this section into aircraft generating equipment, it was felt by some, including Mr P. J. Daglish, the Divisional Manager, that the requirements of the aircraft manufacturers not only for increased capacity but also lighter weights and better reliability would not be resolved by an extension of the same concepts to higher voltages. They suggested that further developments could only be made using alternating current techniques; indeed, this trend had already begun to emerge in the United States.

At this time, also in the United States, there was a significant invention – the constant-speed drive. As was explained above, the main problem in moving to a.c. generators arose because the generators themselves are driven by the aircraft engines. The speed of the engines varies considerably between the idling position on the runway and its top speed. Therefore, before any thought of a.c. generation could be undertaken, the problem of a constant-speed drive had to be solved. This problem had produced many answers, but the innovation of the Sunstrand Corporation at Rickford, Illinois, was the first device to be commercially successful. English Electric went to Sunstrand and successfully negotiated a licence agreement. In this venture they were backed by the then Ministry of Supply who were also interested in providing generating equipment for aircraft.

Although the constant-speed gearbox solved the problem of providing constant-frequency a.c. at sufficiently high voltage to meet modern aircraft requirements, both a.c. and d.c. systems remained unreliable in high-altitude conditions. English Electric began a frontal attack on this problem in 1956. An additional impetus was given to the development of reliable a.c. systems when a Valiant bomber in 1957 and a Vulcan 1 in 1958 crashed as a consequence of the failure of their 112-volt d.c. systems.

In 1957–8, semiconductor diodes with suitable power ratings became available commercially. As a result of this and other developments in the technology of brushless equipment, the A.E.D. made a decision to close the whole of its existing interests in aircraft generators and to initiate a fresh programme of development which had as its aims:

(i) To develop a.c. and d.c. brushless generators and associated control gear using semiconductors to eliminate all the electromechanical devices such as brushes, commutators and relays. Doing this would eliminate the equipment's sensitivity to high temperatures and the low pressures encountered at high altitudes.

(ii) To reduce the weight of electrical generating systems.

In addition, English Electric decided to invest £500,000 in research and development laboratories at Bradford. These were primarily 'environmental' laboratories for testing the generators under conditions of low pressures and extremes of temperatures. The investment was financed entirely by private venture capital. A programme of environmental testing on bearings, insulation and semiconductors was undertaken. The semiconductor problem was rather difficult because of the extremes of high temperature and centrifugal force to which the semiconductors were subjected. Manufacturers of semiconductors, it seems, were unable to meet the specifications set by English Electric. The first development programme was undertaken by English Electric in conjunction with the Ministry of Supply on the performance of semiconductors under extreme environmental conditions; it lasted some two and a half years.

(c) Comments

Although there are many factors which contributed to the commercial success of English Electric's line of brushless generating equipment, one seems to stand out over the others. The company seems to have an active policy of trying to interpret customer requirements in advance of the customer specification. English Electric prefer to develop a range of prototypes and offer it to the customer with most of his problems solved. Clearly, a high degree of anticipation is needed for this approach, and it necessarily involves close contact between marketing, production, engineering design and development.

Basically, the process of 'anticipation in innovation' starts in the marketing section where, through a study of trends in aircraft development and a vast network of contacts, English Electric engineers try to anticipate the next types of generator that will be required by their customers.

Parallel with this, there is a blanket development programme covering all the major areas of generator design: heat transfer, electromagnetic design, insulation selection, rectification and mechanical design. Each of these areas is developing in its own right – of course with a view to ultimately incorporating the results in a generator. Thus when the marketing section 'anticipates' the type of generator which will be required, the latest developments in the development section are combined to make the best possible prototype consistent with customer needs. As a result, when a specific contract appears, the commercial people are ready with a working prototype which contains at least some elements of the customer's requirements. Of course, once the prototype is assembled, there is nothing to preclude further development in points of detail.

Another important factor in stimulating the innovation seems to have been Daglish's alacrity in switching to a.c. generating systems and, further, in obtaining quickly the Sunstrand licence. The other firms in the field stayed longer with their current lines of development and went on to produce 94- and 112-volt d.c. generators. By contrast, in the United States, these two developments were by-passed in favour of the a.c. system and, in this respect, it may be said that English Electric followed the Americans. Daglish had started his career with Parsons after obtaining a B.Sc. in heavy electrical engineering from Durham University. He came to English Electric in the early 1950s and joined the staff of the Directorate of Engineering Office. This office has been described as the place 'where all the bright lads go'. Its function is to carry overall responsibility for co-ordination of engineering activities. In 1953 Daglish was appointed Divisional Manager of the A.E.D. He was in charge there until 1960 and later became Managing Director of English Electric Diesels.

References

1. J. R. Gledhill, 'Recent Developments in Electric Power Generating Equipment for Modern Aircraft', *English Electric Journal*, **21** 6 (1967); see also **22** 1 (1968).
2. The author would like to thank Mr Paul Drath for his assistance in preparing this section.
3. P. Strassmann, *Risk and Technological Innovation* (Wiley, New York, 1967) p. 158,
4. Interviews with Mr N. Cole, Sales Manager, Mr Lynch, Commercial Manager, Mr Hart, Chief Engineer, English Electric, Aircraft Equipment Division, Bradford, 1968.

12 ENGLISH ELECTRIC: FUSES FOR SEMICONDUCTOR DEVICES

The development of fusegear technology has been pre-dominantly 'need-oriented'. The requirements of the electricity supply industry for protection equipment have resulted, among other things, in a continuing supply of more accurate and more reliable fuses opera-ting at both low and high currents and voltage ratings. With the advent of solid-state rectifiers in the early 1960s, a need arose for the appropriate fuses to protect diodes and thyristors. This case study provides a good example of technology transfer 'on the hoof', in that the techniques used to manufacture the fuses were brought into English Electric principally through the transfer of an individual from the electronics industry.

(a) Fuse technology[1]

Fuses have to be designed to perform both passive and active functions: to carry varying loads throughout their life and to interrupt fault currents. From the designers' point of view these two requirements are conflicting, because a large cross-sectional area of conductor or element is desirable for cool running, whereas a small cross-section is preferred for rapid interruption.

The high rupturing capacity (h.r.c.) fuse, as a method of obtaining reliable short-circuit protection at high values of fault current, was pioneered in Britain in the 1920s and developments of the basic principle continue today at an undiminished rate.

All designs of h.r.c. fuses utilise, in various ways, a pheno-menon which occurs when a heavy current is passed through a conductor of relatively small cross-section. If the current is just sufficient to melt the wire, an appreciable time is required for the conductor actually to melt and break. An arc is formed at the break and it elongates until the system voltage can no longer sustain it and the circuit is finally opened. On the other hand, when a similar wire carries a very large current, it is rapidly and uniformly heated along its whole length because there is no time for heat dissipation to occur. The wire, however, does not melt uniformly along its length but breaks into a number of globules and arcs are formed between them. This series of arcs produces a sudden increase in the electrical resistance of the circuit and the current is quickly quenched. In fact, these multiple arcs quickly merge into a single one, but by this time the current has been reduced to a very low value.

The phenomenon whereby a melting wire breaks up into multiple arcs is due to the 'pinch effect' and may be briefly explained as follows. The skin of the wire melts before the core owing to the effects of self-inductance which at high current densities causes the current to concentrate on the skin of the wire rather than at its centre. This high current density also produces electromagnetic forces which apply a constricting force on the wire causing it to break into a number of globules long before sufficient heat has been generated to melt the complete volume of the wire.

It is upon this mode of operation that the design of all h.r.c. fuses is based. The fuse element may be fabricated from many metals, but silver is usually used because of its very low resistivity and specific heat. It is cut from a ribbon of high-purity silver to resemble a concertina. The shape of the silver element is designed to achieve two objectives: the necks (reduced sections) ensure rapid operation under short-circuit conditions while the wider sections help to minimise total resistance and thus reduce the power loss of the fuse when operating normally.

Under short-circuit conditions the element melts at the necks and the bulk of the arcing is confined to these regions. The total length of the necks governs the arc voltage of the fuse mainly by means of the pinch effect described above. 'Arc voltage' is defined as the extra voltage required to maintain current flow through the series of arcs which have been formed, creating a high resistance path. This extra voltage can only be obtained from the stored energy in the circuit due to a change in current. The result is that the current ceases to rise and reduces to zero at a rate governed by the magnitude of the fuse arc voltage.

There are, of course, other factors which have to be considered in successful fuse design, such as the thermal properties of the medium in which the fuse element is immersed, but for the purposes of this discussion they may be ignored as they represent fairly constant factors in all designs. It is sufficient to keep in mind that the performance and rating of the fuse is a critical function of the shape of the element. Since these elements are required in large numbers, they are usually punched out by drawing the ribbon automatically over a die.

In certain applications the accuracy required at the necks of the fuse element may be such that the conventional punching process is not sufficiently accurate. One case where this is found is in the design of fuses for the protection of semiconductor devices.

The outstanding characteristics of semiconductor rectifier diodes are their large power output and small size. This makes them inherently difficult to protect because their low mass involves low thermal capacity and low thermal inertia. They are therefore susceptible to damage from over-currents even for very short times. A further complication is that these semiconductors have a limited ability to withstand over-voltages. To be suitable for protection, a fuse must generate only a small voltage during fault operation, otherwise the reverse voltage capability of the diode would be damaged.[2]

(b) The English Electric innovation[3]

One of the innovations for which English Electric won a Queen's Award was concerned with the production of fuses for the protection of semiconductor devices. The basic principles of the h.r.c. fuse have been used but the method by which the fuse elements are manufactured was new.

There appear to be two main factors influencing this innovation: a more scientific approach to problems of fuse design in general and the introduction of a novel technique. The driving force behind a more scientific approach to fuse design seems to have been Mr E. Jacks, then Chief Engineer of the company's Fusegear Division near Liverpool. Jacks came to English Electric in 1954 when fusegear practice, at least at English Electric, was based on a 'suck it and see' philosophy. He sought to set up facilities so that fuses could be tested in an 'orderly manner' by studying them in a 'properly controlled environment'. In fact, he discovered that the fuses were sound enough but design was governed more by experience than by scientific knowledge.

None the less, Jacks seems to have realised that as requirements for fusegear became more complex, a more scientific
determination to the parameters of importance in fuse design
would be necessary. This approach bore fruit in the design of
fuses for semiconductor devices because with these devices 'the
degree of protection depends not only on the performance of the
fuse and the device it protects but also upon the degree to
which the performance can be expressed in terms that are
mutually meaningful'.[4]

It is possible that Jacks's familiarity with the multiple interaction of a fuse with its environment explains why he adopted
the policy of hiring his staff from a broad base of engineering
skills. This attitude turned out to be of particular importance
in the late 1950s when the Fusegear Division was beginning to
look for a solution to the problem of fuses for semiconductor
devices. It was at this time that Jacks originally conceived the
idea of using printed circuit techniques. The real incentive
here was to overcome the difficulty of having to manufacture
mechanical tools for punching out silver strip.

> In fuse development it is necessary to produce a large number
> of prototype samples in order to get a good statistical spread in
> the testing [of the fuse elements] and it is not practicable to
> manufacture elements by hand. Even where the elements can
> be designed mathematically, this results in a large number of
> alternatives and it is not economic to make punching tools for
> large numbers of alternatives in this way. The photofabrica-
> tion method overcomes this problem and this was its first
> attraction to me.[5]

The first attempts failed because the printed circuit techniques
then available were not accurate enough and it was realised
that the inaccuracies lay mainly in the photographic process.
It was for this reason that Jacks began looking for a person
skilled in industrial photography and

> it was a matter of sheer luck that one day a young electrical
> engineer applied to me for a job . . . and mentioned casually
> that he had a qualification in industrial photography which
> he pursued as a hobby. I employed him and teamed him up
> with two other people who were skilled in fuse design and
> we carried on from there.[5]

The job was not a trivial one as different etching techniques were necessary when working with silver rather than copper. In addition, the whole process had to be more accurately controlled since photofabrication was required in three dimensions instead of the two which usually suffice for laying copper wires on circuit boards. The successful development of photofabrication techniques appears to have benefited greatly from close collaboration with the Kodak Company and was completed within two years.

(c) Comments

The original development team numbered three, bringing together photographic technology, etching processes and fuse technology. None of the initial team members possessed university degrees but all possessed H.N.C. qualifications. The speed with which the development was completed has been partly attributed to the skills of the initial team and partly to the attitude of Jacks who, according to some, 'gave them a fairly free hand which allowed them to get on with the job'. In addition to keeping in close touch with the development team, Jacks personally undertook the task of probing the market, both at home and overseas, to find out directly from the semiconductor manufacturers exactly what their requirements were because the progress of the physical design of the fuse elements was crucially dependent upon accurate evaluation of the various situations in which the fuses would be used.

It appears as if Jacks's hiring policy has met with resistance from time to time. Some of his colleagues felt that, as there was only a limited budget with which to finance fuse development, it might be best utilised by hiring specialists in heavy current engineering. On the other hand, Jacks held that it was possible to design modern fuses on mathematical and scientific bases and, consequently, he needed people with these kinds of skills rather than the conventional ones. In addition, he has always insisted that his staff travel to conferences and take positions in various professional organisations.

References

1. I. Feenan, 'The Protection of Electrical Apparatus with H.R.C. Fuses', *Electrical Journal*, July 1961.
2. E. Jacks, 'The Role of the H.R.C. Fuse in the Protection of Low and Medium Voltage Systems', *English Electric Journal*, **17**, no. 3 (Sep 1961).
3. Interviews with E. Jacks, Commercial Director, and W. Clarke, engineering staff, English Electric Fusegear Division, 1968.
4. E. Jacks, 'The Fundamental Behaviour of High Speed Fuses for Protecting Diodes and Thyristors', Institute of Electrical and Electronic Engineers International Convention (New York, 1968).
5. E. Jacks, private communication, 1970.

13 ENGLISH ELECTRIC/MARCONI: THE MARK VII COLOUR TELEVISION CAMERA

Many interacting technologies often contribute to modern products having a high technological content. It is sometimes more appropriate to speak of the next 'generation' of equipment rather than of a new product. The emergence of a new generation of equipment usually implies that recent advances in each of several contributing technologies have been employed in the latest model. Thus, technological innovation in the context of continuing development may mean that new techniques have been applied in a situation which is relatively well understood and that certain improvements in such parameters as reliability, performance or costs have been achieved.

The Marconi Company at Chelmsford, Essex, has been concerned with the design and development of colour television systems since the early 1950s. The new Mark VII colour television camera provides an interesting example of the dependence of a new generation of colour TV cameras on both technological advances and on developments in overseas markets.

(a) *The Mark VII colour television camera*[1]

Most present colour television systems accept the same form of input from the camera, their differences beginning with the manner in which the signal is processed before transmission. Thus, the Marconi Mark VII camera makes the same standards of picture quality available to users of the American (N.T.S.C.), the French (S.E.C.A.M.) or the German (P.A.L.) transmission systems.

The Mark VII is a four-tube camera, having red, blue and green colour tubes plus a separate fourth tube for the black-and-white luminance transmission. The red, blue and green components of the image are sent over a separate narrow band-width colour (chrominance) channel. This configuration is important to picture quality as judged by the human eye because fine detail is seen in black and white whereas colour is seen with lower resolution. From the operating point of view, it is also a practical approach to colour transmission because the use of a separate camera tube for the black-and-white channel greatly reduces the dependence of the colour picture on extremely accurate alignment of the pictures from the three colour tubes. If the colour alignment system begins to drift because of a temperature fluctuation, for example, the black and white picture is unaffected while a defect in a colour tube blemishes the colour picture only where, for reasons associated with the human eye, faults are less easily perceptible. In addition, colour scenes often include black and white subjects such as newspapers and in these cases the additional black-and-white channel is an advantage.[2]

The heart of the television camera is the television tube. In most tubes, an optical image is converted into a pattern of electrostatic charges which is scanned by an electron beam to produce electric signals. The camera tubes in most general use at the present are the image orthicon, the vidicon and, more recently, the plumbicon. The image orthicon uses photo-emission and secondary emission to form the charge pattern while the vidicon and plumbicon are based on photoconduction. Each of the tubes has advantages and disadvantages and a wide variety of considerations must be examined before a given tube is chosen.[3]

A compact, lightweight camera, stable in operation, has been achieved in the Mark VII by the use of the plumbicon photoconductive camera tube developed by Philips at Eind-hoven, Holland, during the 1950s. The plumbicon is a new television camera pick-up tube similar in many respects to the vidicon but having a lead oxide photosensitive target. Its advantages over the vidicon, besides enhanced temperature stability, are freedom from 'lag' which causes the ghost outline of rapidly moving objects, and the absence of 'dark current' (a residual output from the vidicon even when not illuminated) which tends to make reproduction of pictures in low-level lighting difficult.

All the camera electronic circuits are transistorised except the television camera and viewfinder tubes. For 'hands off' operation, high stability is essential and, to assist in obtaining this characteristic, the Mark VII utilises over sixty thin film circuits. The thin film technique is used here for stability and reliability rather than for miniaturisation which is usually the main consideration.

The most important technological innovations embodied in the Mark VII are the use of four photoconductive plumbicons rather than the more common three-tube configuration and the adaptation of thin film circuits for increased stability. Beside these, the complete television camera has required develop-ments in many other fields as well. Some of these include: a tilting viewfinder and adaptability for zoom lens applications; the 'screened yokes' for protection of the camera tube against unwanted magnetic fields; and enhanced sensitivity and increased operational range (both of which are important for outdoor applications).

(b) The Marconi innovation[4]

The Marconi Company has been closely associated with the development of television services since their inception in 1936, when they provided transmitters and aerials for the first public television service opened in London by the British Broadcasting Corporation (B.B.C.). In the years immediately following the Second World War, the company developed its first television studio equipment and, in common with most international manufacturers, its cameras used the 3-in. image orthicon pick-up tube.

Marconi moved into colour TV around 1952 when it supplied colour equipment for 'operation flower shop' (so called because the building in which it was carried out was at one time a flower shop) at the request of the Postmaster-General's Advisory Committee. The Advisory Committee was facing the general question, 'Should Britain go for colour?', and Marconi, because of its experience with television in general, was asked to give a demonstration of some colour TV facilities.

In 1954 Marconi introduced the first of its range of black-and-white television cameras employing the $4\frac{1}{2}$-in. image orthicon tube. This technological innovation was an important one for Marconi's future as a television equipment supplier and it helped greatly to establish the company's reputation in the field. It is worthy noting that Marconi received in 1961 an award from the National Academy of Television Arts and Sciences in recognition of its work in pioneering the $4\frac{1}{2}$-in. image orthicon. This was the first occasion that an award was made to a company outside the United States. During 1953 Marconi, under the terms of its licence agreement with the Radio Corporation of America (R.C.A.), built a colour TV camera for the B.B.C. colour demonstrations. The basic R.C.A. design was used, but it was built in Britain using British components. Some cameras of this type were sold to the B.B.C. in 1955. In addition, small numbers were built using the 3-in. image orthicon, but the cost was about £65,000 per camera and since, at that time, there was no sign of a home market developing, there seemed little justification for designing a new camera from scratch.

Because of the uncertainties in the American colour TV market, Marconi appears to have held back any major investment in colour television equipment. The breakthrough, or the initial stimulus, came when the Americans themselves decided that 'at last the colour TV market looked like it was going to open up'.[5] As mentioned above, Marconi had established their reputation in black-and-white television equipment markets and when the American market began to expand, Marconi – in the absence of any home demand – began to invest heavily in colour TV development for sale abroad.

In breaking into colour TV, Marconi realised that a brand-new design was required. According to some, 'the R.C.A. design was dead and had not been significantly altered for some years'. The 'usual market assessments' were made into what type of equipment would be most suitable. The most difficult decision arose in February 1965 in connection with the most appropriate camera tube to employ. At that time a 3- or $4\frac{1}{2}$-in. image orthicon or the vidicon were available. The vidicon was not considered entirely suitable because of the intense illumination that it required.

Marconi were, however, aware that for some thirteen years Philips in Holland had been developing a new tube of the vidicon type, called a plumbicon, especially for colour applications. They apparently knew from the research data that had already been published that Philips were 'about due for a breakthrough'. As a result, Mr N. N. Parker-Smith, Development Manager of the Broadcasting Division, and Mr T. Mayer, the then Director of Microelectronics, together decided to embark on a 'four-plumbicon design'. This decision was made at a time when no plumbicons were actually available. The decision to go ahead with the full development of the Mark VII colour TV camera appears to have been made mainly by Parker-Smith, Mayer and Mr R. G. Williams, who undertook the co-ordination of the project from design through production and sales using various Programme Evaluation and Research Techniques (PERT).

One of the most serious development problems encountered was concerned with registration – that is, keeping the information from the red, blue and green colour tubes aligned. Technically, this involved the design of electrical circuits which were highly stable to temperature fluctuations. Thin films – made by deposition of resistors and capacitors on a glass substrate – proved to be the answer. These were developed by Marconi in the research laboratories and production was carried out in the microelectronics factory.

The company's previous skills in television technology were co-ordinated by Williams and the first demonstration of the new camera given to an invited audience in December 1965. In March 1966 it was shown to the National Association of Broadcasters and the first model went to a customer in August 1966.

(c) Comments

Bringing out the camera in such a short time was, perhaps, as much a production achievement as a development one. This product involved 'almost totally'[5] the assembly of components whose characteristics were well understood and Marconi was successful because it brought all these technologies together quickly into a marketable product. The main risk involved the availability of the plumbicon, and this seems to have been a calculated risk. One should not conclude from this that the design and assembly operations are simple or trivial. On the contrary, it was precisely Marconi's skill and experience in fitting the component technologies together that accounts for their success.

Marconi's development work on colour television since the early 1950s has been largely subsidised from the world-wide success of its black-and-white equipment. Even in 1969, the colour television requirements in the United Kingdom are small and consequently the United States has provided the largest potential market. Between the beginning of 1966 and the end of 1967, some 222 cameras were reserved by customers. Of these, 191 are for export of which 184, valued at nearly $10 million, are for North America. It is worth noting that these sales have been made in the virtual absence of any home demand.

References

1. The author would like to thank Mr Paul Drath for his assistance in the preparation of this section.
2. 'Colour Television Camera', *The Engineer*, 24 Dec 1965.
3. N. N. Parker-Smith, 'Colour TV Cameras – the Designer's Choice', *Sound and Vision Broadcasting*, **7**, no. 1 (1966).
4. A. G. van Doorn, 'The Plumbicon Compared with Other Television Camera Tubes', *Philips Technical Review*, **27**, no. 1 (1966).
5. Interviews with R. G. Williams, Manager, Broadcasting Division, and R. L. Murphy, Assistant to the Commercial Manager, English Electric/Marconi Company Ltd, 1968.

14 ETHICON LTD AND H. S. MARSH LTD: STERILISATION BY IRRADIATION

Two case studies are combined together here; Ethicon's Award in 1966 was for 'technological innovation in sterilisation of surgical materials by irradiation' and H. S. Marsh Ltd was cited in 1967 for 'technological innovation in the design and construction of plant for the sterilisation of medical equipment by gamma radiation'.

The effect of various forms of radiation in killing bacteria has been known since the last century but commercial application did not become possible until the late 1950s when the Americans developed the use of fast electron beams for sterilisation and the United Kingdom Atomic Energy Authority (U.K.A.E.A.) developed the use of cobalt 60 for irradiation sterilisation. The U.K.A.E.A. did not confine themselves to establishing a principle; they built an irradiation plant capable of sterilising industrial packages on a contract basis so that commercial organisations could establish the success of this method without having to risk capital in building their own plant.

Ethicon Ltd and H. S. Marsh Ltd are two of the firms that have benefited from the U.K.A.E.A. work, the former by becoming the first company in the world to operate the irradiation of surgical sutures on a commercial basis and the latter by being responsible for the construction of several irradiation plants. Marsh's first plant was commissioned by Johnson's Ethical Plastics Ltd who pioneered the production of disposable plastic syringes.

(a) Sterilisation by irradiation

In 1877, Downes and Blount[1] reported to the Royal Society that radiation in the form of sunlight could have a sterilising effect, and when Röntgen discovered X-rays in 1895 many workers investigated the effect of this new radiation on living organisms. Rieder in 1898[2] was the first to observe a bactericidal effect with X-rays and this was confirmed by several workers, although other workers found X-rays to have no effect on certain species. In the late 1920s gamma radiation and fast beams of electrons were found to have a bactericidal effect.[3,4,5] Fast electrons are produced by linear accelerators and gamma radiation can be emitted by certain radioactive isotopes. Commercial use of these forms of radiation was not possible until the 1950s, when suitable sources of radiation became available and 2·5 Mrad had become accepted as the level of the sterilising dose. (There is one radiation-resistant organism *Micrococcus radiodurans*, which will survive 6 Mrad, but this is unique and there are several bacterial spore formers which can survive conventional heat sterilisation.[6])

The first recorded sterilisation of a commercial product by gamma radiation in Britain took place in 1951 as described in the next section, but at that time there was still no suitable source available for industrial development.

In 1954 the Technological Irradiation Group of the U.K.A.E.A. was set up at Harwell from where it moved to Wantage. The Group was set up to find a use for the radio-activity available from spent fuel rods, but experimental work used cobalt 60. (Cobalt 60 is an 'artificial' isotope manufactured by placing rods of the natural isotope, cobalt 59, into a nuclear reactor where some of the cobalt 59 is converted into cobalt 60, a radioactive isotope which emits gamma rays.) The first gamma irradiation plant in the world was the package irradiation plant at Wantage;[7] it used cobalt 60 as its source of radiation and provided the U.K.A.E.A. with knowledge that was applied in the construction of the world's first commercial gamma irradiation plant. This was in Australia and was used to sterilise goat hair for carpets. Previously, sterilisation had been done in Holland before shipment to Australia.[8,9]

The Wantage Irradiation Plant is sometimes referred to as a pilot-scale or experimental plant but in fact it is larger than the first commercial plants. The commercial use of the techniques developed at Wantage required development work by industry, not so much a scaling-up operation as a process of simplification and cost reduction. The first commercial irradiation plant in this country was built at Slough for Johnson's Ethical Plastics, who pioneered the production of disposable plastic syringes. Both Johnson's Ethical Plastics Ltd and Ethicon Ltd, whose plant for the sterilisation of sutures was completed three months later, are subsidiaries of Johnson and Johnson, an American company which prefers to be regarded as an international organisation with its headquarters in the United States. H. S. Marsh Ltd were responsible for the design of the Johnson's Ethical Plastics plant. They then went on to build several plants, including the first gamma irradiation plants in Germany, Holland, Sweden and Italy.

(b) Ethicon Ltd

Ethicon Ltd of Edinburgh are manufacturers of surgical sutures and ligatures. They are associated with companies in the United States, Australia, Canada and Germany and they have one director nominated by the American Ethicon Company which is part of the Johnson and Johnson International Group of Companies. In 1963 Ethicon opened the first commercial plant in the world for the sterilisation of surgical sutures by gamma irradiation from a cobalt 60 source. Sterilisation by irradiation offers several advantages over the traditional means of sterilisation by heat or chemical attack, notably the elimination of the aseptic filling and sealing process which was always a potential hazard. Sutures can now be sterilised after they have been packed in foil sachets instead of being sterilised before packing and sealing into glass containers which were themselves a hazard to the nurse who had to open them.

In 1951 J. Owen Dawson, the Technical Director of Ethicon Ltd, became interested in new methods of sterilisation. He was aware of the need for a method of sterilisation that could be carried out in the final sealed container and he noticed reports of work being carried out on radiation sterilisation at the Massachusetts Institute of Technology.[10] Dawson approached the U.K.A.E.A. at Harwell where Dr A. Charlesby arranged for a few tubes of catgut to be irradiated. This work was mainly of academic interest as at that time no economic source of radiation was available for commercial sterilisation. Dawson also approached Metropolitan Vickers who were developing 4-MeV linear accelerators capable of producing electron beams which could be used for sterilisation and several experiments were carried out to determine a satisfactory dose for sterilisation. The American Ethicon Company co-operated in this work and they eventually adopted the electron beam method for sterilising their sutures. In the United Kingdom, however, this process was not economical at that time as throughput has to be in multiples of a machine's capacity. The American market was sufficient for Ethicon's associates in the United States to install two electron beam machines, but the United Kingdom market was not large enough to justify the installation of even one machine.[10] One advantage of the cobalt 60 process is that capacity can be increased gradually by using more cobalt 60, though the initial capacity must be sufficient to justify the cost of constructing the concrete chamber which acts as a biological shield.

Meanwhile the U.K.A.E.A. was looking for markets for sources of radiation which it was expected would be available from the British nuclear power programme and research on sterilisation by gamma radiation was carried out. The original intention was to find a use for caesium 137, obtained from spent fuel rods. Cobalt 60 was used for experimental work as it is a more convenient source of irradiation, but it was not intended that this material should be developed as a source.[17] The Canadian Atomic Energy Commission, however, realised that cobalt 60 could be used industrially as a source of gamma radiation if it could be produced at a reasonable price, which they did. In 1959 the U.K.A.E.A. followed the Canadian example and took serious steps to promote the sales of cobalt 60 by reducing its price and building a package irradiation plant at the Wantage Research Laboratory.[8]

The availability of cobalt 60 made Ethicon reconsider their position. Throughput was then sufficiently high to justify the installation of an electron beam sterilisation system but they preferred to go for gamma radiation from cobalt 60. In 1961 government approval under the Therapeutic Substances Regulations was received for the sale of surgical catgut sterilised by gamma radiation and Ethicon went into production using the irradiation plant at Wantage until one could be built for their own use. Ethicon's own irradiation plant was built by Nuclear Chemical Plant Ltd, using U.K.A.E.A. experience and licences, plus improvements of their own, and it was opened in April 1963, shortly after Johnson's Ethical Plastics irradiation plant for the sterilisation of hypodermic syringes had been opened. This plant was designed by H. S. Marsh Ltd.

(c) H. S. Marsh Ltd

H. S. Marsh Ltd was established in 1906 and carried out steam-engine and boiler maintenance. In 1936 it was bought by H. A. Edwards, A.M.I.Mech.E., whose son, Eric Edwards, and grandson, Rex Edwards, joined the firm in 1946 and 1958 respectively. H. S. Marsh Ltd is still a private company owned by the Edwards family.

Eric Edwards gained a wide experience in bulk and package mechanical handling equipment before joining the family firm and used this knowledge to expand the firm's interest. In addition to the main lines of business represented by the subsidiaries, H. S. Marsh Heating and Ventilating Ltd and H. S. Marsh Mechanical Handling Ltd, the Edwards family have always been interested in developing specialised machinery. For example, they manufacture a machine for cutting Thermalite building blocks. Another novel piece of machinery which was developed in response to a problem is a root pruner which cuts the tap roots of young trees without the trees having to be dug up.[11]

Over a long period, hundreds of 'one off' items of equipment have been developed for the U.K.A.E.A. and, through a chance meeting at a cocktail party, Eric Edwards was able to interest the Wantage Research Laboratory in the possibility of H. S. Marsh solving the problem of providing an irradiation plant at a low capital cost. The firm which had built the large experimental plant at the Wantage Research Laboratory did not wish to attempt the commercial development of this plant. Vickers were interested but the cost of an automatic handling plant for irradiation was estimated to be about £100,000, excluding the cobalt 60, and this was too high to encourage commercial use.

Rex Edwards enrolled on a special course at the Wantage Isotope School and both Eric and Rex Edwards attended international symposia held to discuss radiation processing. They were able to state that the figure of £100,000 for an irradiation plant could be reduced considerably. The first commercial gamma radiation plant to be commissioned in this country (and the second in the world) was built for Johnson's Ethical Plastics Ltd who were producing disposable plastic syringes, having convinced the National Health Service that it was cheaper to buy a disposable syringe than to sterilise a glass one and that cobalt 60 sterilisation was safer than the alternative chemical sterilisation using ethylene oxide.

Johnson's had been using the Wantage plant to sterilise their syringes, but the market began to grow and the U.K.A.E.A. persuaded Johnson's to put up their own plant.[11] By simplifying the mechanical details and by not charging for their time, E. & R. Edwards were able to submit a quotation which was considerably less than other quotations received by Johnson's and in 1961 the plant was completed by H. S. Marsh Ltd ahead of schedule. The plant was designed in co-operation with the U.K.A.E.A. Wantage Laboratories whose patents were made use of and there was an administrative controversy before H. S. Marsh obtained a licence to use the patents.

The second United Kingdom plant was Ethicon's plant mentioned previously. H. S. Marsh, the Canadian Atomic Energy Authority and Nuclear Chemical Plant Ltd all competed for the Ethicon contract which went to Nuclear Chemical Plant.

The designs of these first two plants have been published[12, 13] and can be compared. The basic design problems are to ensure safety, reliability and efficiency in usage of the gamma radiation. There are certain differences between the two plants and Marsh's appears to be superior. In both cases the source consists of a frame which supports rods of cobalt 60. The frame can be lowered into a pit where it is 'safe'. Previous pits consisted of a tank about 20 ft deep lined in stainless steel and containing demineralised and filtered water. The plant designed by H. S. Marsh Ltd uses a dry pit which is claimed to be cheaper and more convenient. The dry pit is close-fitting and lined with steel and concrete. The top half of the source frame consists of steel plates so that, when the frame is lowered into the pit, the source in the lower half of the frame is sealed off. This novel construction resulted from design collaboration between the U.K.A.E.A. Wantage Laboratories and H. S. Marsh.[13]

The methods of moving packages round the source are also different. The Marsh packages move 'along and round' the cobalt 60 source rather than 'up, over and down'. This results in higher source utilisation because packages moving past the centre of the source receive more radiation than packages over the top. The packages go past the source more than once which enables a redistribution to take place so that packages that were near the centre during one pass are at the top or bottom on another pass. For example, the packages can move past the source in stacks of six. If one considers the package nearest the ground to be numbered '1' and the package at the top to be numbered '6', then packages 3 and 4 receive the most radiation. At the next pass, packages 4 and 6 have been removed, packages 1 to 3 are now where 4 and 6 were previously and three new packages have been placed in the positions formerly occupied by 1 to 3. Package 3, which received a maximum dose on the first pass, then receives a minimum dose on the second pass. This novel arrangement was devised and patented by the Wantage Laboratories and is simpler and safer to operate than the 'up and over' technique. It enables all hydraulic equipment for the movement of the packages to be located outside the radiation cell.[11]

R. Glasson, the driving force behind Johnson's Ethical Plastics, considered the Marsh system to be superior to others. He states:[14]

it is interesting to note that Vickers had an arrangement for a number of years with the U.K.A.E.A. but were not successful in their designs or prices and it is clear that the design produced by Eric and Rex Edwards was far in advance of any other company and the cost was reasonable; the efficiency better . . . .

It would seem that Edwards, co-operating with the Wantage team led by Jefferson, created a better design than Nuclear Chemical Plant's Ethicon design. When Ethicon (U.S.A.) asked for tenders, the contract went to Atomic Energy Canada, not Nuclear Chemical Plant, and when Ethicon's German company wanted a plant, the contract went to H. S. Marsh Ltd. In fact Nuclear Chemical Plant have not built another irradiation plant. The German plant for Ethicon G.m.b.H. which was completed in 1966 was H. S. Marsh's third plant.

I

Their second, which was almost a copy of the first, was built for Gillette Industries Ltd, who, like H. S. Marsh, are in Reading. Following the German contract, H. S. Marsh have built plants in Holland, Sweden and Italy, each of these three being the first such installations built in these countries. Export contracts have the added advantage for the trade gap that they include sales of cobalt 60 which will continue, as the cobalt 60 decays at a rate of 12 per cent per year.

At the time of the Queen's Award H. S. Marsh Ltd could claim to have sold more automatic irradiation plants than any other company in the world, but they have now been overtaken by Atomic Energy of Canada Ltd. They remain a small company with about 100 employees. H. S. Marsh Nuclear Energy Ltd is the subsidiary company responsible for irradiation plants and H. S. Marsh has taken over another company, Nuclear Engineering Ltd, selling small irradiators for research outlets.

The expansion of H. S. Marsh Ltd into the field of irradiation has not been particularly profitable for the company, though there could still be a financial gain to be made out of irradiation plants. The Dutch plant was for the sterilisation of food and, if this concept gains acceptance, there could be considerably more orders.

(d) Comments

Small firms. This case study illustrates that, while small firms can be successful innovators, they can have special problems. H. S. Marsh were not troubled by lack of capital because they had a specific order to carry out the construction of a plant for Johnson's Ethical Plastics. Even the construction of the trial mechanical handling system at Marsh's factory was financed by Johnson's. From the point of view of company profits, it might be that the company who built the package irradiation plant for the U.K.A.E.A. made a correct decision in not attempting to commercialise this plant. However, from the point of view of the country's interests it seems fortunate that the small firm of H. S. Marsh were able to simplify the plant design and sell at a price that has enabled this country to export plant and cobalt 60 and the National Health Service to improve the sterility of its syringes and also to save money. The contribution to the national good has been recognised also by the award to Eric Edwards of an M.B.E. for technical innovation.

The Edwards family like, working on interesting problems and are not too intent on making large profits. Such freedom is one of the advantages possessed by an innovative private company that does not have to satisfy shareholders every year. They claim that the taxation, system is unfair to small enterprising firms and they are particularly worried about estate duty which, with three generations involved, could be very troublesome.

One of the motivating forces behind the Edwards family would seem to be a desire by the son and grandson to excel in a field different from that of their father's success. Eric Edwards introduced mechanical handling expertise into his father's firm. Eric's son, Rex, did not intend to join the family firm. He served an engineering apprenticeship and worked in the Merchant Navy before joining Eric when co-operation with the U.K.A.E.A. offered the opportunity to develop mechanical handling techniques for an entirely new area of activity. Eric is now Chairman of the firm with Rex as Managing Director.

Johnson's Ethical Plastics, who gave H. S. Marsh their first contract for an irradiation plant, were also a small firm formed when Hughes Brushes took over Bailey's Sterile Syringes. Hughes Brushes had been the smallest member of the Johnson International Group. Under the management of Roy Glasson, the concept of manufacturing disposable plastic syringes was developed. The provision of manufacturing and sterilising facilities required venture capital which small firms cannot always obtain. However, Glasson was able to persuade Johnson and Johnson International to provide an interest-free loan of £75,000. In spite of this initial loan, Johnson's Ethical Plastics were later prevented by lack of capital from taking full advantage of their success in persuading the National Health Service to adopt disposable syringes. Expansion could not keep pace with the growing market and Gillette were able to move in, buy a sterilisation plant from H. S. Marsh, drop the price to the Health Service and capture the major share of the market. Glasson considers that the American management was responsible for Johnson's losing their 80 per cent share of the market.[16]

His fascinating account of the role of Johnsons in the development of irradiation is given in his memoirs.[15]

Innovation not a linear process. The development of sterilisation by gamma radiation illustrates the difficulty of attempting to describe the process of innovation in linear terms with a clearly defined start, followed by a succession of events leading to commercial success. From the point of view of those who see technology as 'applied' science, irradiation sterilisation is an application of scientific discoveries made in the 1930s or earlier, but from the point of view of the U.K.A.E.A., who had to repeat some of the previous work, sterilisation was an outcome of work aimed at finding commercial outlets for isotopes from the atomic energy programme. As far as Johnson's Ethical Plastics were concerned, they were trying to produce disposable plastic syringes which had to be sterilised. If they had not read about the facilities available at Wantage they would probably have developed chemical processes of sterilisation. To the Edwards family, simplifying the U.K.A.E.A.'s Wantage plant for Johnson's was an exercise in mechanical handling; designing a system for moving packages past a radioactive source was an extension of their previous experience which started when Eric Edwards became interested in mechanical handling. From the point of view of Ethicon, the process starts with the realisation of a need for an improved method of sterilising surgical sutures. If the U.K.A.E.A. had not developed the use of cobalt 60 for sterilisation, then Ethicon would have used electron beam sterilisation.

Thus an overall view of the process of innovation, as it happened in this case, has to take into account a variety of different objectives which produced innovation when they happened to coincide. The necessity for different objectives to coincide is often overlooked by those who see innovation in terms of a discovery and its application. People responsible for discoveries often complain about other people's lack of enthusiasm for application, forgetting that their discoveries will only be used when someone has a need for them. Thus Jefferson, the head of the team that developed irradiation sterilisation at Wantage, wrote to Glasson in 1962:[15]

I feel that this is an appropriate moment to express my

appreciation of the initiative which you have shown in taking up gamma radiation processing. Those of us who have worked on the topic for many years sometimes feel that industry is slow in taking advantage of the fruits of our labours, but we do realise that it takes exceptional foresight for anyone to invest large sums of money in such a new field.

Glasson's foresight, however, lay in seeing the market for disposable plastic syringes. His objective was not to make sure that Jefferson's research was made use of but to find a suitable way of sterilising syringes at a time when he was short of capital. The construction of an irradiation plant at Wantage made it possible for these two objectives to coincide.

The innovation gap. It is often claimed that there is an innovation gap between Europe and the United States. This has been explained in terms of the larger home market of America, the difference in management attitudes and the role of the American Government in funding research within industry. In this particular case, the United Kingdom was ahead of the United States in that the plants for Johnson's Ethical Plastics and Ethicon went into operation in 1962 and 1963 whereas the first American gamma radiation plant was opened in 1965.

It is possible to see reasons why Britain led in this instance. One is the earlier use of electron beam sterilisation in America. Work on this method of sterilisation had been carried out in both countries but, in the case of surgical sutures, the American market justified the necessary investment of commercialised electron beam sterilisation and the British market did not. Another method, the use of ethylene oxide for chemical sterilisation, had also been developed in the United States. There was, therefore, a smaller incentive for the Americans to develop the use of gamma radiation from cobalt 60.

Another reason for Britain's lead over America was the provision of an irradiation plant by the U.K.A.E.A. This plant enabled potential users to try out the method and test market reactions without risking capital in a plant of their own. Perhaps this type of development work with government finance could be carried out more often.

The role of 'pure' science. One possible justification for government expenditure in the area of 'pure' or 'curiosity-oriented research is that such research can lead to the formation of new industries in the future. Sterilisation by irradiation might be seen as an example of this. Before the use of a gamma radiation-producing isotope for sterilisation could be developed, two fundamental discoveries were needed: the discovery of 'artificial' sources of radioactivity and the bactericidal effect of gamma radiation. The latter was discovered in 1925, some nine years before Irène Curie and Frédéric Joliot announced the discovery of 'a new type of radioactivity'.

The potential applications for radioactivity were already known; Dominus, for example, in 1910 had shown that cancerous cells were destroyed more readily than healthy cells when subjected to nuclear radiations. Following the discovery of 'artificial' radioactivity, some 200 new radioactive isotopes were produced by 1937 and by 1939 dozens of medical and biological papers had appeared describing work with these new sources of radiation. Whether this work should be described as 'curiosity-oriented' or 'mission-oriented' is open to debate. Potential applications were obvious but so were opportunities for scientific prestige.

Before commercial sterilisation could be developed, an economic source had to be made available and the minimum dose of radiation necessary to make radiation sterilisation at least as safe as heat sterilisation had to be ascertained. The economic source, cobalt 60, came about as a by-product of the nuclear power programme, which can be regarded as a product of 'pure' science but was also, of course, a product of the Second World War and an enormous amount of mission-oriented research.

Once a phenomenon such as radiation sterilisation is known to have commercial applications, academic research into the phenomenon cannot be 100 per cent 'curiosity-oriented'; instead it can perhaps be described as 'understanding-oriented'. Although a considerable amount of scientific work has been carried out aimed at understanding and describing the effects of radiation, the question of what constituted a sterilising dose of radiation for medical products was first answered by largely empirical tests. Ethicon Inc., for example, before using fast electron sterilisation, carried out a study in which 150 different organisms were inoculated into a number of items which were then irradiated at different dose levels and tested for sterility. The lowest dose at which no survivors are found is termed the 'inactivation dose'. Ethicon chose a dose of 2·5 Mrad for their commercial operation as this was 40 per cent above the inactivation dose for the most resistant species tested. This work could not have been carried out without scientists trained in certain techniques, but it is not dependent on an understanding of how the phenomenon of sterilisation occurs.

An understanding of sterilisation was, however, important in a different context. Before surgical catgut sterilised by gamma radiation could be used in this country, approval had to be given by the Ministry of Health under the Therapeutic Substances Regulations. Without some understanding of the issues involved, it would have been difficult to approve a new method of sterilisation. In a related field, the sterilisation of food, approval has not yet been given as there are still uncertainties about possible side effects on food sterilised by irradiation. The reduction of uncertainty about the possible effects of new products and new processes might be one benefit from 'pure' science that has been underestimated.

References

1. Downes and Blount, *Proc. Roy. Soc.*, **28** 488 (1877).
2. Rieder, *Münch. Med. Wschr.*, **24** 101 (1898).
3. Bruynoghe and Mund, *C.R. Soc. Biol. Paris*, **92** 211 (1925).
4. Lacassagne and Paulin, *C.R. Soc. Biol. Paris*, **92** 61 (1925).
5. Wyckoff and Rivers, *J. Exp. Med.*, **51**, 921 (1930).
6. Bridges, *Prog. Indust. Microbiol.*, **5** 283 (1964).
7. *The Engineer*, 9 Oct 1959, p. 395.
8. Jefferson, Rogers and Murray, in E. Glueckanf (ed.), *Atomic Energy Waste: Its Nature, Use and Disposal* (Butterworth, London, 1961) chap. 5.
9. Murray, *Nucleonics*, **20**, no. 12, 50 (1962).
10. J. O. Dawson, personal communication, 1967.
11. R. Edwards, interview, 1968.
12. Ethicon Ltd, 'Cobalt 60 Irradiation Plant for Sterilisation of Surgical Sutures', technical literature (1963).
13. *The Engineer*, 23 Nov 1962.
14. R. Glasson, personal communication, 1968.
15. R. Glasson, *Never Look Backwards* (privately printed memoirs, 1968).
16. R. Glasson, personal communication, 1969.
17. J. O. Dawson, personal communication, 1969.

15 FERRANTI: LIGHTWEIGHT
INERTIAL PLATFORMS

Government contract work is often mentioned as an important stimulus to innovation. The development of miniature, lightweight inertial platforms by Ferranti's Inertial Systems Department, Edinburgh, illustrates the difficulties that may confront a firm when political changes make such contract work uncertain. In addition, this case study shows how heavily the design of sophisticated systems may depend on developments in component technologies.

(a) Aircraft inertial navigation

Ideally, an aircraft navigation system should present an accurate indication of the position and velocity of an aircraft at every point on its flight path from take-off to landing. For most purposes, such equipment may be regarded as composed of three main systems: a detector, a computer and a display unit. The detector is a device which senses the various forces (i.e. accelerations) acting on the aircraft. These measurements are then transformed with the aid of a computer into aircraft velocities and positions which are subsequently displayed for the navigator to compare with the original flight plan. The Ferranti innovation concerns the development of the detector, although the company does market the computer and display equipment as well.

The basic physical law on which the detector is based is that matter resists changes in its motion – that is, it has inertia. For this reason the term 'inertial navigation' is used. The simplest aspect of inertial navigation is the use of an accelerometer for measurement of aircraft acceleration. This acceleration may be integrated once to yield velocity or twice if a measure of the distance travelled is required. The direction along which the accelerometer is lying must be kept fixed in space, and gyroscopes are used to provide this stable attitude reference.

In most systems, the detector instruments (accelerometers

and gyroscopes) are mounted on a rigid platform fixing their orientation relative to one another. The platform itself is mounted on gimbals (pivots) to isolate it from changes in attitude of the aircraft in which it is carried. The platform is kept in position (i.e. at a fixed attitude) by motors acting about the gimbal axes. These motors respond to signals from the gyroscopes which sense any platform rotations about three perpendicular axes. The whole structure forms an 'inertial platform'.[1]

The two main types of instruments mounted on an inertial platform are accelerometers and gyroscopes. Gyroscopes were installed by the Royal Navy as early as 1913 to stabilise ships' compasses. Since then, they have been much improved. Most of the 'difficulties of the gyroscope have been overcome and it now emerges as a fully developed precision instrument for sensing extremely small angular movements over quite long periods of time and distance'.[2]

Accelerometers are also inertial instruments. If a known mass is constrained to move at the same velocity as the aircraft, the force necessary to move the mass is proportional to the acceleration of the mass and hence of the aircraft. This is the principle employed on most accelerometers for inertial navigation use. Various types of accelerometers differ primarily in the methods used for producing and measuring the force. The output of an accelerometer is normally integrated to give a measure of velocity. To avoid the need for a separate integrator, accelerometers have been designed which incorporate an integration stage and provide velocity change directly as their output. One type of integrating accelerometer is the pendulous gyroscope accelerometer which, though it does not appear as such in the Ferranti innovation, forms part of the background to it. It is a device which has an unbalanced gimbal that requires a torque to maintain its attitude when subject to linear acceleration. This torque is balanced by a gyroscope precessional torque and hence a rate of angular displacement is obtained which is proportional to the acceleration. The angle through which the gyroscope is turned is a measure of velocity change. This type of instrument was used in the German V2 rockets and it was from this source that, after the Second World War, it entered British and American practice.

I 2

When, after the war, the United States embarked on a vast programme of rocket development, inertial equipment of the highest accuracy became an absolute necessity and American industry began a massive government-funded research and development programme to produce gyros, accelerometers and stable platforms which would meet the demands of aircraft, missile and space systems. A major contribution to high-accuracy gyros came in 1946. Dr C. S. Draper and a team at the Instrumentation Laboratory of the Aeronautical Engineering Department of Massachusetts Institute of Technology (M.I.T.) developed a 'floated' gyro. This idea of suspending a gyro in a bath of viscous oil is attributed to Schuler and Anschutz in 1911[2] and has found the widest application in inertial guidance and navigation. By departing from the previously adopted ball-bearing support technique, it makes higher accuracy possible.

The inertial platform developed by Ferranti weighs about 20 lb. and uses 'modular' construction to reduce assembly and repair difficulties. It is designed to accommodate a range of accelerometers and gyroscopes to meet a variety of customer specifications. Contrary to some American practice, the platform utilises inexpensive materials and machining techniques.

(b) Earlier work in Ferranti

British industry and the Royal Aircraft Establishment (R.A.E.) at Farnborough were also aware after the war of the tremendous potentialities of inertial techniques. So, after keeping a watching brief on the developments in the United States, R.A.E embarked on an extensive programme in the early 1950s. A survey was carried out of the state of the art in the United States and of manufacturing possibilities in the United Kingdom. It was apparent even before the survey began that there was little advantage to be gained from duplicating the work already carried out in the United States and the decisions concerning the British inertial guidance programme really concerned which types of gyro and accelerometer should be produced in Britain. Eventually, to supplement work already being done in this country on inertial navigation system and its gyroscopes and accelerometers, it was decided that an integrating pendulous gyro accelerometer should be developed and a little later that the manufacture of a single-axis floated gyroscope based on the design of the American firm Kearfott should be undertaken here.[3] The inertial guidance programme was the responsibility of R.A.E. through which the Ministry of Supply subcontracted some of the development work to British industry.

The involvement of Ferranti, Edinburgh, in inertial problems had begun during the Second World War. The Instruments and Fire-Control Department at Edinburgh was set up to produce automatic gunsights as part of the war effort. After the war, when this work ceased, Ferranti faced the question of closing down the Division altogether. Having decided against this, the firm in 1946 embarked on a number of development programmes among which was one embracing vertical and rate gyroscopes. At this time Mr M. P. Powley was head of the Instruments and Fire-Control Department. He had previously been an employee of R.A.E. In 1947 Mr J. Drury joined Ferranti from the Admiralty, where he had been working on the development of magnetic measuring systems connected with protection of shipping against magnetic mines. At that time he was responsible for gyroscopes and other component development. A few years later, in 1950, Mr K. Brown and Mr R. Dawson joined Ferranti and they began work on stabilisation of radar antenna platforms and gyroscope development respectively.

During the early 1950s it became apparent that the growth of sales of gyroscopes and conventional gyroscopic equipment could only be maintained if the Instrument and Fire-Control Department embarked on a programme of work aimed at inertial navigation systems. It was decided in the first instance that the concentration should be on the most difficult aspect of this work, the development of inertial quality gyroscopes, and in 1954 the company received from R.A.E. a contract to develop a pendulous integrating gyro accelerometer as part of the 'Blue Streak' guided missile programme. The Sperry Company were already engaged on the development of the complete inertial navigation system for the 'Blue Streak' missile and the contract which Ferranti received for the development of a pendulous integrating gyro accelerometer was an insurance policy using a different principle altogether to that from the Sperry Company in this most exacting work. As a result of work centred around making a useful measuring instrument out of the pendulous gyro, Ferranti acquired extensive experience in gyroscope technology and control engineering techniques. As a result of the experience gained in this work, Ferranti were asked by the Ministry to undertake the manufacture under licence and the subsequent development of the Kearfott (U.S.A.) type 2502 single-axis rate integrating gyroscope. As a result of the co-operation with the Kearfott Company, Ferranti were later able to make use of another Kearfott-developed gyroscope of smaller size and superior performance. This was eventually used in the Ferranti inertial navigation system.

(c) *The Ferranti innovation*[4]

There was an awareness within Ferranti that work on the pendulous integrating accelerometer was much too specialised to provide a large amount of future business and that once the problems of the components had been mastered, the emphasis of the work should be changed to the development of a complete navigation system. This type of thinking led Drury and Brown, Manager and Chief Engineer respectively of Inertial Systems, to make an extended visit to the United States in 1959 with financial help from the Ministry of Supply. The purpose of this trip was to assess the state of the art with regard to gyroscopes and inertial platforms, and it enabled Brown and Drury to identify some important ideas in connection with lightweight inertial platforms. Three important factors were noted:

(i) The Kearfott Company, with whom they had worked previously, had succeeded in building a miniature plat-form which was much smaller than any which had previously been built.

(ii) Mr W. Reichel of Norden had designed an inertial platform which was built from a group of sub-assemblies. The design simplified manufacture and replacement of parts. This was very much in line with current American ideas on modular construction. Prior to this period with Norden, Reichel had been a leading light in the Engineering Department at Kearfott Company.

(iii) The designs being developed at M.I.T. were using materials like beryllium which, in addition to being expensive, involves complex machining techniques.

It was partly from the first two pieces of information that the Ferranti lightweight inertial platform developed. The Ferranti engineers seem to have regarded the use of beryllium as 'over-sophistication'.

The first opportunity to apply these new ideas to a complete system was provided by Hawker Siddeley Aviation Ltd. H.S.A. were interested in developing a complete weapons delivery system for use in their vertical take-off and landing aircraft and they asked Ferranti to do a feasibility study in connection with this. The Inertial Systems Department took the offer mainly because they needed the business.

The cancellation of 'Blue Streak' in 1960 presented Ferranti with a problem similar to the one that they had faced after the war: highly trained technical staff and very little for them to do. Indeed, Ferranti realised that recurring crises such as these would undermine both profits and the morale of their technical staff and, in addition, make future recruitment difficult. For these and possibly many other reasons Ferranti decided to implement their policy decision of the early 1950s to go 'into the complete system business' and exploit their existing expertise fully. The amount of money involved in the inertial navigation system eventually amounted to more than £200,000.

Encouraged by this, and utilising the experience that had been gained from the American visit, the Inertial Systems Group set about designing a small inertial platform weighing about 20 lb. It is important to note that at this time, in another department of Ferranti Ltd, an R.A.E./Ferranti design of inertial platform was already being developed for use in the T.S.R.2 aircraft. This particular platform was aimed at a fairly specific requirement and a number of features were considered undesirable for the next generation of equipment. These factors and many other aspects of the problem were revealed in many discussions in both this country and the United States, and were taken into account before the Inertial Systems Group started on the final design of the lightweight platform and its system. In particular, a comprehensive survey of the gyros and accelero-meters most suitable for use in an inertial platform of optimum size for future systems was carried out by Brown.

A particularly important aspect of the platform design is the manner in which the gimbal pivot bearings contribute to the

stiffness of the complete gimbal assembly and the way in which
these bearings are given self-aligning properties which give
improved performance and simplified assembly over conventional designs. This design was apparently suggested by Brown
but actually developed by Mr P. Johnson. Johnson was
also responsible for the casing design, as well as its structural
stressing and testing. Other factors contributing to the
successful development of the platform were the use of modular
construction and wire-bound (instead of soldered) pin
assembly techniques and of low-cost and easily machined
materials.

After Ferranti had demonstrated the performance of the
system, built by private venture funds, on the bench, in road
vehicles and in aircraft, the Ministry of Technology bought two
similar systems in 1963 for service evaluation. About a year
later the next opportunity for selling this system was presented
when the Ministry of Aviation called for competitive bids for
the supply of an integrated inertial navigation and weapon
aiming system for the Hawker Siddeley P.1154 vertical take-off
and landing aircraft. Ferranti won this contract in the face
of competition from Elliot Automation, Sperry and Litton.
Another Government cut-back in defence expenditure led to the
cancellation of this contract, but about a year after the competition for the P.1154, the smaller P.1127 aircraft was funded
for operational service. Ferranti was again successful in gaining
the contract for the inertial navigation and weapon aiming
system for this aircraft and the supply of development models
started in 1966. Ferranti systems have also been produced for
the European Launcher Development Organisation (ELDO)
missile launcher, and the 'Black Arrow' missile. In the civil
field, inertial equipment for the Concorde supersonic aircraft
has been supplied.

(d) Comments

The very short development time of eighteen months for such a
complex piece of equipment deserves some comment. According to Mr K. Brown, it was due to the urgency of the situation
which Ferranti faced. They had been given a closely controlled
budget to develop a working system within two years and they

were aware that if they did not succeed in this the team would have to be disbanded.

Closely associated with the tight budget was the necessity of insisting on simplicity as a vital requirement for successful design. According to Drury, 'a tight budget is one of the most powerful disciplines one can apply to a development team'. In addition, the team enjoyed a high degree of insulation from external interference; once the money was allotted they were left to get on with the job.

On the technical side the modular construction technique, adopted originally to ensure ease of maintenance by the airlines, provided a bonus in the sense that each module could be developed in parallel and there was no hold-up of the whole assembly due to difficulties in one module. Also, the design targets seem to have been realistically chosen to ensure that no unnecessary delay would be caused by the natural tendency to over-complexity which sometimes makes itself felt in designing large systems. This in turn may be attributed to the extensive market survey to discover what types of accelerometers and gyroscopes a customer might specify.

The use of the 'inside-out' approach is an important design innovation. In this method the platform is built around the gyros and accelerometers instead of building a frame and then mounting components on it. It seems that Mr K. Brown was a motivating force behind this approach. He brought a very important skill with him when he came to the project, having had experience as a control systems engineer, while most of the Ferranti staff were either specialist electrical or mechanical engineers. According to Brown's colleagues, the fact that he was experienced in systems engineering meant that he kept a clear picture of what the system was intended to do. 'You must make the components the slave of the system, not the system the slave of the component.' This sort of philosophy can have a marked effect on the time allotted to the development of each component and may result in the extensive use of existing technology rather than expending effort on developing new devices.

References

1. J. A. Lee, 'Aircraft Inertial Guidance', *Aircraft Engineering*, **36** 2 (1964).
2. W. F. Hilton, 'Gyros and Guidance', *Engineering*, **196** 226 (1964).

3. J. A. Lee, 'Inertial Guidance–the British Scene', *Aircraft Engineering*, **36** 10 (1964).
4. Interviews with Mr R. Dawson, Sales Manager, Inertial Systems; Mr K. Brown, Project Engineer, Inertial Systems; and Mr J. Drury, Manager, Electronics Department, Ferranti Ltd, Edinburgh, June 1968.

16 FERRANTI: NUMERICAL CONTROL EQUIPMENT

The Queen's Award for technological innovation awarded to Ferranti Ltd of Dalkeith in Scotland covers a range of numerical control equipment for machine tools. The case study illustrates some of the problems confronting a firm when launching a new series of products as well as the relationships between an innovating firm, the industry it supplies and a government establishment.

As the name suggests, numerically controlled machine tools are the result of two interacting technologies – metal machining and computers. The machine-tool industry has a history reaching back as far as the industrial revolution, while computer technology is of more recent origin. The disparity in the backgrounds and interests of these industries goes a long way towards explaining why there have been very few manufacturers of numerical controlled machine tools as such. More often, the numerical controls are supplied by the electronics industry for application by the machine-tool industry. This is the background against which Ferranti interest in numerical controls for machine tools has evolved; they do not make machine tools but supply electronic control equipment to that industry.

(a) Ferranti numerical controls

The main purpose of automation is to remove the need for human skill. There is little point in 'replacing the skilled machinist by a more intelligent and highly specialised trained planner'.[1] The planner should, with a minimum of training, be able to process data for a variety of numerically controlled machines operated by relatively *unskilled* labour. This simplicity of operation is all-important and can only be achieved through the use of complex electronics and computers.

For this reason, the general philosophy behind the Ferranti numerical controls has been to separate any complicated data-processing from the machine tool and keep the machine-tool electronics controls to a minimum. The data-processing is carried out at a computer centre shared by a large number of machines. The service is provided by Ferranti and in this way the capital cost per machine is kept low while the absence of complex electronics at the machine tool makes maintenance by relatively unskilled operators possible. Magnetic tape is used to link the machine to the computing centre. The large storage capacity of this medium allows all the information for controlling the machine tool to be recorded, thus obviating the need for additional calculating equipment at the machine tool.[1]

Numerical control of machine tools takes advantage of the fact that the information on an engineering drawing can be reduced to a set of numbers which specify the important points in a component. In the Ferranti 'continuous path' system, the planner fills in a form which details in sequence the various cutting operations to be performed. In this process the various profiles of the component are subdivided into simple curves (i.e. straight lines, circles, parabolas, etc.). A punched paper tape of this *planning sheet* is prepared and fed to the *general-purpose computer* for processing. The computer transforms the simple planning instructions about these elementary curves into their mathematical constants. The curve constants are fed to a *curve generator* which records continuous information on *magnetic tape* in the form of electronic signals. At the same time, a drawing of the path that the machine tool will follow is prepared on the *plotting* table of the curve generator to give the planner a visual check of his work. The magnetic tape is then brought to a *control console* near the *machine tool* where it is replayed. Information from this console provides 'command' information to the machine tool – that is, information specifying where the cutting surfaces of the machine tool should be. But in order to exercise accurate control, there must be a similar supply of information about the *actual* positions of the cutting surfaces. At any instant in time, the actual and desired positions may then be compared and the difference (the error signal) used as a basis of corrective action.[2]

It is extremely difficult, if not impossible, to measure the surface cut by a milling tool and the next best position to measure is the table on which the specimen to be cut is placed. The accurate determination of the table position has been one of the salient features of Ferranti control systems.

Ferranti realised from the start the importance of having a complete numerical control system. In the early stages they pioneered development in these very important areas. From the work on software for numerically controlled machine tools developed the very simple planning system which is widely used in the United Kingdom called Profiledata. Ferranti were one of the first to pioneer the use of digital differential analyser techniques for the generation of continuous information for controlling machine tools. These developments produced their curve generators and copaths for recording the phase magnetic tape. Although initially a small, high-performance 400-cycle electric servo was used, the full potential of hydraulic servos was soon appreciated, and much original work on hydraulic servos as applied to numerically controlled machines was carried out by the Ferranti team. The case study presented here concerns itself mainly with the development of the new measuring system, but equally big if not bigger teams were deployed in such other areas as the preparation of data for the control system, the conversion of these data into a continuous digital form for control, and the use of high-performance servos to give good following performances for cutting profiles.

(b) The Ferranti continuous-path control system[3]

Towards the end of the Second World War, Ferranti, Edinburgh, had to reconsider its future development. Although there seemed to be no doubt at the time (1946–7) that government spending would remain high for a few years, the company was in doubt about establishing a permanent base in Scotland; the Edinburgh works had been set up as part of the war effort to work on gyro-operated gunsights. After it was agreed that

Ferranti should continue to operate in Scotland, there was a general investigation into what type of work to undertake. A good deal of electronics manpower had come to Edinburgh during the war and it is perhaps not surprising that, with considerable support from the Government, Ferranti turned to radar development with particular emphasis on guidance systems for missiles (cf. p. 253).

Ferranti were, however, concerned about becoming too dependent on government support because changes in the political situation are both difficult to anticipate and could involve a major reallocation of resources. A Director of Research was appointed whose particular interest was in a group known as the Applications Lab. This undertook a variety of investigations which it was thought could possibly lead to viable products. The group of engineers gathered for this purpose included several who played prominent parts in the development of numerical control and in other aspects of Ferranti activities at a later date. This group was subsequently hived off to a Victorian mansion near the main factory to consider the possibilities of numerical control for machine tools.

The first problem for this team of pioneers in numerical control was an 'in-house' one. The company had experienced difficulty in machining wave guides economically and a shortage of skilled men had caused bottlenecks in the machine shops.[2] The difficulty was that waveguide cross-sections must be uniform to a high accuracy. However, it proved very difficult to maintain uniformity when the waveguide was bent, as it had to be for certain applications. The suggestion seems to have arisen that, as there are certain sections of a waveguide which carry no current, one could take a solid block of aluminium, slice it open along these neutral axes and mill out of the solid whatever shapes were required. This is known as the 'milled block' technique. The problem was to make a machine tool capable of milling to the required accuracy. This proved to be a good problem for the newly formed group to start on because they could try out various electronic controls on the machine tools already at Ferranti. Here was an application which did not require negotiation with machine-tool manufacturers and would provide a good deal of experience. In fact the numerically controlled machine tool developed for this application was very successful and the original machine was still in operation in 1968.

In attempting to solve the problem of accuracy in the machining of aluminium waveguides, several difficulties were immediately evident: considerable demands would be placed on servo-motor performance; there would be difficulties in converting the information on the engineering drawing into a form suited to the control system; and a new type of measuring system would have to be found.

Most of the early starters in numerical control circum-navigated the last difficulty by placing a simple rotary device on the leadscrew of the machine tool. Ferranti decided not to be dependent on the inaccuracy inherent in rotary devices or the functional characteristics of leadscrew drive. Indeed, it seems that one of the earliest decisions was to look for a measuring system which would sense the tool displacements directly. In itself this was a major problem.[2] At that time, electrical methods of length measurement were restricted to very small ranges (a few thousands of an inch). Direct sensing of the tool position demanded accuracies of the order 0·001 in. over distances of several feet.

Because of the high accuracy required, Ferranti decided against using analogue computer techniques in favour of digital ones. But, as most of the then available short-range position measuring devices were based on an analogue technique, the decision meant that the machine tool would have to yield positional information which could be digitised. Most of the techniques then available were electromechanical and it appeared unlikely that they would provide the necessary accuracy.

After reviewing the possibilities exhaustively, it was decided that the most attractive proposition was the use of the moiré fringe patterns which are produced when two optical gratings are laid one over the other. These interference fringes were first described by Lord Rayleigh in 1874 and J. L. Baird seems to have taken a passing interest in the phenomenon in 1927. The moiré fringe technique provides a light output whose intensity varies sinusoidally. This light then falls on a series of photoelectric devices which convert the light into an electrical signal also having a sinusoidal variation. The sine waves may be easily converted into the pulses necessary for digital computation. Mr D. T. N. Williamson and Mr A. T. Shepherd, both members of the numerical control group, began looking for optical gratings. There were available some rather coarse gratings (100–200 lines per inch) of small size which were used in printing technology. Kodak also appears to have made some small gratings with up to 1,000 lines per inch. These gratings formed the basis of the first trials that led to the development of the moiré fringe technique which overcame the optical difficulties of viewing a single line of the grating and of methods of ensuring that accurate count could be maintained during reversal and vibration of the machine tool.

A search was made for gratings of higher accuracy, resolution and length than could be attained from printers' screens. It was at this stage (1952) that Ferranti came into contact with the National Physical Laboratory (N.P.L.). The Ferranti engineers had envisaged some experiments in optics and applied to the Department of Scientific and Industrial Research for an interferometer. The D.S.I.R. replied that there was none available on loan but that there was such a machine at the N.P.L. A team subsequently visited the N.P.L. where they further learned that Dr L. Sayce and his colleagues were doing some fundamental work on large gratings for infra-red spectroscopy; the Light Division of the N.P.L. was in the process of developing a suggestion by Sir Thomas Merton for making plastic gratings by replication from a master. These gratings were of sufficient length and the correct order of pitch (1,000 to 10,000 lines per inch) but their line structure was naturally suited to spectroscopy, this being the N.P.L.'s interest.[4,5,6]

Close collaboration was established between the two organisations, the N.P.L. developing gratings suitable for measurement purposes and advising on suitable optical arrangements for producing high-contrast fringes while the Ferranti group developed transducers for production, incorporating a direction-sensitive photocell arrangement. The ability to distinguish direction was a cardinal point in the development of a measuring system based on digital pulses. A counter was provided for N.P.L. in January 1953 and in May 1954 a complete grating measuring system was displayed by Ferranti at the N.P.L. Open Days. In the following months the first complete patent specification was filed,[7] followed by an improved system in 1955.[8] The electronics associated with the grating measuring system was developed primarily by Mr G. S. Walker and is mainly concerned with distinguishing the direction of motion of the machine tool from the pulse information from the gratings.

The account given so far covers the development of the Mark I continuous-path control system for machine tools. It was primarily an experimental system and was completed in 1954. About this time photo-transistors became available and these replaced the photo-multipliers of the older measuring system, resulting in considerable simplification. Other refinements, to eliminate the effects of temperature and voltage change, were incorporated into a Mark II system but the only models built were those for the company's own works at Crewe.

A Mark III version of the system was exhibited in 1956 at the Olympia Machine Tool Exhibition. Although it sold for two years, there were disadvantages which Ferranti wished to eliminate as better ancillary equipment became available.[2]

A choice between improving the Mark III and developing a new series was made in favour of the latter and it was decided to transistorise completely, changing also from electrical to hydraulic servos in order to give more versatility in dealing with different sizes of machine tool. The most radical change was the conversion of a digital system to one using phase modulation.[1] The immediate twofold advantage of this was the ability to use much coarser gratings (100 instead of 2,500 lines per inch) and the fact that the replay speed of the tape was reduced from 15 m/sec. to 3·75 m/sec. There were, in addition, considerable gains in simplicity of circuitry.[2] The decision to adopt hydraulic technology followed from the fact that Ferranti had learned a great deal from a co-operative venture with Fairey Aviation which produced the Fairey–Ferranti skin milling machine.

The Mark IV system in fact represents a very long stride towards a wholly satisfactory system and is the basis of all the current developments. In 1958 it was fitted to a wide range of machine tools by different manufacturers. The massive effort which was put into this system was crucial in demonstrating that numerical control was in fact capable of a wide range of applications. In addition, it laid the foundation for the range of practical knowledge of the use of numerical controlled tools which is considered to be one of the strongest elements in Ferranti know-how. The production of numerical control systems is thought to account for about 10 per cent of Ferranti's total business in Scotland. Ferranti estimates that nearly half of the 1,800 numerically controlled machine tools in the United Kingdom are equipped with its system.[9]

(c) Inspection machines

The development of control techniques which would make machine tools fully automated occupied the numerical control group from 1951 to 1963. In 1959 the Ferranti work on numerical controls was extended to include inspection machines. In principle, inspection machines are much simpler to design. Being open-loop systems, they require none of the elaborate control equipment required for the fully automated machine tool. An inspection machine checks the components produced by the machine tool. Previous to the Ferranti development, inspection seems to have been done mostly by hand by checking the component against an engineering drawing, and was very time-consuming.

Opinions vary as to the origin of the idea for the inspection machines. The earliest suggestion seems to have come around 1954 from the engineering laboratory where the suggestion was made, 'Why don't we run the machine backwards?' This statement was meant to imply that instead of using the control equipment as a closed-loop system to machine a component, it could also be used to find the location of a hole or the length of a cut by having a human operator close the loop and push the machine-tool head (which would now be a 'pointer') from one position to another on the component. In this way the accuracy of the machining could be checked.

A machine was in fact run backwards to plot an error contour in 1957 but, to a team preoccupied with the development of numerical control, the development of inspection machines seemed secondary. However, in spite of the considerable effort which had been put into demonstrating the potential of numerical control, general acceptance by the machine-tool industry and its customers was rather slow and the need was felt for diversification which would create a more rapid rise in turnover. Against this background, the idea of inspection machines was given active consideration. Although there is no doubt that there have been several expressions of opinion and suggestions of methods for carrying out the inspection function, the practical realisation of the first machines lay with Mr H. Ogden, some of whose ideas about kinematic mounting of sliding members were very adaptable to the problems of inspection machines. Mr A. Lodge, who had come from a production department into the sales side, was able to appreciate the value of inspection machines and enthusiastically promoted the new machines. As it has turned out, the much simpler inspection machines have enjoyed at least as great a success in terms of turnover as the sophisticated numerical control equipment and have provided products, complete in themselves, whose production is entirely in the hands of the company.

(d) Management

Sir John Toothill played an important part in the establishing of a Ferranti numerical control group. After the war, Toothill, then Managing Director of Ferranti, Edinburgh, was directly involved in trying to find new product lines which would have a high growth potential. Toothill seems to have believed that the application of computer techniques to machine tools was one such market. How he arrived at this hunch is not fully clear, but it would appear that it was at least partly based on knowledge of recent American work on numerically controlled machine tools, together with an awareness of the potential success of the Ferranti Mark I computer being developed at Manchester University by Professor F. C. Williams. The presence of somebody at the top of the management spiral who was personally committed to the development of numerical controls for machine tools seems to have contributed greatly to the confidence of the original team.

Toothill was not alone in supporting numerical control. In the early 1950s M. K. Taylor, who had been appointed Director of Research, visited the United States and, acting upon information about what was going on there plus his own knowledge of computer techniques, recommended that a small numerical control interest be started. Taylor seems to have had a 'flair for getting things started and a keen sense for picking out interesting things to do'. His sense of technical adventure also helped to gather a team which included Mr D. T. N. Williamson, who, together with A. T. Shepherd and G. S. Walker, laid the technical foundations of numerical control. M. K. Taylor has since joined Ferranti Packard in Toronto and Mr Williamson later became the Director of Research and Development of Molins Machine Ltd, London. Before he left, the team had been joined by D. F. Walker who has been responsible for many of the significant advances, including the important elements of the Mark IV system.

It appears that it was from his early work with aluminium when developing the 'milled block' technique that Williamson gained the experience necessary for him to apply the properties of light alloys to the System 24 which Molins have developed to improve the manufacture of cigarette making and packing machinery.[10]

(e) Some personnel

The original numerical control group numbered about six and over fifteen years has grown to include some fifty engineers, nearly all of whom are graduates, and thirty technicians. Short histories of some of the original development team are given below.

A. T. Shepherd studied physics at Edinburgh University. He moved to Ferranti, Edinburgh, from Ultra Electric in 1948 and worked on electronic instrumentation for a prototype steam catapult on H.M.S. Perseus. In 1951 he joined the numerical control group and was responsible for measuring systems and data recording. In 1963 he moved to the numerical control department at Dalkeith to work on digital measurement applied to precision inspection and lathe setting.

M. K. Taylor seems to have been an 'ideas man'. He was Director of Research when numerical control work started, though it was shortly afterwards that he moved to Ferranti Packard. He attended Cambridge University but did not graduate. He was in charge of TV development in Ferranti,

Manchester, and moved to Edinburgh after the war when the company was looking for new products.

D. T. N. Williamson originally worked with Ferranti on the steam catapult for H.M.S. *Perseus*. Later, under M. K. Taylor, he was transferred to work on inertial guidance systems, but he put in a strong claim for continuing the work on numerical control and this was agreed to by Toothill. He is generally regarded as an 'ideas man who could carry things through to completion'. He has since left Ferranti to become Research Director of Molins Machine Company. He took the full war-time course in electronic engineering at Edinburgh University but did not graduate.

H. Ogden left Halifax Municipal Technical College in 1944 and subsequently worked with A. V. Roe and English Electric on aircraft stress analysis. From 1948 to 1954 he was with the National Gas Turbine Establishment where he did extra-mural studies on electronic instruments and servo control. He joined the numerical control group of Ferranti in 1954 as Chief Mechanical Engineer and was responsible for much of the successful application work on numerical control of machine tools. He started work in 1956 on the inspection machines which became important products in 1959.

A. Lodge was originally a production engineer and was brought from there to the sales department to become Sales Manager of the numerical control department. His transfer from production to sales seems to have been important in the introduction of the Ferranti range of inspection machines.

References

1. D. F. Walker, 'The Ferranti Mark IV Continuous Path Phase Control System', *International Journal of Machine Tool Design*, **3** 61 (1963).
2. 'A Review of Ferranti Continuous Path Control Equipment', *Ferranti International News*, no. 1 (1965).
3. Interviews with Mr G. I. Thomas, Chief Engineer, Measurement Department, Ferranti, Edinburgh, 1968, and Mr A. T. Shepherd, Project Engineer, Numerical Control Department, Ferranti, Edinburgh, 1968.
4. C. D. Dew and L. A. Sayce, *Proceedings of the Royal Society*, **207A** 278 (1951).
5. R. G. N. Hall and L. A. Sayce, *Proceedings of the Royal Society*, **215A** 536 (1952).
6. G. D. Dew, *Journal of Scientific Instruments*, **33** 348 (1956).
7. D. T. N. Williamson, A. T. Shepherd and G. S. Walker, British Patent No. 760,321 (1953).
8. A. T. Shepherd and G. S. Walker, British Patent No. 810,478 (1955).
9. *Financial Times*, 16 Jan 1969.

10. D. T. N. Williamson, 'The Pattern of Batch Manufacture and its Influence on Machine Tool Design', *Proc. Inst. Mech. Eng.*, **182** (1) (1968).

17 FERRANTI:
MONOLITHIC MICROCIRCUITS

The Ferranti innovation in logic microcircuits is the result of close co-operation between the Automation Systems Division (A.S.D.) and the Electronics Department. The case study illustrates how important clearly defined product specifications may be in the innovative process. The technology of monolithic circuits had been known for some time to the Ferranti Electronics Department, but it was not until the process control interests of the company, the A.S.D., requested a fast logic circuit that a development programme was initiated. The A.S.D. specification was the result of a decision to use knowledge of computers developed for military purposes in civil applications.

(a) *Monolithic circuit technology*[1,2,3]

A monolithic circuit is a silicon chip about one millimetre square having all the components of a conventional circuit fashioned within the chip. Each area of the chip contains a particular functional element (e.g. transistor or resistor or capacitor) whose properties depend on the way in which that region has been treated. It is possible to reproduce the majority of conventional component functions by diffusing layers of impurities in different amounts and types into the silicon substrate. The layers are insulated from one another by the high impedance of a suitably biased silicon junction.

In constructing a monolithic circuit, one difficulty is to diffuse the correct amount of impurity in exactly the right place (which may be an area as small as 10^{-5} sq. in.). The technique used is known as oxide masking. Areas of silicon into which it is not desired to diffuse impurity atoms are protected by an oxide layer (SiO_2) and diffusion can occur only in the exposed silicon. The process of fabricating a typical monolithic circuit may be divided into three stages:

(i) growing an oxide layer over the silicon crystal;
(ii) removing the oxide from the areas which are to be exposed to diffusion of impurity atoms;
(iii) diffusing impurities into the silicon crystal.

The first stage is performed by baking the crystal in a furnace at about 900°C. To remove the oxide from certain areas of the silicon crystal a technique known as *photoetch* is employed. The oxide layer, which now covers the whole silicon crystal, is coated with a photographic emulsion called *photoresist* which polymerises when exposed to light. A photographic *mask* is prepared (see below) to cover the crystal so that no light can reach the areas where diffusion is to take place. The silicon crystal and the mask are then exposed to ultra-violet light and the crystal washed to remove unpolymerised photoresist. The photoresist which has been polymerised is insoluble in the developer and thus remains as a shield over the oxide layer. Where no polymerisation has occurred, the photoresist is washed off and the oxide exposed.

The crystal is then etched in an acid bath to remove the exposed oxide, leaving clean surfaces of silicon into which impurity atoms may be diffused. The remaining photoresist is then removed. Diffusion occurs much more quickly in the silicon than in the oxide and consequently only the preselected (clean) areas are affected. The actual diffusion operation is accomplished by passing gas containing the impurities over the crystal at a high temperature.

The construction of each photographic mask is itself a precision operation as the masks have to be small and accurately drawn. They are made by designing and cutting out patterns of the appropriate shape, but about 10 to 1,000 times the required size, from dark celluloid. Such large-scale drawings are used to minimise errors. The cut-outs are reduced in scale by photographic techniques and after two or three reductions they are the correct size for the etching process. A step-and-repeat camera is then used to produce a mask which consists of one pattern repeated many times. For any one type of circuit a number of different masks may be required. A fundamental feature of this technology is that many circuits are being made simultaneously – that is, it is a batch process.

It is possible to diffuse more than one impurity layer using the oxide-masking technique with different masks. A typical transistor complex may require two or three isolated layers of impurity atoms.

(b) *The Ferranti innovation in monolithic microcircuits*[4]

The Micronor II range of microcircuits originated from a specification of the Automation Systems Division for a 'range of fast, low dissipation, and reliable digital integrated circuits for use in the Argus 400 and 500 process control computers and a number of similar products'.[5] To understand this innovation, it is necessary to consider the interaction of two technologies:

(i) process control technology and the A.S.D.'s early involvement with the Argus family of computers;

(ii) monolithic microcircuit technology and the Electronics Department's previous experience with transistor and integrated circuit techniques.

Automation Systems Division. This Division of the company has, for some time, been involved with the Argus computer. The original Argus machine was built under government contract and was used as the guidance control computer for the Bloodhound missile whose development began in the early 1950s. The first industrial prototype, Argus 100, utilised discrete devices (i.e. conventional circuit components) and has been improved by successive generations of new technology. Later models, the Argus 400 and 500 systems, incorporate the Micronor II monolithic microcircuits and are among the first integrated circuits for digital computers to be used in Europe.

In the late 1950s, within A.S.D. itself, there seemed to exist two distinct but complementary philosophies about how the Division should develop. On the one hand, there was pressure to look for possible civil applications for computer systems originally designed for military purposes. The Sales Department, on the other hand, was primarily concerned with trying to survey the civilian market demand and then make process control equipment to supply this demand. The two views were reconciled and in the early 1960s a decision was made to opt for a commercial model of the guidance control computer of the Bloodhound missile.

Much of the design philosophy incorporated in the military system has been carried through to the commercial machines. For instance, plug-in semiconductor packages of uniform design are used. Rationalised design of circuits minimises the number of types of component used. The more complex logic configurations are obtained by arrangements of standard circuits rather than by introducing special-purpose units. This leads to improved reliability since only tried and tested components and circuits are used. Similarly, the specialised input–output equipment required to link the computer to the process is built up of standard circuits.[6]

The reliability of the latest designs of the Ferranti process control computers seems to be due to the fact that these systems use solid-state microcircuits. Applying microcircuits to process control computers provides the solution to two problems simultaneously. Microcircuits make computers cheaper as standard circuits produced in this form cost less than the equivalent circuits utilising conventional components interconnected with conventional wiring. In addition, the number of exposed interconnections is reduced compared to conventional electronic assemblies and customer fear of computer breakdown lessened.

Electronics Department. In the early 1960s the A.S.D. initiated enquiries with the Electronics Department about the possibilities of using microcircuit techniques on the components of the Argus 400. In 1962 Ferranti had introduced what appear to be Britain's first range of digital microcircuits – Micronor I. These microcircuits were subsequently used in both the Marconi 'Myriad' computer and the 'Priam' computer developed for the Admiralty. But, prior to 1962, the Micronor I range had not been used in any of the process control computers.

There were, however, some problems with Micronor I. This series used a 'multi-chip' assembly approach. In this technique, overall yields may be quite high because individual parts of the circuit are tested before assembly, which ensures that if any particular chip is defective in some way, it will not appreciably affect the overall production of finished devices. However, weighed against this are the advantages of the monolithic (single chip) approach: easy testing and marking by multiprobe equipment, fewer interconnections around the periphery of each chip, general amenability to mass production and consequent lower costs. Most manufacturers regarded multi-chip circuits as a stop-gap until the technology was advanced sufficiently to make monolithic circuits. There also seems to have been the feeling among some customers that the multi-chip circuits were somewhat complicated and required specialist skills in using the circuits properly.

In 1963 Ferranti were trying to develop a range of logic circuits. Ferranti 'were very conscious that the design of circuits required skills both in semiconductor technology and in circuit design. The co-operative venture between Electronics and A.S.D. married these two skills together and enabled us quickly to form a team with the right qualities.'[7] The A.S.D. seem to have provided not only a customer for the Electronics Department but also to have laid down in concrete terms exactly what was required. Having a clear specification by no means reduces the development task to a trivial one, but it does remove a great deal of the guesswork about the right road to follow. Working in close collaboration with A.S.D., the Ferranti Electronics Department produced a completely new range of digital microcircuits within three years.

One of the basic design requirements was that the switching takes place in a time of the order of 10^{-8} seconds. Ferranti were aware that the fastest discrete transistor in existence was the R.C.A. 2N2475 and, in fact, had been manufacturing this device under licence for some years. The Ferranti engineers used this as a model for design. Even though it could not be copied directly, they knew that any fast switching device would in many ways resemble the R.C.A. design. Thus 'it was decided to base the design on this type (2N2475) of transistor which has a gold diffusion to give very low storage and fast switching times'.[5]

The three years that it took to develop the new range of microcircuits was considered a very short time in 1965, considering that each circuit design has a two-month cycle from draughting of the original masks to production and testing of the first monolithic circuits. At that time, a two-month cycle was exceptionally short. There seem to be three main reasons for the rapid development:

(i) The product (i.e. the microcircuit) was well defined.

(ii) Close liaison between A.S.D. and Electronics Department.

(iii) The presence of a competent and enthusiastic engineer in the A.S.D. who, though he had a junior position, took the initiative to get things done. According to several of his colleagues, 'he is impossible to work with but his enthusiasm carried the project'.

(c) *Comments*

Close collaboration between A.S.D. and the Electronics Department at Gem Mill, Manchester, has existed for some time. In fact, the Electronics Department's R. & D. staff are permanently resident alongside A.S.D. near Wythenshawe, Manchester. An important factor in this innovation is design. The overall design specifications for the Micronor II microcircuits were drawn up by the A.S.D. systems engineers, but the precision of the components to meet the specification was the responsibility of the specialist circuit and semiconductor engineers in the Electronics Department. This interaction seems to have been particularly helpful to the Electronics Department who felt that they were able to achieve rapid development because the path they were to follow was indicated by the specification.

In microcircuit technology, it is necessary to have good integration between development and production. Often, one of the determining factors of a good design is its suitability for mass production, and as a result a particular production process may limit the number of possible designs. On the other hand, a single innovation in the production process often seems to involve a complete redesign of the basic circuit.

The size of the group who were initially concerned with Micronor II numbered about five. The team comprised one scientist with Ph.D. qualifications, three to four engineers, and two to three management personnel who were about head of department level. In addition, there was an ancillary technical staff of about twenty.

American domination of the microcircuit market has been a constant worry to the Ferranti Company. According to the company Chairman, Sebastian de Ferranti, 'unless we choose to discriminate against the Americans, there soon won't be a British semiconductor business'.[8] The situation is apparently acute in microelectronics. Ferranti estimate that 37 per cent of the integrated circuits sold in Britain are wholly imported from the United States and another 19 per cent are imported in a part-finished condition for the technologically much less demanding assembly process to be completed in the United Kingdom.

In 1967 Ferranti had a 46·4 per cent share of United Kingdom microelectronics production and 20·4 per cent of the United Kingdom microelectronics market. Other British companies in the field include Elliott-Marconi and Plessey. The largest share (23·3 per cent) of the United Kingdom market is held by the American-based Texas Instruments Company who have only a 5·3 per cent share in United Kingdom production. Ferranti's market research people estimate that it now takes £800,000 per year of research and development expenditure to keep an independent microelectronics manufacturer fully competitive in technology with its American competitors.

Although the total microcircuit industry in 1967 accounted for only £4 million of an estimated £35 million British semiconductor market, these components are essential for the operation of a multiplicity of types of equipment worth many times that amount. Microcircuit development itself is heading towards large-scale integration which simply means integrating more and more circuit elements on the same silicon chip. Ferranti plans to incorporate some advanced large-scale integration in a later version of the Argus process control computer.

References

1. H. C. Lin, *Integrated Electronics*, (Holden-Day, San Francisco, 1967).
2. R. M. Warner and J. N. Fordemwalt, *Integrated Circuits*, (McGraw-Hill, New York, for Motorola Co., 1965).

K

3. The author would like to thank Mr Anthony Thornley for his help in the preparation of this section.
4. Interviews with I. Breingan and D. Grundy, Electronics Department, Ferranti Company, Gem Mill, Manchester, 30 August 1968.
5. P. J. Bagnall, 'The Micronor II Exercise', *Electronic Components*, Dec 1966.
6. 'A Review of the Ferranti Automation Systems Division', *Ferranti International News*, no. 3 (1965).
7. R. I. Walker, Chief Engineer, Electronics Department, private communication.
8. *Sunday Times*, 12 Nov 1967.

18 FERRANTI: ELECTRONIC SUMMATION METERING

Two previously unrelated technologies were brought together in the Ferranti, Moston (Manchester), innovation in electronic summation metering: metering techniques and electronic circuitry. This case study provides an interesting illustration of technology transfer and in addition gives some insight into the operational problem that confronts a firm when the technology on which its products are based is changed.

(a) *Electronic measurements of power consumption*

Since its adoption by the then Central Electricity Board (C.E.B.) in the late 1920s, summation metering has become firmly established in the field of electrical metering and few large schemes can now be completed without it. The initiative for the development of summation metering seems to have originated with the C.E.B. when it was confronted with new problems in the addition and subtraction of large amounts of power from a multiplicity of locations often supplied at different voltages.

Although the idea of summation metering arose initially in response to the problem of measuring the total power delivered to the national grid, the private consumer has recently been impelled to take an interest. Factories and other large undertakings are increasingly being supplied at two or more points. Although the total energy consumption can be determined simply by adding meter readings, the maximum demand for power over a given period cannot. This is important because the maximum demand for power has a very large bearing upon the total amounts which the consumer has to pay, and a continuous record of the amounts of power consumed becomes essential.

Over the years, the equipment used for summation metering has been of the electromechanical type. These devices required frequent and specialised maintenance and, in addition, imposed a load on the circuit meters which adversely affected their accuracy. The logical operations which could be performed by mechanical summators such as pulse rate division, subtraction and the provision of sub-totals were limited and equipment had to be custom-made for each application.

For many years these limitations had been recognised by meter engineers and equipment designers, but it was not until the widespread availability of the transistor that a practical solution using electronic summation became possible.

Most present electronic summation metering systems make use of well-known configurations of transistors such as 'gate' and 'flip-flop' circuits. The electronic summation systems developed by Ferranti embody a temporary store of information on each power source being monitored and some form of 'scanner' which empties these stores in a predetermined sequence. In this way it is possible to separate into an orderly sequence power consumption information (usually in the form of pulses) which may have arrived simultaneously from several power sources. These pulses may be divided, subtracted and totalised as required.

(b) *The Ferranti innovation*[1] [2]

Ferranti have been involved in measuring power consumption for national, industrial and domestic users since the 1920s. The approach to summation metering was determined by the technology available at the time. Since customer requirements were expressed previously in terms of accuracy and reliability, mechanical and electromechanical types of design utilising differentials, gears, relays, etc., were required. Ferranti entered the market with equipment of this kind which was very successful and through which they acquired experience concerning the requirements of the electricity supply industry.

In the late 1950s the Ferranti Instrument Department was aware that the existing range of equipment design was reaching the end of its useful life. In addition, summation metering involving mechanical techniques was inflexible in that it was a major task to extend the capacity of the system. On the other hand the equipment was reliable and enjoyed a good reputation among consumers. The question which faced the Instrument Department in 1958 was whether to replace existing equipment with similar designs based on mechanical operating principles or to opt for fully electronic metering systems. Valves and other thermionic devices had been considered previously but were rejected because of their poor reliability. The estimated lifetime of a metering system is twenty years, during which there should be no major replacements; valves would have to be changed two or three times during this period. Two development paths seemed possible: more sophisticated electromechanical devices or solid-state circuitry.

In 1959 several meetings were held to discuss the issue. Although the use of solid-state circuitry seemed attractive, the state of the art was rather primitive at the time and it was changing rapidly. It was not until the early 1960s that the devices became reliable enough and prices low enough to make solid-state circuitry a reasonable alternative.

Despite these difficulties, a decision was made in 1959 within the Instrument Department to look into solid-state devices and a few meters using germanium semiconductors were designed and built. At this time there does not seem to have been any other firm attempting to market new designs in summation meters, and with the Ferranti customers apparently satisfied with their equipment, the Instrument Department were able to obtain little guidance as to what sort of system might be desirable.

One of the first steps was to consult the Ferranti 'computer boys' who were working on the Pegasus computer in an adjacent laboratory. They were asked to submit some designs but, having only a slight acquaintance with metering technology, their circuits tended to be too complex and too expensive for the need.

In 1958 Ferranti had engaged Mr N. Mascarenhas, an electrical engineering graduate from Manchester University. From the very beginning of his appointment to the Instrument

Department he had made it quite clear that he was less interested in the meter field than in electronics, and as a result he spent most of his time devising circuits for the various electronic applications which the Instrument Department required.

In 1961 Mr M. L. Done, then senior engineer in charge of metering, realised that Mascarenhas might usefully work with him on the development of the necessary electronics for the summation metering system.

This alliance seems to have produced results in a short time. The first production summator was designed in December 1961 and delivered in June 1962. According to Mr Done, it was the result of the 'right skills in the right place at the right time and neither one could have done it alone'. The young electronics expert without knowledge of customer needs and the operational requirements of the electricity supply industry would have been just as ineffective as a highly experienced meter engineer without detailed knowledge of the possibilities and limitations of solid-state electronics. Each influenced the other in arriving at the final design.

(c) Sales promotion

Mr H. L. Harrison, then Sales Manager of the Instrument Department, realised that there would be several problems in presenting novel electronic devices to an industry whose engineers and technical staff were brought up in the electro-mechanical tradition. To overcome this natural resistance, the new equipment was designed in such a way that it could be used and effectively maintained without knowledge of the new electronics. In addition, Ferranti's Instrument Department gave lectures and courses to all technical levels of those companies likely to use their equipment. This seems to have been an important factor in creating customer acceptance of the new equipment when it appeared.

Ferranti sell this equipment extensively abroad. As there is no patent on the electronics involved, it is natural to question why the Americans have shown no interest. No doubt many reasons can be put forward, but the Ferranti Instrument Department themselves seem to feel that so much of America's technological effort is directed at space and military programmes that there is not enough manpower left over 'to develop the more mundane fields'.

(d) Some key personnel

The number of people who were directly involved in the development of the new electronic summation metering equipment is about eight. They include the following:

N. Mascarenhas holds a degree in electrical engineering from the University of Manchester. He graduated in 1952 and joined the company in 1958. He left Ferranti in 1967.

M. L. Done holds a degree in electrical engineering from Birmingham University. He joined the company in 1950. At the time of the meter development he was a senior development engineer. He is now Chief Systems Engineer in charge of metering and control systems design in the Instrument Department.

H. L. Harrison is not a graduate but holds a Higher National Certificate in Draughting. He joined Ferranti before the war in the drawing office. In 1960 he was an outside representative for the company and in 1961–2 was Sales Manager of the Instrument Department. In 1966 he was appointed Manager of the Aircraft Equipment Department.

J. D. Carter also holds an H.N.C. He joined the company in 1937. After a break during the 1939–45 war he became Manager of the Instrument Department and was responsible for the decision to develop along electronic rather than electromechanical lines. In 1967 he was appointed Manager of the Meter Department.[3]

(e) Comments

Ferranti's main metering customer is the Central Electricity Generating Board (C.E.G.B.). While the C.E.G.B. have suppliers other than Ferranti, they do not seem to have exerted any pressure on the supplying companies to modernise their equipment. The stimulus seems to have arisen within Ferranti. Although the equipment was in the end developed by Done and Mascarenhas, it is difficult to assess the effect of the computer people who for some years occupied an adjacent laboratory.

While there were no formal links between the two groups, there do seem to have been many informal conversations and meetings about the application of electronics to metering technology.

In the transfer of electric circuit technology to metering, the electronics seems to have required little development. This is indicated by the fact that the circuits which were used were all standard ones in computer technology, and as a result no new patents were taken out.

References

1. M. L. Done, 'Electronic Summation Metering', Ferranti Company, private communication.
2. Interviews with A. K. N. Thomas, Sales Manager, Instruments Department, and M. L. Done, Chief Systems Engineer, Instrument Department, Ferranti, Moston, 1968.
3. A. K. N. Thomas, private communication, 1970.

19 FREEMAN FOX AND PARTNERS: SEVERN BRIDGE DESIGN

The Severn bridge was officially opened in 1966 and so followed the opening of the Forth bridge by two years. Although the two bridges were designed by largely the same team of engineers, they represent two radically different approaches to the problem of building large suspension bridges. The Forth bridge is built in the 'classical' mode, with an open-truss type, 27 ft-deep deck structure, whereas the Severn bridge is provided with a distinctive streamlined aerofoil-type deck only 10 ft deep that presents the minimum resistance to the wind. The conception, design and significance of the Severn bridge is commented on here with regard to the Queen's Award for technological innovation received by the consulting engineers – Freeman Fox and Partners of London.

(a) *The Severn bridge*

The southern industrial region of Wales, centred on Newport and Cardiff, is separated from England by the Severn river

estuary. The river crossing at the English Stones was described
by Thomas Telford, the famous engineer and bridge-builder
of the nineteenth century, as 'one of the most forbidding places
at which an important ferry was ever established – a succession
of cataracts formed in a rocky channel exposed to the rapid rush
of a tide which has scarcely an equal on any other coast'.[1] The
passage, subject to tidal variations of over 40 ft with currents
of up to nine knots, thus presented an effective barrier to
commerce between England and Wales. A ferry service operated
between Aust on the English side of the river and Beachley on
the Welsh side at a point where the river was narrowest; the
alternative surface route was a fifty-four-mile drive through the
city of Gloucester.

With the development of road transport and the extension of
industrialisation in the South-west, the provision of an adequate
bridge over the Severn became a matter of urgency. So in
November 1935 a joint committee was set up by the Gloucestershire and Monmouthshire county councils to 'take all such steps
as they might consider necessary to promote a [Parliamentary]
Bill for the purpose of constructing a road bridge over the river
Severn at or near the English Stones and for other purposes
incidental thereto'. The committee sponsored a Bill in 1936
for the proposed bridge but it was rejected and it was not until
1943 that

> the then County Surveyor of Gloucestershire, Mr E. C. Boyce,
> after reviewing the position of the pre-war schemes, advocated
> a high-level bridge, 6,000 ft in length, on the Aust–Beachley
> site, a scheme which was supported by the Government.
> It was not, however, until May 1945 that the Ministry of
> Transport informed the County Council that the Severn
> crossing proposal would be taken over as a Government
> scheme. Messrs Mott, Hay and Anderson, in association with
> Messrs Freeman Fox and Partners, were appointed the
> Consulting Engineers and Sir Percy Thomas the Consulting
> Architect and an order fixing the line of the bridge was
> confirmed by the Minister on 22 July 1947.[1]

The efforts to bridge the Severn were paralleled by other
efforts being made to span the Firth of Forth at Queensferry.
In January 1926 the Ministry of Transport had appointed

Messrs Mott, Hay and Anderson to make a survey of possible sites for the Forth road bridge. Various investigations were made but it was not until 1947 that the scheme was brought to the contract-letting stage.

It was first thought that a bridge over the Severn was the prior need, and aerodynamic investigations were therefore instigated by the Ministry of Transport for the Severn bridge in the knowledge that the data so obtained would be of equal use for the Forth. To this end the Ministry of Transport built the Severn Bridge wind tunnel at Thurleigh near Bedford in 1947; using this wind tunnel, the Aerodynamics Division of the National Physical Laboratories carried out aerodynamic tests under the direction of Dr R. A. Frazer and Mr C. Scruton in collaboration with the Engineers, Sir David Anderson and Sir Ralph Freeman. At the time, this wind tunnel was probably unique in having a working chamber 60 ft square in plan and 8 ft high, large enough to take a 1/100 scale model of the whole bridge (3,240 ft main span with two side spans of 1,000 ft each) which could be slewed to allow wind effects to be studied athwart or along the bridge or at any intermediate angle.[2]

The study of aerodynamics as a major design factor with regard to bridges had been thrown sharply into focus in 1941 when the Tacoma Narrows bridge collapsed after only four months' service. The bridge had a main span of 2,800 ft, was only 39 ft wide and was stiffened along its sides by solid plate girders.

During the four months it was in service, vertical bending oscillations occurred in certain winds. These motions were not very large (records show a maximum throw of about 5 ft) and they did not prevent traffic from using the bridge. On the day of the failure the bridge was destroyed within an hour by the single-noded torsional oscillations which started suddenly in a wind of about 42 miles per hour.[3]

Extensive investigations followed the failure in order to determine the relationship between the vibration of the bridge and the airflow past it. This problem had in fact been noted previously in 1839 with regard to Telford's suspension bridge

K 2

over the Menai Straits when the bridge deck was seen to oscillate in waves 16 ft high[4] and in 1836 when the Brighton chain pier bridge collapsed due to torsional oscillation of the deck in high winds.[3] But a theoretical understanding of the phenomenon had eluded bridge-builders until, with the advent of wind-tunnel testing methods, it became possible to relate the form of a bridge to its in-service behaviour by studying models in varying wind and load conditions. Two factors were clearly of importance – the degree of resistance which the bridge structure presents to the wind and the rigidity of the bridge in resisting both torsional and bending movements. The Tacoma bridge had been built with a large solid surface area exposed to the wind and it was this in combination with its lightness which led to its collapse. The answer as adopted for the second Tacoma bridge and for the first Severn bridge model was to build an open structure with many air passages between the structural members, so as to reduce as far as possible the formation of large eddy currents, and to adopt a deep cross-braced rigid framework for the deck. In

a demonstration of the effect of wind on the model at Thurleigh it was shown that with small rectangles of paper covering the deck girders and other minor changes, violent movements [of the deck] took place at relatively low wind speeds, up to about 60 m.p.h. full scale. In these conditions the bridge was intended to resemble the Tacoma Narrows bridge and it behaved similarly, the movements at first being of the vertical or 'galloping' type, giving way later to torsional motion of the deck. A succeeding test was made with the paper removed and the model withstood a wind equivalent to 105 m.p.h. without any sign of periodic motion.[5]

A cine-film study was made of the model under the various wind conditions likely to be met in service.[6] This film showed quite clearly the oscillatory motion of the vertical hangers from which the deck was suspended under adverse wind-induced vibrations. The loosening of one of these hangers on the main cable had immediately preceded the failure of the Tacoma bridge and was clearly of interest to the engineers since the film record drew particular attention to the hanger's motion.

Four main designs of deck structure were evolved for the

bridge that was to be built first at the Severn and later at the Forth The final choice was a four-carriageway structure with substantial air-gaps between each carriageway and cross-bracing across the top and the bottom levels of the stiffening struts.[7, 8, 9] All components for the structure were to be prefabricated and a diagonally cross-braced design was selected for the 492-ft-high towers.

Owing to the economic crisis of the late 1940s the construction of the two bridges was postponed until in 1956 'the Forth road bridge got the go-ahead, when originally it had been behind the Severn in the Government's priorities queue. So we took the Severn bridge design and built it in Scotland instead of Wales. The Welsh were as mad as hell that the Scots had beaten them to it, so to pacify them the Government said the Severn would follow in two and a half years'.[10]

(b) The innovation

The principal elements of a suspension bridge are the deck, main cables, hanging cables, towers, foundations and anchorages. Each of these is intimately related to the others. Thus, a lighter deck structure will permit savings in the size of hanger cables, towers, foundations and anchorages and it is an obvious design objective to build the lightest possible deck structure consistent with its service loads.

Deck structure. The Forth road bridge had been constructed with a deck structure $27\frac{1}{2}$ ft deep and with an independent deck surface that was not intended to contribute to the bridge bending strength. Steel girders in a triangulated arrangement were used to form the open-truss type design. To ensure that the new Severn bridge was as light and economical as possible, a new design was suggested based on the experience gained with the Forth bridge. This retained the open-truss type of construction but it had a deck structure depth of only 14 ft and continuous deck plating to take its share of the bending loads on the bridge.

This design was fully developed, and a scale model of it was being tested in the wind tunnel to check its aerodynamic stability when an accident provided an opportunity to test a completely new concept. The model of the truss design broke loose in the tunnel and was destroyed completely, and to occupy the time before a new model could be built it was suggested that the National Physical Laboratory should make up and test some simple alternative shapes in timber to represent a possible alternative design, namely, box section. [As in the monocoque construction form used for cars, strength in the box-section bridge is provided by the continuity of the plates forming the structural surface.] A model was made of two sheets of plywood with interchangeable edge pieces of various shapes. These tests were sufficient to indicate another approach to design for aerodynamic stability besides the dispersal of wind eddies by the use of trusses and openings in the deck, that is by streamlining the section so that virtually no eddies are created.[11]

The reduction in wind drag on the Severn bridge to one-fifth of that experienced by the Forth bridge in similar conditions was of considerable economic importance. With a suspension bridge, approximately 70 per cent of the wind force on the deck is sustained as a lateral reaction on the top of the towers. Further advantage is derived from the plated box design because of the torsional rigidity conferred on the deck by this form of construction which at the same time offers a fuller surface area for roadways and reduced maintenance costs.[12,13]

On the occasion of the announcement that the design team responsible for the Severn bridge had been granted the MacRobert Award, it was stated that M. F. Parsons of Freeman Fox and Partners 'proposed the feasibility of the shallow box suspended structure and showed its adequacy for strength and stiffness'.[14] Parsons had studied civil engineering at Bristol University and on graduating in 1949 he joined Freeman Fox and Partners and started working on the early structural analyses for the Severn bridge. Subsequently he was 'responsible for all the Severn bridge design calculations and . . . his long experience of suspension bridge analysis enabled him to assess the costs of alternative structural proposals with great speed'.[14]

Following the promising results of the preliminary tests a 7-ft long detailed sectional model was tested in another facility, the industrial aerodynamics wind tunnel at the National

Physical Laboratory at Teddington. This work confirmed that the aerodynamic characteristics of the design were satisfactory for all likely operational conditions up to 110-m.p.h. winds, except for a slight movement in a narrow range of wind speed around 15 m.p.h. This would not be considered significant in the more usual type of structure with riveted or bolted joints, where movement results in the dissipation of energy at the joints with a consequent tendency to 'dampen' out any vibration which may be excited in the structure. With a completely welded box-section structure, however, no such slip can occur and so the structure tends to 'ring' like a tuning-fork.

Hangers. It was at this point that the second major development in suspension bridge practice was made. Sir Gilbert Roberts, a senior partner in Freeman, Fox and Partners, was principally responsible. He

had had a wide range of experience in civil engineering. After serving in the Royal Flying Corps in 1918 he studied civil engineering at City and Guilds College (Imperial College of Science and Technology), London. In 1923 he joined Sir Douglas Fox and Partners (now Freeman Fox and Partners). From 1925 to 1936 he was with the bridge department of Dorman Long and Co., and from 1936 to 1948 with Sir William Arrol and Co. as a Director and Chief Engineer. He rejoined Freeman Fox and Partners in 1949 as partner in charge of work on the Severn bridge for the joint Engineers, Mott, Hay and Anderson and Freeman Fox and Partners.[14]

The slight tendency of the new design of bridge section to oscillate at low wind velocities induced a movement in the hangers similar to that which had been studied with the aid of the cine film of the bridge motion. If the hangers could be restrained from oscillating, or if the motion of the hangers could be utilised to dissipate energy in the structure, then the new box type of deck could be used. So, instead of following the conventional practice of hanging the deck from the main suspension cables by means of vertical hanger ropes, he suggested that the hangers be inclined to form an inverted vee formation with the apex of the vee attached to the main suspension cables.

Thus, any movement of the deck relative to the main cables would stretch one of the hangers and relax the other.[15] In this process of alternate stretching and relaxing, energy is dissipated – energy that would otherwise have been used to sustain the bridge oscillation. Parsons 'demonstrated that the inclined hanger concept's capacity for energy absorption by hysteresis could contribute significantly to the bridge's aerodynamic stability'[14]. In this, he collaborated with Dr T. A. Wyatt, who until 1962 worked with Sir Gilbert Roberts 'on a wide variety of projects including problems of the effect of wind on large structures, and this led to his specialising in that field, particularly on the aerodynamic oscillation of suspension bridges'.[14] In 1962 Wyatt took up a teaching post in Imperial College, University of London. He liaised between the designers and the aerodynamicists at the National Physical Laboratory and 'he also analysed the action of the inclined hanger system, with particular reference to its contribution to the stability of the bridge, and made tests with an elastic model to confirm his calculations'.[14] The model tests had to be confirmed, however:

> The absorption of energy by wire ropes subjected to cyclic stress does not appear to have received much attention hitherto from the rope makers but at Sir Gilbert's instigation tests on small 'model' ropes and on full-sized ropes and strands were made by Bruntons (Musselburgh) Ltd and British Ropes Ltd specifically for this purpose. For the first anti-symmetric torsion mode of oscillation, which has proved the most damaging to suspension bridge structures, 7 per cent of the energy of oscillation is stored by the hangers at the instant of maximum displacement, and at least 75 per cent of that energy will be dissipated in each half-cycle.[11]

Explaining why this method of damping could not have been used for bridges such as the Forth, Roberts wrote:

> It is impossible safely to use triangulated hangers when the deck is of truss construction because in this case the lower ends of the hangers must be attached to the deck at the points of intersection of the members of the truss to enable the deck to withstand the stresses imposed upon it. These points are normally equally spaced and while they may enable triangulated hangers connected to high points of the cables near the

tower to extend at included angles of less than 35° (the limit imposed by safety factors), they necessarily require the hangers near the centre of the span and connected to the low portions of the cables to have included angles greatly in excess of 35°.[15]

By adopting the box-section form of deck construction, the hangers could be attached to any points on the deck and thus spaced so as to comply with the 35° limitation. The design assumptions 'appear to have been justified, because the bridge has shown no tendency to oscillate at either high or low wind speeds since it has been completed'.[11]

Other developments.[19,20] With the confirmation of the box-section deck and inclined hangers as the principal features of the bridge, several further developments were made on the original design. The main suspension cables were reduced in size to take advantage of the lighter deck construction and the sag:span ratio was decreased to 1:12 (Forth road bridge 1:11) to compensate for the reduced stiffness of the cable system:

New thinking applied to the towers resulted in the adoption of a design in which each leg is a simple rectangular tube formed from four stiffened plates. The use of this logical form, facilitated by the great reduction in lateral wind forces to be carried, is far more efficient in use of material than the cellular construction previously used in suspension bridge towers. The Severn towers are no more than half the weight that would normally be expected with this span and loading.[11]

The detailed design of the box-section deck for the Severn bridge was the responsibility of Dr W. C. Brown of Freeman Fox and Partners. Brown graduated in civil engineering from Southampton University in 1948 and undertook three years' research at Imperial College, London, on the design of stiffened plates.

In 1951 he joined Freeman Fox and Partners and worked under Sir Gilbert Roberts on the design and supervision of construction of large bridges and other steel structures including cranes and radio telescopes. Dr Brown was responsible for all detailed design of the Severn bridge superstructure and the adjoining Wye bridge. Following his experience with earlier proposed box girder designs he initiated the concept of the shallow box deck structure, and the highly effective single cell tower leg design was also his conception. The economy of both these main parts of the bridge is due to the advanced stiffening techniques employed, the basis of which was Brown's own Ph.D. degree research work.[14]

Brown commented that the research project had been of benefit largely in terms of background reading on the stiffening of web and plate girders in aluminium. But his familiarity with the problem of plate stiffening, combined with the expertise available within Freeman Fox on the design and construction of large steel structures, provided the necessary range of conditions to permit them to advocate the new design for the deck with confidence.[16]

Brief reference has already been made to the fact that the bridge was to be of welded construction, and it is worth noting that this decision had a major influence on the economic practicability of the design. Unless the prefabricated sections of the bridge could be easily assembled on site with virtually no corrections to be made for errors in the machining of the plates, etc., the design would not have proved economical. Site difficulties could have been avoided by specifying small tolerances on plate dimensions, but this would have been prohibitively expensive and certainly beyond the 'shipyard' quality of the work available. Again, the solution to the problem was simple and interesting. The major sub-assemblies of the bridge (such as the 60-ft-long deck sections) were assembled on a slipway at Chepstow, starting with the two centre sections of the bridge and working towards the towers. Each section was temporarily built onto its predecessor in order to ensure that the plates and joints matched each other exactly. Once the fabrication of the sub-assembly was complete the temporary welds were broken and the unit could be floated under the suspension cables and winched into position beside the unit on to which it had first been built. Thus, the sub-assemblies were always matched and could be welded directly with the minimum of remedial measures. This simple and effective method of assembly of the deck sections was the outcome of Sir Gilbert Roberts's wide experience in the design of welded structures and methods of erection. This experience included the design of dock gates, landing craft and cranes, all structures dependent on the welding of large steel plates that would rarely be exactly the right size.

(c) Comments

The Severn bridge study is interesting because it shows how simple changes in design philosophy can have profound effects on a major technological project such as the Severn bridge. The decision to make the Severn bridge structure streamlined so that it would generate little turbulence in the airflow around it, and the use of inclined hangers to provide controlled damping in the structure, radically altered the original design.

> Figures quoted by the [Severn bridge] contractors on the basis of a negotiated target contract showed a saving of about £800,000 in favour of the box-type bridge. To this could be added a considerable sum to represent the savings in maintenance costs owing to the reduced area of exposed steelwork and better access for repainting. The saving on foundations and anchorages if designed from the beginning for the box structure would have been about £100,000.[11]

These savings may be considered in the light of an overall cost of £6 million for the supply and erection of the bridge superstructure and a £2 million cost for the foundations and Aust approach viaduct.

Neither of the two major design changes – the use of box sections and inclined hangers – required a 'breakthrough' in engineering knowledge. Rather, they required a clear understanding of the problem to be solved and of the range of solutions then available. The damping properties of wires had been used for many years; typical applications were the control of electric power line oscillations and the damping of gas-turbine compressor blade vibration.[21] The application of wire rope damping to the Severn bridge structure using a vee-form cable arrangement was an eminently simple and successful way of controlling a potentially hazardous situation. Similarly, the use of box sections was based on well-known engineering principles of construction. The welding of large structures was a well-understood art and Sir Gilbert Roberts had had much experience in the use of these techniques for the construction of dock gates, landing craft and cranes.[17] The use of welded fabrication for the superstructure of the bridge was further facilitated by the specialised knowledge of Dr W. C. Brown, Chief Assistant to Sir Gilbert Roberts on the project. Data on structural welding were obtained from the British Welding Research Association who had done tests in the early 1950s to develop a welding code for high tensile steels similar to those used in the bridge. 'The methods of fabrication and erection have to be considered in the early stages of design and, in fact, may have a decisive influence on the basic conception. Into the design of the bridge fabrication techniques went fifteen years' experience in the design of large welded structures.'[17]

Many engineers had contributed to the development of suspension bridge design theory.

The early elastic theory which assumed equal pull in all hangers, i.e. an infinitely stiff hardening girder, was replaced in 1889 by the deflection theory developed for a single span by Melan, and in 1904 extended to three spans by Moissieff for the design of the Manhattan bridge. Since then different mathematical treatments were applied to the theory by various workers in this field, such as Steinman, Timoshenko and in 1939 by Southwell and Atkinson whose theory Crosthwaite modified and perfected in 1947.[18]

C. D. Crosthwaite joined Freeman Fox in 1923, left in 1925 for experience abroad and rejoined in 1934. 'His main contribution to the design of the Severn bridge was his fundamental work in methods of analysis of suspension bridge structures and assessment of their aerodynamic behaviour by the use of sectional models.'[14] Making such specialist expertise available is of course the *raison d'être* of Freeman Fox as consulting engineers. Each of the features of the Severn bridge design would affect the rest and in such a situation a small, tightly knit design team is essential. Brown commented on this factor in particular as having assisted the radical development of earlier practice which the Severn bridge design represents.[16]

To determine the cost of the Severn bridge and its associated works is a relatively simple exercise, but the measurement of the benefits from it is difficult. An indication can, however, be derived from an interim study[22] of traffic flows across the bridge published in 1968 which showed that in its first year of operation nearly 6 million vehicles crossed the bridge, which was 20 per cent more than had been expected. This figure far exceeded the traffic flows on the Tay and Forth road bridges in their opening years. The bridge carried a higher proportion of commercial vehicles than do trunk roads nationally and the volume was growing more rapidly than on trunk roads generally or motorways.

A further significant benefit to flow from the Severn bridge was realised with the appointment of Freeman, Fox and Partners as consultants and designers of the Bosphorus suspension bridge to be built at Istanbul to link Europe with Asia. This bridge will have a centre span of 1,074 metres with two end spans of 240 metres each.

References

1. L. T. C. Rolt, *The Severn Bridge* (Gloucestershire County Council, 1966, for Ministry of Transport).
2. J. K. Anderson and W. Henderson, 'History and Financial Arrangements of the Forth Road Bridge', *Proc. Inst. Civil Eng.*, **32** (1965).
3. C. Scruton, 'An Experimental Investigation of the Aerodynamic Stability of Suspension Bridges with Special Reference to the Proposed Severn Bridge', *Proc. Inst. Civil Eng.*, **1** 189 (1952).
4. W. A. Provis, 'Observations on the Effects produced by Wind on the Suspension Bridge over the Menai Straits', *Trans. Inst. Civil Eng.* **3** 357 (1839).
5. 'Wind Tunnel for Suspension Bridge Models', *The Engineer*, **186** 449 (1948).
6. 'Oscillations of a Model Suspension Bridge in Wind', reference 49/3/16 (National Physical Laboratory, London).
7. R. A. Frazer and C. Scruton, *Summarised Account of the Severn Bridge Investigation*, NPL/AERO/222 (H.M.S.O., London, 1952).
8. D. E. Walshe and D. V. Rayner, *A Further Aerodynamic Investigation for the Proposed River Severn Suspension Bridge*, NPL/AERO/1010 (H.M.S.O., London, 1962).
9. G. Roberts, 'Design of the Forth Bridge', *Proc. Inst. Civil Eng.*, **32** 333 (1965).
10. P. Deeley, 'Interview with Sir Gilbert Roberts', *Observer*, 21 July 1968.
11. G. Roberts, 'Severn Bridge, Design and Contract Arrangements', *Proc. Inst. Civil Eng.*, **41** 1 (1968).
12. G. Roberts, 'Evolution of a Concept', *New Scientist*, 8 Sep 1966, p. 562.
13. G. Roberts, 'The Severn Bridge—a New Principle of Design', Symposium on Suspension Bridges (Lisbon, Nov 1966).
14. MacRoberts Award to Freeman Fox and Partners for the design of the Severn bridge superstructure. Notes by the firm, Jan 1970.

15. G. Roberts, 'Improvements relating to Suspension Bridges'. British Patent No. 911,350, applied for 16 May 1960, published 28 Nov 1962.
16. W. C. Brown, interviewed by W. G. Evans, 20 May 1969.
17. G. Roberts, interviewed by W. G. Evans, 18 Dec 1968.
18. O. A. Kerensky, *The Forth Bridge* (Institution of Civil Engineers, London, 1967) p. 204.
19. G. I. B. Gowring and A. Hardie, 'Severn Bridge, Foundation and Substructure'. *Proc. Inst. Civil Eng.*, **41** 49 (1968).
20. K. E. Hyatt, 'Severn Bridge, Fabrication and Erection', *Proc. Inst. Civil Eng.*, **41** 69 (1968).
21. R. G. Voysey, 'Some Vibration Problems in Gas Turbine Engines', *Proc. Inst. Mech. Eng.*, **153** 483 (1946).
22. *Severn Bridge Traffic*, Welsh and South-West Economic Planning Councils, 10 July 1968.

20 GENERAL ELECTRIC CO.: SEMICONDUCTORED RADIO EQUIPMENT

G.E.C. (Telecommunications) Ltd received the Queen's Award in recognition of technological innovation in telecommunications. Advancing technology and growing markets interacted to influence the design of certain kinds of radio transmission equipment. In order to get an idea of the range of G.E.C.'s activities in telecommunications, it is worth while to summarise the services and equipment they provide. G.E.C.–A.E.I. Telecommunications Ltd (as the company is now known) manufactures equipment ranging from small internal communication systems to large national networks. In addition, a complete service is provided for systems planning, route surveying, site installation and commissioning, customer staff training and after-sales servicing. The equipment includes private and public telephone exchanges, both manual and automatic; a wide range of telephone instruments; carrier multiplexing equipment; line transmission systems for open-wire, balanced-pair and coaxial cables; and radio transmission systems at frequencies from 70 Mc/s to 7,000 Mc/s, to carry telegraph, telephone, television and data signals. The company received the Queen's Award for producing a completely 'semiconductored' range of microwave radio relay equipment.

(a) Microwave transmission systems

The transmission of telephone conversations and television broadcasts over long distances has been achieved in two ways – by coaxial cable and by microwave radio relay. The second method dominates the former because 'for a given investment cost it provides for greater capacity with comparable performance'.[1]

A microwave radio relay network is composed of a chain of repeater stations distinguished by towers carrying aerials; these are spaced some thirty miles apart in such locations as will give line-of-sight clearance over intervening hills and buildings. Each repeater contains equipment to receive signals from adjacent stations, equipment to amplify these signals and equipment to transmit them to the next repeater. At the terminal in the microwave relay network originating the telephone or television signals these signals are modulated on to a very high frequency carrier for transmission within the system; they are demodulated where the information leaves it.

The first major microwave relay system – the TD-2 system – was put into service in 1949 by the American Telephone and Telegraph Company (A.T. & T.) and by 1951 had crossed the North American continent; its microwave signals have frequencies around 4,000 Mc/s. A second system – the TH system – operating around 6,000 Mc/s, was developed in the Bell Telephone Laboratories of A.T. & T. beginning in 1952, and manufacture began in 1958. The purpose of the TH system was to augment the capacity of the TD-2. High-capacity microwave systems are almost invariably of the heterodyne type (this is a standard technique of lowering the frequency that is used in every radio) and the amplification of the signal at the repeater station is provided mainly by an intermediate frequency amplifier. Subsequently the signals are returned to their original microwave frequency band by employing various frequency converters.

While these systems were being developed in North America,

parallel developments were being carried out in the United Kingdom, mainly by Marconi, Standard Telephones and Cables (S.T. & C.) and G.E.C. The American and British systems developed along similar lines as each company capitalised on the latest developments in microwave components and valve technology.

Increasing experience of the reliability of transistors and other semiconductors in carrier and line transmission equipment from about 1958 led to a trend in the early 1960s to replace all valves by solid-state devices. 'The use of all-solid-state techniques has undoubtedly contributed to the improvement of reliability, reduction in size and power consumption and general improvements in system stability'.[2] Some of these developments are illustrated in the table below.

The evolution of microwave repeaters[2]

Year of manufacture	1955	1963	1966	1968
Cost (as a percentage of 1955 cost)	100	64	66	38
Percentage of cost of active devices which are semiconductors	0	5	17·5	100
Volume as a percentage of 1955 volume	100	34	14	6
Power consumption as a percentage of 1955 consumption	100	44	30	12

At the same time the use of these devices has been accompanied by very severe problems of thermal dissipation, and an increase in the complexity of circuit design.[3] Compared with valves, transistors (to mention one solid-state device) have parameters which vary widely and the problem of designing to very tight performance specifications for quantity manufacture and long operating life are considerable. None the less, solid-state devices have made possible many operations which would not be economically considered before their advent. A particular case in point is the variable capacitance diode (varactor) which has made possible the generation of microwave frequency signals at reasonable cost and acceptable power levels.

Coupled with the trend towards solid-state systems has been the demand for systems operating from simple, reliable and economic primary power sources. Since solid-state circuits employ low voltages, batteries are once again becoming a possibility after a period when most equipment was operated off the alternating current mains. In highly developed countries, where mains power supplies are readily available at repeater sites, the amount of power consumed by each repeater station is unimportant. In developing countries, however, where the primary power supply may derive from diesel generator sets whose fuel may have to be transported over long distances and inadequate roads, minimum power consumption is an important consideration.

(b) The G.E.C. innovation[3]

G.E.C. were in the microwave communication field in the late 1940s and in 1949 installed the first permanent television link in the United Kingdom, between London and Birmingham, for the British Post Office. Throughout the 1950s and 1960s G.E.C. developed first 2,000 Mc/s and then 6,000 Mc/s microwave radio relay systems utilising valves. The 4,000 Mc/s band was avoided partly because one of their competitors was already established in this band, but mainly because they believed (in the event, correctly) that a given performance could be matched or surpassed at a lower cost with the higher frequency (and therefore more compact) system.

The company, who were steadily increasing the percentage of their equipment to be exported, experienced continual pressure from overseas customers for enhanced reliability from the equipment. Oddly enough, the most unreliable part of the system was the primary power supply, which had nothing to do with microwave technology. It was here that the potentialities of solid-state systems looked most attractive because their relatively lower power consumption and low d.c. operating potentials meant that batteries could be used to prevent loss of service during temporary failures of diesel or a.c. mains supplies.

By 1960 the use of semiconductors in carrier transmission equipment had revolutionised that part of the transmission business and the management became concerned lest a similar use of semiconductors within microwave equipment should render existing valve equipment obsolete. In the event this state of affairs never arose and mixed valve/semiconductor microwave equipments are still sold. None the less, the management felt that their next systems should be *fully* solid-state as this would be a good selling point, and they authorised development work.

As a result, in 1960, it was decided to study the feasibility of fully transistorised microwave equipment in the 7,000 Mc/s band. It was Mr B. Wilson, at that time Chief Radio Engineer of the Transmission Division, who suggested that experiments be concentrated in this region primarily because it was internationally allocated to 300 channel systems but also because he reasoned that, if they were successful with a 7,000 Mc/s system, subsequent development of lower frequency systems would be relatively easy.

The 'completely semiconductored' equipment developed by G.E.C. was made possible by a change in the state of the art in solid-state devices. The crucial device was the variable capacitance diode (varactor) developed originally at the Bell Telephone Laboratories in 1957 for use in low-level parametric amplifiers, an application well reported in the literature. G.E.C. have a close working relationship with the Bell Laboratories and according to some 'it is useful to keep an eye on the Bell Labs to see what new developments are likely to be'.

However, it was not the amplifying but rather the frequency multiplying and mixing characteristics of the varactor which were of importance in the 7,000 Mc/s experiments. The use of a varactor as an efficient 'high-level mixer' was evolved and patented by Wilson.[4] The mixer is based on the well-known fact that

> each of the sideband components of the modulated signal has only one-half of the amplitude of the [incoming signal] ... and accordingly only one-quarter the power. If, therefore, this principle is made use of in a [mixer or] frequency changer [a device to raise or lower the frequency] in which a filter is used to select one of the modulation sidebands to provide the output, there is an overall loss of 6 dB [i.e. of three-quarters of the power] from purely theoretical considerations.[4]

Because of this loss of power, high-gain microwave amplifiers containing valves were generally used to boost up the power before transmitting the information to the next repeater in the chain.

Wilson and his team sought a solution to this problem and ultimately evolved a

> microwave frequency changer which utilised a non-linear element [varactor] ... that is arranged to modulate a microwave signal [in which] ... a filter ... is arranged to select the signal containing one sideband resulting from the modulation, this filter being so disposed that, during use, at least part of the signal supplied thereto, other than the said sideband, is reflected back to said non-linear element in the appropriate phase to give an overall power loss between microwave signal and selected sideband signal passed by the filter which is less than 5 db.[4]

The basic experiments with semiconductor units for a 7,000 Mc/s equipment proceeded until early in 1963 by which time the practicability of such a system had been demonstrated and the marketing department requested that such a system, for 300 telephone channels, should be made available for sale as quickly as possible. The experimental work had also indicated that a higher transmitter power could be made available at 2,000 Mc/s using similar techniques and development was also started on this system. The resulting systems were the first of their type in the United Kingdom and the 2,000 Mc/s system (for 960 telephone circuits or colour television) was for a number of years the highest capacity all-solid-state microwave system available anywhere in the world.

(c) Comments

Wilson and his electronics teams solved a problem defined by the following constraints: the need to reduce power consumption on microwave repeater stations in developing countries; and the availability of space in the frequency spectrum in the region of 7,000 Mc/s. The technology on which the innovation is based derived mostly from common knowledge of communication electronics, but in particular a recent advance in the state of art – the varactor – was utilised in a novel way to produce a low-loss frequency changer.

A critical element in the success of this innovation is the systems planning approach. In G.E.C. there exists close collaboration between radio development in general and marketing. This is primarily due to the work of E. C. H. Organ who between 1950 and 1952 set up a systems planning unit as a co-ordinating body. This enabled G.E.C. to offer a complete systems service. A large part of the orders for microwave radio relay equipment is for developing countries. To secure a contract in such countries it is often necessary not only to design, manufacture and install the equipment but also to survey, plan stations and networks and to train engineers and technical staff in its proper use and maintenance. In addition, the equipment is often subjected to extremes of environment which in turn are reflected in more stringent design specifications.

The laboratory in which microwave development took place was staffed by about forty persons holding a range of technical and university qualifications; only about six of these were directly involved in the work on the 7,000 Mc/s equipment.

Overseas contracts resulting from this microwave development programme already totalled £6 million by 1967. Equipment has now been supplied to twenty-six countries in all six continents.

References

1. T. A. Marshak, 'Strategy and Organisation in a System Development Project', in National Bureau of Economic Research, *Rate and Direction of Inventive Activity* (Princeton U.P., 1962).
2. D. Davidson, 'High Capacity Systems Improved by Solid-State Techniques', *Electronics Weekly*, 31 July 1968.
3. Interview with Mr E. M. Hickin, Chief Radio Engineer, G.E.C.-A.E.I. Telecommunications Ltd, 1968.
4. British Patent No. 1,099,505 (1968).

21 HAWKER SIDDELEY GROUP (MIRRLEES NATIONAL): LARGE DIESEL ENGINES OF MEDIUM SPEED

Heavy diesel engines in the 1,000-h.p. and over class have had two main areas of application – ship propulsion and power generation. Both these applications have the attendant requirements that the engines operate with maximum fuel efficiency and maximum reliability. The tendency therefore was to design engines with a relatively lower power to weight ratio because of the low stress levels which were specified by the designers. Emphasis was placed on low-speed engines operating at speeds in the region of 150 r.p.m. In the 1950s the medium-speed diesel engine, operating at approximately 700 r.p.m., gained widespread acceptance but the thermal and mechanical loads for which these were designed remained low. In the 1960s a systematic research programme was undertaken to increase the power output of the Mirrlees National engines without increasing their size and whilst retaining the required reliability. A 50 per cent increase in engine performance was achieved by 1967.

(a) The company

Through a complex ancestry, Mirrlees National Ltd derived from a line of companies with a long tradition of work on diesel engines.[1,2,3,4] Indeed, one of its predecessors, Mirrlees, Watson and Yarvan Co. Ltd, had in 1897 produced the first diesel engine to be made in Britain. Mirrlees National was formed when Mirrlees, Bickerton and Day Ltd merged with the National Gas and Oil Engine Co. Ltd following a take-over[5] by the Hawker Siddeley Group in 1958. As a result of this take-over there was rationalisation of product ranges within the Group. The Technical Director of National, Mr J. Smith, moved to Petters to undertake the rationalisation of the small engine interests of the Group, and Dr J. A. Pope was appointed Technical Director and Director of Research of Mirrlees National in 1960 in succession to Mr Brownlow. Prior to his appointment, Dr Pope had been Professor and Head of the Department of Mechanical Engineering at the University of Nottingham and during that time he had acted as consultant to the Hawker Siddeley Group.

By 1963 the rationalisation in the range of engines being produced by the Hawker Siddeley Group had been substantially completed. The Mirrlees National range of engines was limited to the Mirrlees TL, J, K and A range of engines and to the National R and F ranges with production centred on the Stockport site. In 1963 Hawker Siddeley Diesels Ltd was set up and Pope was appointed Chairman of its Technical Co-ordination Committee for large engines. As a holding company, Hawker Siddeley Diesels Ltd has no sales force or technical programme of its own but serve to co-ordinate the policies of its constituent companies, Mirrlees National, Petters Ltd, R. A. Lister and Blackstone.[6]

(b) The innovation

The Mirrlees National Company Ltd of the Hawker Siddeley Group received its Queen's Award in 1967 for technological innovation in the development of its K Major and KV Major engine ranges which have power outputs from 1,200 b.h.p. to 8,000 b.h.p. The company had, in the years 1948–50, developed a similar-sized range of engines with lower power ratings which was designated the K and KV range. The development of these K and KV engines was announced in October 1949 at a joint exhibition organised by the Mirrlees, Bickerton and Day Company Ltd and the British Oilfield Equipment Co. Ltd by the Chairman of the two firms, A. P. Good.[7] By 1966 over 1,000 of these engines had been produced by Mirrlees, approximately 70 per cent of these being for export.

On joining the company in 1960, it fell to Dr Pope to define the areas in which the firm would seek to develop its engine range. To do this, he called upon the Economic Appraisal Department of the Hawker Siddeley Group to carry out an investigation of the market for large diesel engines and of the equipment sizes it would require in the future. The results of this survey indicated that the primary target for engine development should be to achieve a 50 per cent increase in power output from the engine without suffering an increase in overall engine size. A portion of the results of this study was published by Pope in 1963.[8]

To increase the specific power output of the Mirrlees engines it was necessary to increase their thermal and mechanical loading whilst retaining the same degree of reliability as the lower-rated K engines. At the same time it was intended to design the engines so that they would be able to burn cheap residual heavy fuels obtainable anywhere in the world. 'Usually the cost of fuel is between 70 and 80 per cent of the total cost of [power] generation and frequently the price of heavy fuel is about three-quarters that of diesel fuel and quite often as low as two-thirds.'[9]

Valve design. To accomplish these objectives, it was necessary to obtain a precise knowledge of the thermal loadings and limitations of the existing K engines. The focal points of these investigations were the piston and valve assemblies. A thermal test rig to determine optimum piston design was built and, on the basis of test results derived from it and from test running of a prototype engine, a two-piece piston design was developed to replace the one-piece cast-iron piston used in the K engine.[10,11] The crown of the piston was made from a high-tensile steel and the skirt of cast iron. 'The reduction in top ring groove temperature by some 126°F achieved by the new design has made available a wide potential for increase in rating in the future, before any limitation due to lubricating oil break-down is reached.'[11]

Heavy fuel oil usually contains appreciable amounts of vanadium and sodium salts in addition to a high concentration of sulphur. The former can cause severe wear of the valves if the salts build up on the exhaust valve because of the subsequent leakage of high-temperature gas past the valve. The high sulphur content of the fuel requires that care be taken to ensure that sulphuric acid is not deposited on the engine components. The melting points of the principal sodium and vanadium salts likely to be deposited on the engine components were all determined[9] to lie above 550°C and so it was necessary to ensure that the matching faces of the valve and valve seat were maintained at a temperature lower than 550°C. 'More than 60 exhaust valve and cage configurations have been tested, of duration between 300 and 1,300 hours each, to determine the effect of different factors in the design.'[11]

As a result of these tests, a valve/cage design was chosen wherein the valve seat is formed from a proprietary steel ('Stellite') with a cooling groove formed in its rear face. Cooling water is circulated through the groove and by this the valve seat and valve face temperatures have been kept between

450°C and 500°C, which is well within the required range. The valve stem, too, had to be cooled because of the necessity to remove heat from the valve face, and to ensure that the lubricating oil was not subjected to temperatures high enough to cause the formation of hard carbon deposits. By reducing machining tolerances and applying cooling water, it was possible to reduce the valve stem temperature to 77°C.

A secondary problem was thereby induced, however, because the valve stem temperature was low enough to cause condensation of the sulphuric acid vapour from the exhaust gas on the stem and so cause severe corrosion of that component. A possible solution lay in providing a high-pressure lubricating oil feed to the stem which would prevent the ingress of exhaust gas into the stem passage, but this would have meant accepting the penalty of oil leakage and carbon build-up on the stem if any wear occurred. The solution to these interacting problems was simple and novel. It was provided by one of the development engineers employed by Mirrlees – Mr S. Kryzwski. He proposed that the stem be intermittently lubricated with high-pressure oil only during the time in each engine cycle when the high-pressure exhaust gas was present at the entry to the valve stem. Thus, the lubricating oil pressure and gas pressure could be matched so that no flow of gas or oil to exhaust passage or stem took place. To achieve this cyclic variation in oil pressure, Kryzwski proposed that the stem be machined flat at points corresponding to the inlet and outlet oil ports. On each valve stroke the oil flows only when the entry and exit ports are open, i.e. when the high gas pressure exists at the stem. A simple rotator mechanism is mounted on the top of each valve and this ensures that the valve and valve seat are automatically brushed clean during each cycle and a uniform temperature distribution is maintained on the valve.

The importance of these developments in valve design can be gauged from the fact that, whereas the K-type engines required valve overhaul after each 500 hours running time, the K Major engines with water-cooled valve cages give periods between maintenance of over 3,000 hours. For the K-type engines, maintenance costs accounted for 10 per cent of the average cost of 0·781d per electrical unit generated, and of that 10 per cent the major portion was due to the frequent replacement of the valve assemblies. Fuel costs typically amounted to 79 per cent of the running cost, with a further 11 per cent contributed by lubricating oil, storage, water and operating personnel cost.[9]

Fuel injector nozzles. A similar problem to that met in the design of the exhaust valve cages was the design of the fuel injector nozzles to ensure that no carbon build-up, cold corrosion or softening of the nozzle valve occurred. The solution of this problem was to control thermostatically the flow of cooling water to each fuel injector nozzle so as to maintain it within the permissible temperature limits. Considerable development effort was necessary to optimise the performance of the fuel injection phase of the engine cycle. This involved not only the design of the injection nozzles themselves but also the design of the cams operating the fuel pump. Use was made of a computer to determine the necessary cam profile to optimise the rates of opening of the valves without exceeding the permissible limits of valve acceleration. The development of this system was the result of a joint investigation by Mirrlees National Ltd and Bryce Berger Ltd, manufacturers of fuel line equipment. Extensive use was made of test rigs on which the effects of the many parameters to be considered could be assessed.

Connecting rod. One of the largest components in a medium-speed diesel engine such as the K Major is the connecting rod between piston and crankshaft. This, besides being one of the most highly loaded components in the engine, is also one of the most vital ones. A failure of the connecting rod would normally result in the complete disintegration of the engine and so it is necessary to design these units for maximum strength whilst reducing their weight and size as much as possible. To optimise the connecting rod proportions, tests were carried out

in a full-scale static rig, in which gas loads and inertia loads are simulated by hydraulic pressure and the resulting stresses measured by strain gauges attached to the connecting rod. It was thus possible to reduce the weight of the connecting rod by 15 per cent from that of the original K rod so that, even at the increased operating speed and loads [of the K

Major engines], the connecting rod stresses are lower than in the original design.[11]

(c) *Comments*

The Mirrlees National case study reflects the importance of relating market requirements to research and development targets. A world-wide study of the competitive position of diesel engines for base-load power generation showed that to achieve a significant increase in the company's share of the power-generating market it would be necessary to achieve a 50 per cent increase in power output per cylinder over its existing range of engines. The responsibility for setting the targets to be achieved rested with the newly elected Director of Research and Technical Director of Mirrlees, Dr J. A. Pope. He had been recruited into the firm to lay down a research policy, build up a research team and develop a range of diesel engines which would be in a position to capture a large sector of the world market.

To obtain the necessary increase in specific power output of the company's engines, Pope sanctioned and directed a series of developments aimed at taking large engine design out of the 'blacksmith era' and into the 'scientific era'. The problems having been divided into thermal and mechanical ones, a series of test rigs was used to determine the optimum configuration for each component. The techniques used in these tests were well proven and standard, but provided a real information basis on which to design the up-rated engine.

The most significant development made in the K major engine design was the introduction of long-life exhaust valve cages. An increase in the time between overhauls for the exhaust valve assemblies from 500 hours to 3,000 hours was thus achieved. The methods used to do this were to adopt a water-cooled valve seat and valve stem design and to use a novel oil control unit. Water-cooling of valves for high-duty application was already standard practice, as was the use of 'Stellite' valve seats. A textbook[12] published in 1945 referred to the fact that 'exhaust valves of high-speed engines are made of special heat-resisting alloys, such as chrome-nickel steel, and for severe service, as in aircraft engines, the contact cone is surfaced by welding on a "Stellite" layer'. Two novel aspects of the Mirrlees design, however, are the provision of a cooling-water groove behind the 'Stellite' ring so that precise control of valve seat temperature may be maintained, and the use of electron beam welding to secure the ring in position. Because of the size of the components being welded (the rings are approximately 5 in. in diameter) and because of the precision with which the welding operation had to be performed, electron beam welding proved to be the ideal choice. The welding unit used in this operation was designed for Mirrlees, at the suggestion of Pope, by Hawker Siddeley Dynamics Ltd on the basis of an original design by the Hamilton Corporation of the United States with which Pope was familiar. Its use reflects the value of recruiting personnel into an organisation where their knowledge of the latest developments in a particular field of research activity may be immediately applicable to products under development.

The K Major engines make extensive use of cast iron, both in the main frame and subsidiary components. Pope had carried out a 'very thorough theoretical and experimental investigation for the British Shipbuilding Research Association into the causes of failure of pistons, liners and cylinder heads in marine oil engines and it was not therefore surprising to see the attention which Pope and Lowe had given to thermal and pressure-induced stresses in the design of the two-piece, oil-cooled piston'.[13] The concept of a two-piece piston was well established when the K Major engines were being designed,[14] but the important feature was the use of well-developed experimental techniques to determine the optimum form which such a piston should have. Such was the pattern of innovation for all but one of the developments previously mentioned – a pattern of sophisticated development rather than of novel design.

The overall design of the K Major and KV Major engines was largely the responsibility of Mr N. Fletcher as Senior Designer for the programme, with Mr J. M. Radford as Senior Development Engineer responsible for the thermal analyses of the design. These were backed by a Research and Development Division staffed by seventy people, of whom approximately half were of graduate or equivalent academic standard. Over two thousand people were employed on the site in all capacities. In addition to its own development effort, Mirrlees made use of

L

the specialist experience and facilities of D. Napier and Sons Ltd, Bryce Berger Ltd and the David Brown Corporation to develop specialist items such as turbo-chargers, fuel pumps and lubricating pumps for use on the K Major series.

References

1. Mirrlees National Ltd, Publication No. 2001 (1967).
2. 'A Pioneer Called Diesel', *Esso Journal* (autumn 1967).
3. *The Engineer*, **187** 697 (1949).
4. *The Engineer*, **189** 271 (1950).
5. Interview with Mr J. Smith, Technical Director, Petters Ltd, Staines, Middlesex, 30 Oct 1968.
6. Interview with Dr J. A. Pope, Technical Director and Director of Research, Mirrlees National Ltd, Stockport, Cheshire, 28 Nov 1968.
7. 'British Oilfield Equipment', *The Engineer*, **188** 531 (1949).
8. J. A. Pope, 'The Economics of Large Diesel Engines for Electrical Power Generation', *Proc. Inst. Mech. Eng.*, **177** 1075 (1963).
9. R. Greenhalgh, 'Turbocharged-Diesel Generating Plant Burning Residual Fuels', *Hawker Siddeley Technical Review*, **1**, no. 3 (1964) 23, and *Proc. Inst. Mech. Eng.*, **178**(3K) 74 (1964).
10. R. A. Dennis and J. M. Radford, Symposium on Thermal Loading of Diesel Engines, 'Piston Stresses – Theoretical and Experimental Developments', *Proc. Inst. Mech. Eng.*, **179**(3C) 19 (1964–5).
11. J. A. Pope and W. Lowe, 'The Development of a Highly-rated, Medium-speed Diesel Engine of 7,000–9,000 h.p. for Marine Propulsion', *Trans. Inst. Marine Eng.*, **78** 325 (1966).
12. V. L. Maleev, *Internal Combustion Engines: Theory and Design*, 2nd ed. (McGraw-Hill, London, 1945) p. 326.
13. R. Cook, contribution to the discussion of ref. 11, p. 339.
14. Maleev, *Internal Combustion Engines*, view of Nordberg oil engine piston, p. 19.

22 IMPERIAL CHEMICAL INDUSTRIES: PROCION REACTIVE DYES

This case study deals with a discovery made and exploited within the large research and development organisation belonging to the Dyestuffs Division of I.C.I. The discovery was made in 1953 and was described thirteen years later by the *I.C.I. Magazine*[1] in the following terms:

For more than 60 years, research workers sought a practical process for dyeing cellulosic fibres – cotton, linen and viscose rayon – by chemically reacting the dyestuff with the fibre. The breakthrough came at Blackley in the 1950s when the work of Dr W. E. Stephen, who is now retired, and Mr I. D. Rattee, now Professor of Colour Chemistry and Dyeing at Leeds University, led to the manufacture of the first three 'Procion' dyes – a yellow, a red and a blue. This completely new type of dyestuff could be reacted with any cellulosic fibre to give economical, bright shades with high fastness to washing. It was a revolutionary event.

(a) Dyes

The commercial value of a dyestuff is not completely predictable in terms of its chemistry and potential dyes have to be tested empirically. Early synthesisers of coloured chemicals, like Perkin who synthesised his famous mauve in 1856, used dyers to test dyeing behaviour, but the testing function eventually came to be transferred to the dye manufacturers who set up their own dyeing facilities. At the headquarters of I.C.I. Dyestuffs Division at Blackley in Manchester, where offices and laboratories share a large site with one of the Division's factories for the production of dyestuffs and other chemicals, the testing of dyes was carried out in a department known until 1965 as the Dyehouse.

This provision of a separate department with a technological rather than a scientific base provided the model for research into rubber, pharmaceuticals, plastics, fibres, etc., which also grew out of organic chemical research. It is of particular significance in the discovery of Procion dyes as the two men, Stephen and Rattee, who were credited with the discovery were members respectively of the Research Department and the Dyehouse.

For a coloured substance to be a dye there must be a suitable method for attaching it to the fibres of the material being dyed and, once attached, it must resist subsequent removal. The technical term for resistance to removal is 'fastness'. One type of dye molecule contains a chromophore, i.e. a chemical group responsible for the colour, and a group which makes the dye soluble or dispersible in water. After the dye molecule has been

absorbed by a fibre from a solution of the dye, chemical treatment of the group responsible for water solubility can produce
an insoluble molecule which is not removed by washing. For
example, the I.C.I. Alcian Blue dyes are made soluble by the
presence of a cationic quaternary ammonium type of group
which after dyeing is hydrolysed by alkali to leave an insoluble
dye. Another method of developing wash-fastness is to use two
soluble chemicals which can be made to react with each other
after they have been absorbed by the fibre. Large molecules
can be built up in this way and they are then trapped in the fibre
structure. A third method of dyeing is to react a dye chemically
with the fibre. To a chemist this might seem an obvious sort of
process to attempt, but in fact the Procions were the first
commercially successful dyes which could be attached to a fibre
by chemical reaction.

Attempts to colour cellulose, the main constituent of cotton, by
chemical reaction date back to the last century. In 1895 Cross
and Bevan,[2] as part of their extensive research into cellulose,
attached colour-producing molecules to alkali-impregnated
cellulose. Several separate steps were used and no practical
dyeing process was developed.

Cellulose contains abundant hydroxyl (–OH) groups and in
theory it should not be difficult to produce coloured substances
that would react directly with these groups. One explanation
as to why this approach was not tried has been given by Vickerstaff,[3] a former Director of Dyestuffs Division:

> The reason ... was the belief that the hydroxyl groups of
> cellulose were relatively inert so that drastic conditions
> (leading to fibre degradation) were needed to promote
> reaction. The alternative approach of using a highly reactive
> grouping in the coloured molecule was similarly considered
> impracticable unless a non-aqueous system could be used,
> since a grouping highly reactive towards hydroxyl groups
> would probably be hydrolysed by water.

Reactive groups in non-aqueous solvents could be used, but
most practical dyeing processes use water (though this might
change if present trends in the direction of water shortage
continue).

(b) Cyanuric chloride

The chemical reactivity of the Procion dyes to the hydroxyl groups of cellulose is made possible by the use of cyanuric chloride $(CNCl)_3$. The cyanuric chloride molecule consists of a ring structure with three chlorine atoms on the outside. If, for example, one of the chlorine atoms is made to react with a cellulose hydroxyl group and another chlorine with a colour-producing molecule, then the colour-producing molecule has been attached to the cellulose by a chemical bridge.

Cyanuric chloride has been known since 1828 when it was discovered by a pupil of Liebig.[4] Its composition, $(CNCl)_x$, was determined by Liebig, who prepared it by passing chlorine over dry potassium thiocyanate. Its cyclic nature with three chlorine atoms per molecule (and hence its possible use in 'bridging') was widely accepted[5] by 1883. The reaction of cyanuric chloride with ammonia was described by Liebig and its reactions with hydroxyl, amino and other groups were known in the last century; for instance, Fries[6] described the reaction with amines in 1886. Thus the Procion dyes can hardly be considered as the product of recent discoveries in pure organic chemistry (which is not the same as saying that a knowledge of chemistry played no part in their discovery).

Cyanuric chloride was used extensively in dyestuff research in the 1920s by the Society of Chemical Industry in Basle (now known as CIBA). Some 200 Swiss patents indicate the extent of the research into the use of cyanuric chloride in producing dyestuff intermediates and derived azo dyes. For example, Chlorantine Fast Green BLL was produced by using cyanuric chloride as a chemical bridge to couple together a blue component and a yellow component to give a very good green dye.[7]

There are two possible ways of using cyanuric chloride as a bridge between a colour-producing molecule and cellulose. One is first to treat the cellulose with cyanuric chloride and then add the colour-producing component; the alternative is the reverse process of reacting the cyanuric chloride with the colour-producing component and then using the resultant product to dye cellulose. The latter approach is used in the Procion dyes; the former approach was attempted by CIBA in a process patented in Britain[8] in 1931. The CIBA process was, however, based on the Cross and Bevan series of reactions using cyanuric chloride (or bromide) in solution in xylene and it was not developed for commercial use. The first step of the I.C.I. process was also carried out by CIBA who prepared many dyestuff intermediaries by reacting cyanuric chloride with a colour-producing component. They did not, however, attempt the reaction of these intermediaries with cellulose. In a 1923 CIBA patent[9] one of the intermediaries described is chemically identical to one of the first Procion dyes and the existence of this patent helped to prevent I.C.I. from gaining a patent-protected monopoly of reactive dyes.

CIBA used intermediaries that were similar to the Procion dyes as a means of coupling together two coloured components or a dye and some other chemical aimed at modifying its dyeing properties. That the use of these intermediaries for reaction with cellulose was not considered may have been due to the accepted theory of cellulose activity or it may have been that no advantage was seen in dyeing cellulose in this way.

With the advantage of hindsight, it can be seen that all the important facts necessary for the discovery of Procion dyes for cellulose were available many years before the discovery took place. Even the belief that a group reactive enough to attack cellulose hydroxyl groups must be unstable in water could have been questioned in 1937 when a paper appeared[7] describing the resistance to alkaline hydrolysis of some of CIBA's intermediaries. However, in 1952, when Warren *et al.*[10] published some work on the reaction of cyanuric chloride with cellulose, they still thought it necessary to follow in the Cross and Bevan tradition by first impregnating the cellulose with alkali and then reacting with cyanuric chloride in xylene solution.

(c) The Procion dyes

In 1953 Dr W. E. Stephen, working in the Research Department at Blackley, considered the possibility of using cyanuric chloride as a bridge between a coloured molecule and the wool fibre. In his own words, Stephen was 'always searching the literature' and he was aware of the work carried out by CIBA some twenty years previously. He was also aware of previous work at Blackley aimed at using labile chlorine-containing compounds to aid the fixation of dyes to wool.[11]

However, at that time work aimed at chemically reactive dyes was not part of the annual research plan and Dr Piggott, the Associate Research Manager, was not keen on the idea of reactive dyes. Stephen was aware of the high reactivity of cyanuric chloride and there is no doubt that it was Stephen alone who, on his own initiative, conceived the idea of reacting a coloured molecule with one of the three chlorine atoms contained in the cyanuric chloride molecule, leaving two chlorine atoms which might be able to react with a fibre molecule. Stephen was aware of the reactivity of the chlorine atoms in dichlorotriazinyl compounds (compounds formed by reacting cyanuric chloride with amino dyes) and he was also aware from his long experience and familiarity with the literature that such compounds would react with amino groups on other dyes in solution in water. Heating for four hours at 40°C, for example, would give a good yield.[11] Wool fibres contain amino, carboxyl and hydroxyl groups and Stephen's inventive step was to consider the possibility of reaction in water between a dichlorotriazinyl compound and the wool fibre instead of with another amino dye.

He reacted an amino-azo dye with cyanuric chloride and in September 1953 he handed the resultant dichlorotriazinyl amino compound to Mr I. D. Rattee, who was working in the Wool Section of the Dyehouse, where it was found that the new compound did not give very good results as a wool dye. However, through information circulated from the literature-searching activities of the Division's library, Rattee read the paper by Warren et al. referred to above, and without consulting Stephen he decided to try Stephen's sample as a cotton dye.[12] Rattee found that Stephen's dichlorotriazinyl amino dye could be used in aqueous solution to dye 'alkali cellulose'. Thus the xylene used in Warren's work was not necessary. He also found that the strongly alkaline conditions of all the previous work were not necessary and that relatively dilute caustic soda could be used if the solution was saturated with common salt.[13]

Rattee discussed his results with Stephen who then prepared more dichlorotriazinyl amino dyes. Rattee's section leader in the Wool Section, E. Waters, told him to press on with his investigation even though dyes for cotton were supposed to be examined by the Cotton Section and not the Wool Section.

The final stage in Rattee's experimental dyeing of cotton fabric with Stephen's samples consisted of washing off unreacted dye under the cold water tap. On one occasion Rattee, who carried out most of his experiments unassisted, washed the dyed fabric under the hot water tap by mistake and he found that much better 'fixation' of the dye resulted.[12] He showed his results to the Cotton Section where a Technical Officer, D. Weston, took up Rattee's heat fixation discovery, investigated the effect of temperature and developed a practical dyeing process involving steam treatment.

G. S. J. White, who was Dyestuffs Division Technical Service Director, gave enthusiastic support to the discovery, as did J. D. Rose, the Division's Research Director, and the two of them were able 'to ginger the whole development machinery into an unusual activity'.[14] This short-circuiting of the usual procedure started about July 1954. Stephen's first sample had been prepared in September 1953, a laboratory dyeing process had been developed by Rattee in October and five of the dyes that were eventually to be marketed had been prepared by Stephen and examined by Rattee in the Wool Section before the Cotton Section were involved.[12]

Considerable effort was devoted to the development of the discovery. In the Research Department, work started by Stephen on the stabilisation of the dyes by buffers was enlarged, more samples were prepared, manufacturing processes developed and work to counter competitive patent activity was started. In the Dyehouse, development spread to other sections. The Textile Printing Section, for example, found that sodium bicarbonate could be used instead of caustic soda and successful printing techniques were developed.

In 1955 I.C.I.'s sales organisation began to press for the release of some of the new dyes as they claimed to have heard rumours that competitive reactive dyes were being tested. The first three Procions were marketed in April 1956 and they were to be followed by many others. In June 1957 CIBA entered the reactive dye market with ten 'Cibacron' dyes. These were monochlorotriazinyl dyes in which only one reactive chlorine atom remains, making them less reactive but more stable than the dichloro derivatives. I.C.I. just managed to beat CIBA in the marketing of the monochloro dyes by introducing two 'H' dyes in May 1957, and nine more 'Procion H' brand dyes appeared in June 1957.

By 1966 I.C.I., with more than seventy Procions being marketed, could claim that it sold more Procion dyes than any other class of dye,[1] although other firms such as Bayer, Geigy and Sandoz had also introduced reactive dye ranges. Successful exploitation of foreign markets earned a Queen's Award for exports as well as for innovation.

(d) Comments

The discovery of the 'Procions' illustrates several points about the process of innovation.

(i) *The relationship between science and technology.* The chemistry of the Procion dyes is an example of the fact that many technological developments in so-called science-based industries depend not on the recent discoveries of 'curiosity-oriented' research but on the presence of a person (in this case Stephen) who has been educated and trained in the theories, knowledge and techniques of a particular science so that, when a technological problem requires it, he can make use of the relevant science which may have originated many years before and possibly in a different country.

The type of relationship between science and technology illustrated by this case study is in marked contrast to the origin of the synthetic dyestuff industry. The chemical structure of Perkin's Mauve was not known at the time of its discovery in 1856, but its preparation would not have been possible without the availability of aniline, which had been prepared only three years earlier by Béchamp from benzene via nitrobenzene. Benzene, itself the basis of a new branch of organic chemistry, had been isolated only thirty years before and Kekulé's cyclic structure, which was to provide a rational basis for reactions in this branch of chemistry, was not formulated until 1865. Another example of the closeness between chemical discoveries and the manufacture of new dyes in the last century is provided by Griess and the azo dyes. In 1859 Griess discovered a new chemical reaction involving the coupling of two benzene rings

L 2

by an —N=N— group. The first commercial dye based on this reaction was Aniline Yellow,[15] sold in London in 1863. This closeness between science and technology was one of the factors which influenced the British Government in setting up the Department of Scientific and Industrial Research in 1917.

If, however, Procion dyes are typical of modern advances, then it would seem that the universities' role in supplying manpower has now become more important to industry than the supply of new chemical compounds and new reactions. Many technological advances can be seen to depend on a previous technological discovery rather than a discovery in the field of pure science. It is much more realistic to see reactive dyes as stemming from the technological work of CIBA than from the work of academic chemists.

(ii) *Reasons for delay.* Given that CIBA demonstrated the usefulness of cyanuric chloride as a 'bridge' between dyes and other compounds, and that in 1923 they actually patented the preparation of a compound identical with one of the Procions, the time-lag of thirty years before Rattee thought of using Stephen's dye for cellulose needs some explanation. There are two possible reasons for the delay. One already mentioned is the theory of cellulose reactivity that suggested the need for a drastic reaction of the Cross and Bevan type.

The second possible reason for the delay is that no one saw any point in developing chemically reactive dyes for cellulose. The most obvious attraction of a dye that reacts chemically with a fibre is the resultant wash-fastness. Since wash-fastness had been achieved by other methods, it seemed that there was no need to develop reactive dyes for this purpose. Widmer, CIBA's Dyes Division Research Manager, expressed this point of view as follows:[16]

> Dyers will not necessarily rejoice when told of the existence of dyes that are linked with the fibre chemically. The trade are primarily interested to know whether simple and safe methods are available by which to obtain the fastness properties and shades specified.

Much of what has been written about the delay between discovery and application ignores the fact that industry does not develop a technological concept unless there is some reason for doing it. Rattee has claimed[17] that the reason for the delay in the discovery of reactive dyes is that chemists were set the wrong targets, and it seems likely that a major contributory factor to the delay was the lack of appreciation of some of the benefits that were to be gained from developing reactive dye systems.

(iii) *Factors making for success*. Certain factors can be selected as contributing to the success of the Procions. These are good co-operation, the involvement of senior management and the availability within I.C.I. of large resources capable of exploiting the discovery on a world-wide scale.

The good co-operation in this case was the informal contact that existed between Technical Officers in the Research Department and in the Dyehouse. The Dyehouse was responsible for development work, testing and providing technical service to customers and it employed chemistry graduates like Rattee as well as people with experience in the industries using the Division's products. A tradition existed of workers in the Dyehouse calling at the research laboratories for informal chats. Co-operation between the two departments was further encouraged by the provision of a large number of internal telephones with up-to-date directories. This informal co-operation was certainly of importance in the discovery of the Procions. Stephen in the Research Department and Rattee in the Dyehouse were both Technical Officers, the normal career grade for qualified scientists and technologists and below the level of the sort of co-operation that depends on committees and more formal structures.

Once the discovery had been made and senior management had become interested to the extent of 'gingering the whole development machinery into an unusual activity', the resources available to I.C.I. must have been an important factor in the success of the Procions. No small firm could have developed manufacturing processes and industrial dyeing processes and sold its products to a world market in the face of the competition that soon developed.

It might be said that Dyestuffs Division in this case managed to combine the advantages of a large firm with those of a small firm. The informal approach to research which is often typical of a small firm certainly played a part in the discovery of the

Procions. Stephen's original work was not part of any departmental plan and Rattee's discovery that Stephen's dye could be used on cotton was made in the Wool Section, though the Cotton Section played an important role in developing practical dyeing processes.

(iv) *A piece of luck?* It is common knowledge that research in one field sometimes produces discoveries that turn out by chance to have application in another field. Perkin, for example, was attempting to synthesise the drug quinine when he discovered his first dye. A related phenomenon is the development for a particular market of a product which subsequently becomes a commercial success in a market different from the one first envisaged.

There is some evidence that many of the advantages of the Procion dyes were not entirely clear to all concerned in the early days of their development. A paper by Vickerstaff[3] published in 1957 contains no clear statement of the advantages of the Procion dyes over other dyes for cotton. Similarly, the first publicity material of CIBA[16] for their 'Cibacron' reactive dyes does not contain any convincing reasons for their use in normal dyeing processes.

According to Rattee in his inaugural lecture as Professor of Colour Chemistry and Dyeing,[17] the Procions were first developed for continuous dyeing as opposed to the traditional batchwise method and reactive dyes had not been discovered earlier because chemists had not realised the need for special dyes designed for continuous dyeing.

In fact, the I.C.I. sales organisation was not very interested in dyes for continuous processes and effort was switched to the development of the use of weak alkalis in cold batchwise dyeing. The most successful of the early Procion dyes was Procion Brilliant Blue H 7GS, which was a batch dye with a very attractive turquoise-blue shade. The new shades that were possible with Procions helped in the development of brighter fashions, which in turn actually reduced the amount of continuous dyeing that was carried out in this country as there was less demand for thousands of yards of fabric all dyed in the same shade.

Prior to the advent of the Procions, the improvement of dyeing properties and the introduction of brighter shades had been restricted by the overriding requirement for wash-fastness. The chemical reactivity of the Procions made the choice of shade and the improvement of dyeing properties independent of wash-fastness. The brilliant shades that became available had one unpredictable result, as expressed by an I.C.I. publication:[1] 'The pop-art fashion trend in clothing which has spread from Britain all over the world might never have started but for the work done on Procions at Blackley and imaginatively exploited by I.C.I. customers.'

It is not being critical of I.C.I. to claim that luck played a part somewhere on the road between a speculative dye for wool and a new fashion trend for cotton.

References

1. 'Winning the Queen's Award', *I.C.I. Magazine*, Oct–Nov 1966, p. 152.
2. C. F. Cross and E. J. Bevan, *Researches on Cellulose* (Longmans, London, 1895), p. 34.
3. Vickerstaff, *J. Soc. Dyers and Colourists*, **73** 237 (1957).
4. Serullas, *Ann. Chim. Phys.* **38**(2) 379 (1828).
5. Hofmann, *Ber.*, **16** 2893 (1883).
6. Fries, *Ber.*, **19** 242 (1886).
7. Fierz-David and Matter, *J. Soc. Dyers and Colourists*, **53** 424 (1937).
8. Swiss Patent 144,228; German Patents 554,781 and 560,035; British Patent 363,897–all to CIBA A.G.
9. British Patent 209,723 to CIBA A.G., 1923.
10. Warren *et al.*, *Textile Research Journal*, **22** 584 (1952).
11. Interview with W. E. Stephen, 1968.
12. Interview with I. D. Rattee, 1967.
13. I. D. Rattee, *Procion Dyestuffs in Textile Dyeing* (I.C.I.,1962) chap. 1.
14. G. S. J. White, private communication, 1968.
15. F. W. Gibbs, *Organic Chemistry Today* (Penguin Books, Harmondsworth, 1961).
16. Widmer, *Ciba Review*, no. 120 (1957), p. 5.
17. Rattee, *Inaugural Lecture: Discovery or Invention?* (Leeds U.P., 1964).

23 J. AND S. PUMPS LTD: SEALED MOTOR PUMP UNITS AND TURBINE ALTERNATOR UNITS

In 1968 J. and S. Pumps Ltd employed approximately eighty people in the design, development, testing and sale of special pumps and turbo-alternators. The firm was founded to exploit, in the chemical industry, certain advances in the design of leak-proof pumps made to meet the increasingly rigorous demands of modern power plants. A useful diversification on the basis of highly specialised technology available in the company was made possible by using the pumps 'working backwards' as turbo-alternators to generate electrical energy on off-shore oil and gas wells. For its activities the firm was awarded the Queen's Award to industry for technological innovation in April 1967.

(a) Background to the innovation

Two distinct but interrelated developments in the electrical energy industry served to generate a market for 'canned' motor pump units of the kind for which the J. and S. Pumps Ltd company received its Queen's Award, namely, the increasingly stringent operating conditions of conventional (oil- or coal-fired) power stations and the development of new types of power stations as economic competitors to conventional designs.

In a conventional power station, the working fluid, water, is successively pressurised, heated, converted to high-pressure steam and expanded through turbines to low pressure and temperature, yielding a high proportion of its energy in the process. The thermal cycle efficiency of such a power station is directly related to the difference between the upper and lower operating limits of the working fluid. Very significant economic advantages may be derived from increasing the maximum temperature of the fluid. In 1952 it was stated[1] that there has been a 'consistent upward trend of boiler pressure and temperature; pressures have reached 2,500 lb/sq. in. for commercial operation and temperatures have increased 1,050°F.' By the mid-1960s operating pressures in excess of 3,000 lb/sq. in. were common. These operating pressures made the attractions of a leak-proof pump evident to manufacturers and operators of power plant.

As pressures and temperatures in power plants have risen, troubles have been encountered in gland leakgage in circulating pumps, and it has also been necessary to use large volumes of water for stuffing-box (i.e. shaft seal) cooling purposes. To overcome these difficulties, a totally enclosed pump and motor unit in which there are no glands or rotating seals has been developed by Hayward Tyler and Co. Ltd of Luton. In this boiler circulator the pump and motor are enclosed in a single pressure-tight shell. The shell is completely filled with water which serves to lubricate and cool the motor, and the design is such that the possibility of leakage round the pump shaft is eliminated and there are no glands to be packed.[2]

A range of pumps was available to provide pressures of up to 3,000 lb/sq. in. and motors of up to 2,000 h.p. to drive them. The motor used in the Hayward Tyler pumps was an induction motor with immersed P.V.C.-insulated stator coils driving the rotor and pump impeller; because of the insultant an operating limit of 135°F had to be imposed as the maximum permissible motor temperature to protect the motor windings.

The second major stimulus for the development of leak-proof boiler circulator pumps was provided by the advent of atomic-energy power plant. In 1951 the design studies for Calder Hall, the world's first major atomic power station, had begun and by 1956 the station was supplying power to the British national grid. In America, efforts were directed towards the development of a different type of nuclear energy conversion process. Whereas the Calder Hall power station used circulating carbon dioxide gas to transfer the thermal energy released by the nuclear reaction to the steam-generating plant, the American developments were aimed at producing a pressurised water reactor, the thermal energy being transferred by water prevented from boiling by the pressure imposed on it. The American developments were stimulated by the possibilities which nuclear energy offered for military use,[3] and the *Nautilus* submarine formed an operational test-bed for subsequent American civil reactors.

Following on the *Nautilus*, a civilian programme of development was undertaken in the United States to produce a suitable reactor system for an experimental merchant ship, the *Savannah*, and for the Shippingport commercial atomic power station. In July 1953 the Shippingport project was formally authorised. Work began on the station site in September 1954 and on

23 December 1957 the plant achieved its full output rating of 60 MW. In a description of the plant design philosophy it was noted that

> in the design, construction and operation of the Shippingport plant, every necessary precaution has been taken to guard against hazards to the operating personnel or to the surrounding area. The principal safety problem is to prevent, in the event of any conceivable accident, the release of harmful quantities of fission products from the reactor core. . . . Although the radioactivity of pure water decays rapidly, dissolved impurities, corrosion products, or fission products that have longer periods of radioactivity may be in the coolant water. . . . Therefore it was decided early in the project that the portions of the reactor plant containing coolant at operating temperature and pressure would be, or at least be made capable of being, hermetically sealed. With a sealed system, the plant could more feasibly continue to operate even though the coolant might contain fission products from a number of minor fuel element leaks. This decision for a sealed system dictated use of canned motor main coolant pumps and control rod drive mechanisms. The main coolant pumps are of the vertical, single-stage, centrifugal type. The coolant circulates between the rotor and the stator, lubricating the bearings and cooling the motor. The motor is mounted directly on the pump casing, and can be seal-welded to the casing, forming a leakproof unit. This Westinghouse pump is designed to deliver 18,300 gallons per minute of coolant at 340 ft of head and develops approximately 1,500 h.p.[4]

Similar pumps were used on the discharge and vent system, and the failed element detection and location system.

The *Nautilus* submarine reactor and the Shippingport civil reactor provided a ready market for the development of canned motor pump units. Their advantages for conventional applications, particularly in power stations, were readily evident. In America, Germany and Britain, three principal suppliers undertook to develop a series of canned motor pump units. In America, the principal supplier was Westinghouse with its extensive experience gained during the atomic submarine and power station development programmes. In Germany, the most important manufacturer of the units was the firm of Klein, Schanzlin and Becker (K.S.B.) of Frankenthal/Pfalz. As mentioned above, Hayward Tyler were the principal manufacturers in Britain. K.S.B. produced a pump similar to the Hayward Tyler design with exposed stator windings but with the motor mounted vertically below the pump instead of above it, a practice which had been established in the design of borehole pumps.[5]

It was evident that the range of usefulness of these pumps could be improved if they were given a greater operating temperature range and if the electrical connections and fittings in the pumps could be protected against the corrosive effects of the circulating fluid. The Hayward Tyler design of 1955 had a maximum operating temperature of 135°F because of the problem of maintaining the insulation of the motor windings. The solution to these twin problems of insulation and corrosion was to enclose the stator windings within a double cylinder formed by the motor casing and a corrosion-resistant sheath of stainless steel or other suitable material. Similarly, the rotor of the motor was encased in a thin corrosion-resistant shield, leaving only a thin annulus of fluid between the rotor and stator. By placing a thermal barrier between the pump and motor the heat flow to the motor could be controlled within acceptable limits with the aid of an external heat-exchanger. The German firm of F. Ladendorf produced such a pump expressly designed for pumping corrosive liquids.

(b) The beginnings of J. and S. Pumps Ltd[6,7]

One of the people who opted to leave Hungary after the uprising of 1956 was P. Somlo, a graduate in chemical engineering of the University of Budapest. Somlo's experience had included a period of three years spent at the Central Research Institute on machine tools in Budapest prior to obtaining a position as chief engineer of an industrial concern. When he came to England he worked at the General Motors Frigidaire Division and left to join the firm of Hayward Tyler in 1958. There, Somlo was engaged in designing a range of pumps for service in power stations in association with Mr T. Hetherington, a designer, and Mr G. P. E. Howard, the firm's Technical Director. Hetherington had begun his career as an engineering apprentice with Vickers prior to joining the Sigmund Organisation where he worked on the design of machine guns during the 1939–45 war. After the war he was engaged in the design of pumps and became the company's Chief Designer prior to joining the firm of Hayward Tyler. Howard, an engineering graduate of Cambridge, first went to work for George Kent, the firm of instrument manufacturers, prior to joining Hayward Tyler – the firm his family had founded – where he held a wide range of jobs before becoming Technical Director of the firm. He resigned from Hayward Tyler in 1959 to join J. and S. Engineering Ltd, just as it was about to be reorganised to become J. and S. Engineers Ltd.

J. and S. Engineering Ltd had been formed in 1946 as a partnership between Mr T. J. Johnson and Mr Smith to manufacture a range of engineering products. Shortly afterwards they were joined by Mr R. Palumbo, who was later to become Group Managing Director of J. and S. Engineers Ltd, a holding firm founded in 1960 from J. and S. Engineering Ltd. In rapid succession six subsidiary companies of J. and S. Engineers Ltd were formed to provide specialist services to the engineering industry. These were J. and S. Electronics Ltd, J. and S. Precision Ltd, J. and S. Marine Ltd, Jarrett (Dartford) Ltd, Jastac Ltd and J. and S. Pumps Ltd. The last-named company arose naturally from Howard's previous activity in Hayward Tyler. A service company, J. and S. Engineers (Sales) Ltd, was formed to provide marketing assistance to each of the companies in the group. In 1962 the group's sales amounted to £352,275 and by 1967 these had grown to £821,278 with a total group strength of approximately 370 staff.[8] J. and S. Pumps Ltd, the firm with which we are principally concerned, consisted initially of five people, Johnson and Palumbo, who had been with J. and S. Engineering since its earliest days, and Howard, Somlo and Hetherington who had joined them in 1959. Between them they had a wide range of experience in industry and particularly in the area of pump design. For eight months the group worked to plan its policies and the products it was to manufacture.

The guiding principle of the firm's activities was to offer a high ratio of specialised engineering skill to hardware in pump and motor production. They chose to bring the many technological advances made for (and paid for by) the nuclear engineering industry to the chemical engineering industry. They would offer a range of pumps and motors suitable for specialist use in industry (such as the pumping of liquid chlorine) which would be designed on the basis of the high reliability leak-proof techniques developed for nuclear reactors. They aimed to stay small, with a staff of less than one hundred people, so that they would remain economic in activities where larger firms with design staffs of, say, fifty people and a total of five hundred employees would be unable to operate satisfactorily. It was decided that a firm deliberately kept small would retain its ability to offer specialist designs to industry where the numbers of units involved would be far too small to be attractive to the larger firm. An indication of the limited nature of the market gap towards which J. and S. Pumps aimed its activities is the fact that by 1968 the firm had installed less than thirty chlorine pumps in Europe but this was more than the sum of all their competitors' deliveries.

The fundamental technological features of the J. and S. Pumps Ltd designs were settled in 1960 and manufacture got under way in 1961. The principal product was to be the canned motor pump, with both the rotor and stator separated from the pumped fluid by corrosion-resistant 'cans'. The stator windings would be enclosed in a hermetically sealed casing filled with inert gas and the electrical connections taken through pressure-tight containers. The rotor bearings were to be lubricated by means of the pumped fluid after it had passed through an external heat-exchanger to keep the motor within the required temperature limits. A variety of pump impeller units was to be provided to give a range of capacities and delivery pressures for the units. Initially, the firm would design only medium-pressure low-temperature units and later it would expand both of these capabilities. To make this expansion possible, a range of standardised designs for flanges and similar component interfaces was decided on, so that a change of unit specification would entail the minimum of design alteration and disruption to existing production procedures.

After the initial plans for the company had been completed

in 1960, the founders of the firm set about to recruit additional members of staff. They sought to employ a few selected individuals well skilled in the design and development of motors and pumps, but the rest of the staff was to consist of people with little or no experience of these activities so as 'to prevent inbreeding of designs – a potentially disastrous situation for a small company',[6] which can survive only by being up to date with customer requirements and potential developments.

(c) Design development

The basic product of J. and S. Pumps Ltd – the medium-pressure low-temperature canned motor pump – went into production in 1961 and in the first year 50 of the units were made. The following year production increased to 150. Development of high-pressure high-temperature pump units was planned. To finance these developments and to expand the production facilities, the merchant bankers Hill Samuel were approached to provide additional funding in return for an equity holding in the company. This they provided, and the development of very high-pressure pump units and high-pressure high-temperature units got under way.

A common difficulty to be faced in designing these units was that of providing a stator and rotor can combination that was sufficiently thin to ensure a high electrical efficiency in the motor whilst being adequately strong to resist the fluid pressures being imposed upon it. Because of its limited resources, the company tried to interest its suppliers in jointly developing the solutions to various problems that it encountered. But the assistance offered 'was not beyond the bounds of normal commercial practice'.[6] Accles and Pollock co-operated with J. and S. Pumps in the development of the tubular cans and in the fabricating and the inert gas weld-sealing techniques used on them, but 'J. and S. Pumps Ltd paid for the necessary tooling costs and the development costs were levied on sales of the cans'.[6] Discussions were also held between the firm and Associated Electrical Industries (Rugby) and Midland Silicones Ltd (Bury) on the development of insulators for the motors, 'but no development work was undertaken by them'.[6]

Commenting on the development of the very high-pressure

pumps, Somlo said that 'the need was there and anyone could tackle it. There was no commitment by either the manufacturer or user to make or accept the units and technical co-operation was limited to deciding on the specification'.[6] 'Caustic solution circulators are used on ammonia synthesis plants as part of the carbon dioxide system. The pumps circulate the process liquor round the scrubber column without the pressure having to be dropped to atmospheric. The highest operational pressure so far is 5,500 p.s.i. . . .'.[9]

The high-temperature, high-pressure circulators are designed for

experimental work with water, under conditions which are beyond the previously established properties of heat transfer, corrosion, etc. The extreme conditions are for supercritical steam duties at 3,800 lb/sq. in. pressure and 400–600°C temperature. Multi-stage as well as single-stage pumps have been developed for a test facility at the Reactor Engineering Laboratory of the U.K. Atomic Energy Authority, Risley Establishment. The most advanced unit, the multi-stage pump motor for supercritical steam systems, was previously considered impracticable because of insurmountable technical difficulties.[9]

In this case 'the problem [in development] was to convince the U.K.A.E.A. that a multi-stage unit could be built, and then to persuade it to buy it – but they did not finance its development'.[6] Part of the reason for the U.K.A.E.A. reluctance lay in the fact that the canned motor pump industry in general had concluded that such a unit was impracticable because of the problems involved in:

(i) making it capable of withstanding the instability caused by water to vapour conversion should this occur;
(ii) providing a third support bearing on the rotor; and
(iii) providing an effective heat barrier between the pump and the motor without increasing the stresses in the shaft beyond the permissible limit.

The first problem was solved when J. and S. Pumps Ltd built a small test loop to show that no water/vapour instability would, in fact, occur. Then the hydraulic paths were redesigned to give a very close impeller spacing so that the shaft overhang could be reduced and the third bearing eliminated. Finally 'we did an elementary calculation of the heat transfer in the thermal barrier to show that the operating stresses were not as bad as anticipated'.[6]

A major source of delay was met in a less obvious area – in the design of the main flange from the pump to the motor. This is subjected not only to high temperatures and high pressures but also to high rates of cooling (200°C/hour) and so the danger of cracking the flange was very real. Virtually no work had been done in Britain on this problem – a fact that is perhaps explained by the rather mundane nature of flanges, millions of which are used each year. The published data first used in their flange analysis by J. and S. Pumps subsequently proved to be in error and so they turned to the U.K.A.E.A. for assistance. This was forthcoming and the problem was solved, but only after a delay of eight months had occurred in the programme.

The use of the accumulated experience of the U.K.A.E.A. to help to solve the problem of designing the main flange of the motor/pump unit to withstand thermal cycling is typical of the attitude taken by J. and S. Pumps Ltd to outside assistance. Its willingness to learn from other groups' experience has provided it with ready access to far more experience than it could generate itself. Co-operation with the National Engineering Laboratory provided information on the design of fluid-lubricated bearings, the PAMETRADA Marine Research organisation provided details of high-pressure jointing techniques, and the experience of pump users was used to determine the best materials to withstand the corrosion conditions to which the pumps would be subjected.

(d) The turbo-alternator

In the course of his dealings with British Petroleum on the sale of pumps, Somlo, as Managing Director of J. and S. Pumps Ltd, had come to know of the problems encountered by B.P. in the control of its offshore oil-producing wellheads in the Arabian Gulf. These wellheads were provided with diesel-engine-driven generators to provide power for telemetry and valve actuation purposes. It was desirable to find some means of replacing the diesel sets with a more reliable electrical energy source that could ultimately be placed directly on the sea-bed, thus eliminating the need for a sea platform. Several solutions to the problem were under active investigation, including the possible use of fuel cells and bacteriological cells, when Somlo proposed to use what amounted to a canned motor pump unit operated in reverse.

By using the energy in the fast-flowing oil and gas mixture emerging from the well to drive the impeller (as a turbine), the rotor could be driven at zero cost by an energy source that would exist as long as the mixture was flowing from the well. By attaching permanent magnets to the rotor an electrical supply could be taken from the stator windings and used to drive the valves and telemetry equipment as required. A battery system would provide the necessary energy storage for opening the main flow valve. Thus, the solution to the energy supply problem was based directly on the existing technological base of the company and did not require any significant additions to its manpower or equipment to permit its exploitation. The company could diversify its product base whilst intensifying the use of its resources because it had identified its technological capabilities. 'A number of alternators have been supplied to Abu Dhabi Marine Areas Ltd, through British Petroleum. These are operating on pressures up to 3,000 lb/sq. in. and generate 680 watts. Similar equipment is being (1966) designed for operation in the North Sea, the turbines in this case being driven by gas flowing from the wells.'[9]

In connection with this development, the National Engineering Laboratory provided background information on the design and development of the gas-lubricated rotor bearings, but this was of limited value because of the suspended solids in the operating gas flow from the wellhead. Experience gained in designing the sealed systems of the canned motor pump units could be applied directly to the design of sealed electrical accumulators used for driving the motors which open the gas control valves.

(e) Comments

(i) Although many of the developments made in the field of canned motor pumps were made and patented by Westinghouse, J. and S. Pumps has not had to buy a licence to their designs and has in fact been able to by-pass any features on which this would have been necessary. Nor has J. and S. tried to patent its own canned motor pump designs, believing that this is an expensive process that can only be justified when a master-patent is held. Somlo suggested that the strongest protection available for J. and S. designs arises from the reluctance of other manufacturers of canned motor pump units to use the J. and S features simply because they had been developed elsewhere – it is a practical example of the 'N.I.H.' ('not invented here') factor in operation. Neither are the turbo-alternator units patent-protected.

(ii) Throughout the existence of the company a deliberate policy of 'staying small' has been implemented to ensure that at no time will it be prevented from providing a specialist service to industry through having to keep large production teams operating on long product runs. Product diversification is sought and developed only in so far as it is compatible with the more intensive use of existing technological resources.

(iii) The company was at one time a member of the British pump industry's research association – the British Hydro-dynamics Research Association – but resigned from this because it felt that a small company is penalised on two counts: first, because of its small size, it is less likely to have the cost of membership readily available whereas in a large firm this is readily 'lost' in the general 'overheads'; second, because a small firm is likely to concentrate on some specialist sector of the market – a research programme which would directly interest it is unlikely to be of general use to other member firms.

(iv) Because of the smallness of the company and the high proportion of technologists on the main group's Board of Directors and also on the Board of J. and S. Pumps Ltd, it is usual for proposed development programmes to be completely evaluated by the Boards rather than having this function delegated. Responsibility for determining the market require-ments is accepted by each member of the directorate of the company with Hetherington paying particular attention to motor-pump units and Somlo to turbo-alternators.

(v) Each project undertaken is assigned to a project leader to be his primary responsibility. By the nature of the firm these

teams are usually small, and for the turbo-alternator development programme it consisted of three people with drawing
office and machine-shop support. Heavy emphasis is laid on
teamwork. 'We prefer to operate as a team rather than
emphasise the contribution of individuals however senior they
might be.'[7] The small size of the firm is of course a help in
achieving this.

(vi) At the time of the Award there was a technical design
and development staff of approximately twenty people. Of
these, only three were graduates – Howard, Somlo and the
Contracts Engineer. The rest of the technical staff held H.N.C.
or equivalent qualifications.

(vii) Research and development costs for the group amounted
to 4·75 per cent of its trading profit in 1966 (i.e. £6,400 in
£134,579) and 4·96 per cent in 1967 (£7,853 in £158,299).
Capitalised research and patents written off in 1966 amounted
to £29,099 (zero in 1967), and the group capitalised research
expenditure stood at £78,600 in 1967.[8]

References

1. G. A. Gaffert, *Steam Power Stations* (McGraw-Hill, London, 1952) p. 224.
2. 'Glandless Circulating Pump', *The Engineer*, **199** 774 (1955).
3. C. Blair, *The Atomic Submarine and Admiral Rickover* (Holt, New York, 1954).
4. A. Tammaro (ed.), *The Shippingport Pressurised Water Reactor* (Addison-Wesley, Reading, Mass., 1958).
5. S. Lazarkiewicz, *Impeller Pumps* (Pergamon, Oxford, 1965) p. 525.
6. P. Somlo, Managing Director, J. and S. Pumps Ltd, interviewed by W. G. Evans, 21 Oct 1968.
7. P. Somlo, letter to W. G. Evans, 1 Dec 1969.
8. J. and S. Engineers Ltd, Report and Accounts for the year ended 30 April 1967.
9. J. and S. Pumps Ltd, Statement of Case to the Board of Trade, 1966, in support of application for the Queen's Award to Industry for Technological Innovation, 1967.

24 LINEN INDUSTRY RESEARCH ASSOCIATION, AND McCLEERY AND L'AMIE LTD: THE 'ATOZ' PROCESS

A joint Queen's Award was given to the Linen Industry Research Association (LIRA) and McCleery and
L'Amie Ltd for 'innovation in the production of

knitting yarns'. McCleery and L'Amie Ltd have
successfully manufactured and sold a new type of
synthetic yarn with excellent properties for use in
knitting. This new yarn was made possible by a new
process, the 'ATOZ' process, developed by LIRA.
The ATOZ process is primarily a new concept in yarn
production. 'ATOZ' was originally 'A to Z' and its
name implies that a whole range of properties can be
obtained by the use of a new concept involving the
bulking of a yarn made from fibres of varying shrink
properties. The knitting yarn made by the ATOZ
process was the first commercial application but there
may be other applications.

(a) Wool-like synthetics and high-bulk yarns

When shoppers buy a fabric in the form of a sweater, blanket,
etc., they are influenced by the feel of the fabric. Bulk or low
density produces a soft, light, airy and warm handle characteristic of wool. Woollen yarns are bulky because the wool fibre is
not straight but possesses a crimp. Any process which introduces
kinks or loops in a synthetic filament introduces bulk to a yarn
made from the synthetic and gives a wool-like handle. Many
such processes have been developed, mainly by the synthetic
fibre producers. Crimplene, for example, is bulked Terylene.

The technological ancestor of the ATOZ process is the 'high
bulk' process developed in the early 1950s by Du Pont for the
production of wool-like acrylic yarns.[1] High-bulk yarns are
produced by blending ordinary fibres with fibres that have been
heat-stretched. When the blended yarn is heated, the stretched
fibres tend to return to their original length. The stretched
fibres shrink and form a core to the non-stretched fibres which
have to accommodate themselves to the reduced length by
buckling and forming an airy mass round the core. Fabrics
made from such yarns have a desirably soft feel.

The ATOZ process differs from the high-bulk concept in that
a blend of fibres of various shrinkages is used instead of a
mixture of the two extremes of high-shrinkage stretched fibres
and low shrinkage non-stretched fibres. It was developed at
the Linen Industry Research Association as a logical de-

velopment from lines of thought concerned with ways of blending linen fibres with synthetic fibres.

(b) The Linen Industry Research Association

LIRA is at Lambeg near Belfast, as the linen industry is situated mainly in Northern Ireland, where it was founded by Huguenot families escaping from France in the seventeenth century and bringing with them their knowledge of production of fabric from the long fibres of the flax plant. Linen prospered in the nineteenth century, especially when the American Civil War cut off supplies of cotton, but more recently the market for linen has dropped. In 1966 flax formed only 3 per cent by weight of fibre usage in the United Kingdom.[2]

LIRA was set up in 1919 and attempted to improve the competitive position of linen. One successful outcome of the research was the development of crease-resistant finishes. Under the Directorship of D. A. Derrett-Smith and, more recently, H. A. C. Todd, LIRA has rethought its role. In effect, it has asked itself, 'What business is the linen industry in?' The answer has turned out to be the spinning of flax-like fibres into yarn and the production of fabrics. With the help of the Research Association, the Northern Ireland textile industry has ceased to be an almost one-fibre industry and has now started to use synthetics. In addition to Courtaulds, who have produced viscose rayon in Northern Ireland for several years, various fibre producers have been attracted to the area. I.C.I. make Terylene and Ulstron in Northern Ireland and there are Chemstrand with Acrilan, British Enkalon with Nylon 6 and Polythene Fibres with Spandex.

LIRA has not of course turned its back on flax and has improved the technique of flax harvesting and subsequent processing. Much of the effort in recent years, however, has been in showing how existing flax-based equipment can be used to work the new fibres, either alone or in blends. The ATOZ process is a logical development from this line of thinking. Since the granting of the Award, LIRA has had an even more radical rethink about its role. Since May 1970 the name has been changed and the LI in LIRA now stands for 'Lambeg Industrial' and not 'Linen Industry'.

(c) The development of ATOZ

ATOZ arose 'from a succession of events and thoughts' through 'seemingly haphazard but logical development by people with vast experience of fibres, and the contribution of varying fibre properties to blends'.[3] The success of high-bulk yarns had shown what could be achieved by blending two types of the same fibre. Todd was interested in blending different fibres, flax and man-made, as at the time of the development of ATOZ he was head of the Spinning Division as well as Deputy Director and LIRA was attempting to help flax spinners to diversify.

Under the direction of Todd, J. Nelson Ruddell carried out odd pieces of research that were not part of any formal programme but were aimed at applying the high-bulk concept to blends of different fibres. Thus a blend of flax and acrylic fibres could be treated with caustic soda which would cause the flax but not the acrylic to shrink. Alternatively, a blend of heat-stretched acrylic and flax could be heated to shrink the acrylic but not the flax. Blends of flax and nylon were also examined. All the two-component yarns had one characteristic in common – the shrinkable fibre formed a core inside the non-shrunk bulked fibre. Thought was given to how a yarn could be produced without this core. Experiments outside LIRA had been carried out with novel twisting techniques but this approach seemed to be expensive. It would be possible for a yarn to be constructed from several types of fibre of varying degress of shrinkage, but this would produce a problem in supply and storage of different fibres with the possibility of getting them mixed up.

Todd and Ruddell then arrived at the concept of varying the amount of potential shrinkage in a continuous operation by utilising the fact that when heat-stretched fibres are subsequently reheated in the absence of stress they contract. Acrylic fibre is supplied by the manufacturer in the form of continuous tow. Ruddell developed a machine that passed the tow through a steam box and into double nip rollers. By varying the speed of the rollers, the amount of extension given to the heated fibres could be varied continuously. The tow was then stapled (i.e. cut up into short lengths) and the staples blended and spun into yarn. A yarn was thus available which contained staples of varying degrees of shrinkability without the necessity of having a variety of raw materials. The variety had been built into a single tow. Various yarns were produced by this technique.

Two events then happened which were to lead to a commercial application of this concept. The first event was a visit by John Taylor of Shirley Developments Ltd, who look after the patenting and commercial exploitation of textile research association ideas. In an internal report Todd had written: 'it is also of importance to consider the production of a yarn which uses a nearly continuous range of fibre shrinkages instead of the simple two-component system'. Taylor said 'That is an invention' and in August 1961 patent applications in the name of Todd and Ruddell were made. Taylor was responsible for naming the new method of yarn production 'A to Z', eventually shortened to ATOZ.

The second event was the Acrilan exhibition. As part of their policy to help the introduction of synthetics into the Northern Ireland textile industry, LIRA had organised exhibitions jointly with the fibre producers. 1962 was to be the year for a LIRA–Chemstrand Research and Development Exhibition. (Chemstrand Ltd, the manufacturers of Acrilan, are now Monsanto Textiles Ltd.) A Chemstrand employee, Jack Brown, was shown Acrilan yarns made by the ATOZ process and thought they were unique. The ATOZ process was made a feature of the exhibition and a publicity campaign was mounted. The *New Scientist* published an account of ATOZ,[4] as did various technical journals.

Two young Ulsterman, Tony McCleery and John L'Amie, saw ATOZ-produced yarns at the Chemstrand exhibition and became interested in the possibilities. Todd had originally thought of selling the ATOZ process to the fibre manufacturers but in 1962 they could sell all their output and were not interested in new types of yarns. Todd then encouraged McCleery to go into the production of Acrilan yarns for hand knitting as there was no suitable form of Acrilan for this outlet and it had been found possible to produce knitted garments with outstanding dimensional stability from ATOZ yarns.

Market research was carried out which established that an improved Acrilan knitting yarn produced by ATOZ could be a commercial success. McCleery and L'Amie formed a limited company in March 1963, worked out a licensing agreement with LIRA and Shirley Developments Ltd and set to work to scale up to the ATOZ process.[5] McCleery is reputed to be extremely good in the commercial development of new ideas. Since ATOZ, he has developed a new dyeing technique for carpet yarns and is diversifying into carpet yarn production. L'Amie looks after the organisational side of things, which kept him busy, as the construction of a new factory started in July 1963. With the help of a grant from the Government of Northern Ireland, the most efficient modern machinery was installed to go with the scaled-up ATOZ machine produced by McCleery.

In July 1964 the new factory went into full-scale production of round, soft, resilient yarns of good dimensional stability suitable for both hand and machine knitting. Large sales were built up with Emu Wools Ltd, Great Universal Stores and other customers. By July 1965 it became necessary to double the size of the factory which by 1967 was consuming some $1\frac{1}{2}$ million lb of Acrilan per year.[5] (The total consumption of all man-made fibres for apparel uses in the United Kingdom was 257 million lb.[2])

The success of McCleery and L'Amie may be just the start of the application of ATOZ. An enormous range of properties can be obtained by varying the distribution and amount of the stretch given to the tow in the ATOZ process and of course different fibres can be used. This means that the textile industry is not quite so dependent on the fibre manufacturers, who are not very keen on producing different types of tow. From the interest being shown at present, LIRA are sure that new yarns with advantageous properties will emerge as the possibilities of ATOZ are explored more fully.

(d) Comments

This innovation clearly illustrates some of the difficulties which face current attempts to carry out 'research into research'. Much is being written about project selection and control on the assumption that research can be carried out in clearly defined projects with objectives whose benefits can be at least guessed at. This assumption does not fit the ATOZ case.

LIRA has a series of research projects which are controlled by project committees containing industrialists as well as members of the research staff. This would seem to be an admirable way of making sure that the research is of interest to industry and that results are likely to be used. However, the research which led to ATOZ was not part of this formalised project structure and its initial objective, to use a blend of flax and synthetics to enable existing spinners to diversify, was not achieved. In the eyes of those who see research as consisting of discrete projects, the work on two-component bulked yarns was a 'failed project'; but it is highly unlikely that the ATOZ process would have been produced without this initial research. One of the possible benefits of *any* research is its unforeseen usefulness in a direction different from that at first envisaged. If the probability of this happening is very small and is of a similar order for all research, then it can be ignored in project selection. If, however, certain types of research are much more likely to have unforeseen benefits than others, then much present thinking on project selection and control will have to be revised.

The ATOZ innovation also illustrates factors found in many of the other case studies. The 'person in authority' phenomenon is illustrated by Todd, described by one of his staff as being able to 'think widely'. Because of his position he was able to initiate experiments outside the formal programme and to press for development when the results looked interesting.

In terms of the history of ideas, the ATOZ concept has a clear technological ancestry and cannot be described as the application of a scientific discovery. The technical concept is, however, only one dimension of the innovation and Todd was prompted to think along the lines that he did because of the special needs of the Northern Ireland textile industry. These needs did not affect the chemical companies who had developed earlier bulking techniques to assist the sales of synthetic fibres.

The ATOZ case is one of the two studies reported here in which research associations won Awards, the other being the Chorleywood Bread Process. In both cases the established industry was initially reluctant to make use of the research association's discoveries. Entrepreneurial outsiders were required to scale up the machinery. As the established industry was not at first interested in the results, it seems unlikely that they would have been interested in paying for the research. The present trend to encourage research associations to rely more on contract research and less on block grants from the Government must therefore be viewed with some caution.

References

1. S. J. Davis and P. H. Cohen, *Skinner's Silk and Rayon Record*, **30** 258, 276 (1956).
2. 'The Pattern of Fibre Consumption in the U.K.', *The Shirley Link* (winter 1967–8) p. 7.
3. Interviews with J. N. Ruddell, H. A. C. Todd and other workers at LIRA, 1968.
4. *New Scientist*, **15** 680 (27 Sep 1962).
5. Information supplied by I. T. Hamilton, including the commentary to a film, *The ABC of A TO Z*.

25 JOSEPH LUCAS (ELECTRICAL) LTD: HIGH-VOLTAGE TRANSISTORS

(a) *The company*

Joseph Lucas (Electrical) Ltd (J.L.E.) has for sixty years been prominent in the design, development and manufacture of complete electrical installations for British cars, light commercial vehicles, motor-cycles and agricultural tractors. The products of the company include generators, starters, ignition equipment, lighting and signalling units, batteries, windscreen wipers, horns – in fact, practically every item of electrical equipment on the motor vehicle.

The company is a member of Joseph Lucas (Industries) Ltd, a group which includes C.A.V. Ltd (engaged in similar work for the heavier classes of vehicle, including passenger service vehicles) and Rotax Ltd (whose primary activity is electrical equipment for aircraft) as well as other companies with diverse interests.

(b) *Involvement in the field of semiconductors*

At an early stage in semiconductor technology, it was anticipated that semiconductor devices would play an important part in future developments of electrical equipment for road and air transport. Consequently a research programme was initiated at the Lucas Group Research Centre, an establishment providing member companies of the Lucas Group with facilities for investigating problems requiring research of a more fundamental nature than that normally carried out in the individual companies' laboratories. The research staff at the Centre numbers about 200, divided into departments which cover physics, chemistry, electronics, applied mechanics, fabricating processes, metal shaping and metallurgy.[1]

The Lucas involvement in solid-state physics began in a small way, with quite a small original research team headed by Dr G. C. Williams, but it soon became obvious that to meet the future foreseen for the employment of semiconductor devices in the Group, a large and continuing investment would be necessary. It was therefore decided to set up a semiconductor production division with J.L.E., having its own engineering and design facilities for process and device development based on the fundamental work carried out by the Group Research Centre. The small team which commenced the original research work at the Centre formed the nucleus of this production division, Dr Williams becoming General Manager. Today, the products of the Silicon Semiconductor Division cover a very wide range of devices, both for application to the automobile and aircraft electrical equipment manufactured by the Lucas Group, and for the more general fields of electronics engineering.

An example of the expanding use of semiconductor devices in the electrical equipment of road vehicles is found in the trend towards alternators, replacing the d.c. generators universally employed hitherto. Alternators, by virtue of their design and construction, provide improved performance to meet the generating requirements of modern vehicles. Until the advent of semiconductor rectifiers, alternators had been impracticable for car use because of the problem of rectifying their output for battery charging, although they had been used for some years previously for certain passenger service vehicles where it was possible to accommodate the comparatively large selenium or copper oxide rectifiers then available. The modern alternator employs nine semiconductor diodes in a rectifier pack only some $2 \times 2 \times 1\frac{1}{2}$ in., built into the machine. Then again, control of alternator output is achieved by a solid-state

M

electronic regulator, also housed in the machine and incorporating a semiconductor voltage regulator, quench diodes and field switching transistors.

The need for a high-voltage transistor – now to be discussed in greater detail – stemmed from the requirement for vehicle ignition systems of increased performance, which itself became necessary as the result of advances in internal combustion engine development.

(c) The Queen's Award

The Queen's Award to J.L.E. was granted in 1967 in recognition of pioneer work in the development of high-voltage silicon transistors, which have made possible significant advances in ignition systems for petrol engines.

In this connection, the following comment by Freeman[2] is of interest. Discussing the invasion of Europe by American solid-state technology through licence agreements, he notes:

> There are, it is true, a few component developments in Europe which have been licensed in the U.S.A. For example, the Lucas development work in industrial semiconductors which began in 1954 has resulted in the successful development of high-voltage devices for ignition systems which have been licensed to Delco in the U.S.A.

(d) The requirement for a high-voltage transistor

In the conventional form of ignition equipment employed for many years on petrol engines, an engine-driven contact breaker switches the primary current of an induction coil. Each time the contacts separate, the collapse of current in the primary winding induces a high voltage in the coil secondary winding, which at that instant is connected to the appropriate spark plug through a jump-spark high-tension distributor.

The performance of such a system is adequate for the four- and six-cylinder engines of normal performance that are prevalent in Britain, but mechanical and electrical factors associated with a contact breaker of this type impose a maximum sparking rate of about 400 sparks per second. With the development, particularly in the United States, of more advanced higher-speed engines requiring sparking rates considerably greater than this, the thoughts of ignition engineers turned to the possible use of a transistor as the switching device, triggering being accomplished by a suitable form of timing pulse.

Prior to the development of the high-voltage transistor, power transistors capable of handling currents of the values used in ignition primary circuits (up to 5 amperes) were relatively low-voltage devices, rated for voltages below 60 volts. These were not suitable, since the current collapse in the primary winding of the induction coil gives rise to an induced voltage in that winding, normally in the 100–300 volt range but rising to 500 volts under open circuit conditions.

Further developments in ignition systems of this nature were therefore held up pending the development of a transistor capable of withstanding this primary self-induced voltage. At the time, the Group Research Centre had been looking at the feasability of a high-voltage transistors for future Lucas ignition systems of this type, based on techniques already investigated by Dr Kirkman and Mr M. Searle for high-voltage diodes for Rotax aircraft ignition equipment.

(e) Joseph Lucas (Electrical) and Delco

In 1959, a team of experts from Delco (U.S.A.) were touring Europe, enquiring from firms about the feasibility of transistorised ignition systems. The need for a transistorised ignition system was more pressing in the United States than in the United Kingdom mainly because, in the United States, the larger and more common six- and eight-cylinder engines require sparking rates higher than 400 sparks per second. The extension of traditional techniques was proving inadequate and consequently breakdowns involving high replacement costs were frequent. An earlier survey in the United States had suggested that such a system would require a transistor similar to that described in the previous section, and it seems that the semiconductor manufacturers consulted – concerned mainly with communications and other 'small current' applications of transistor technology – had expressed the view, looking at the problem merely as an extension of current practice, that such a device was not feasible at present.

In the course of this tour, the Delco team visited Lucas. With the Delco and Lucas requirements being virtually identical and with the implication that successful development of a high-voltage transistor would have international application, an impetus was given to further work. Six months later, sample transistors were sent to Delco and, having demonstrated the feasibility of the project, the arrangements for a licence covering the manufacture and selling of high-voltage transistors was drawn up. They were finalised in June 1961.

Although this work had shown that a high-voltage transistor could be made, it was abundantly clear to Lucas and Delco that the device in its original form was too bulky and too expensive to be marketed. Further development work carried out jointly by the two companies over a period of two years resulted in a device which was small, light and less expensive and which could be made by mass-production methods. Subsequently, further progress has been made by Lucas and Delco independently and to some extent in competition. The production facilities developed by J.L.E. have contributed considerably to the knowledge of the process, both as regards the basic conditions required and in the all-important area of yield and cost optimisation to make the high-voltage transistor a commercial proposition.

(f) Development of the high-voltage transistor

As already indicated, the work of Kirkman and Searle at the Lucas Group Research Centre on high-voltage diodes provided the early encouragement to tackle the high-voltage transistor problem. Building on earlier studies by Fuller and Ditzenberger[3,4] of Bell Telephone Laboratories on the diffusion of various materials into silicon, experiments with aluminium as a diffusant achieved some success in handling high voltages, and these diodes were later produced commercially.

The major problem in the development of a high-voltage transistor was the production of a high breakdown voltage at the collector-base junction, at the same time maintaining a narrow base width to produce significant current gain at the required current level. Using aluminium as the p-type base dopant, the diffusion profile of the aluminium was developed to produce a useful compromise in terms of current gain and breakdown voltage.

Based on a simple ignition circuit, the initial high-voltage transistor specification was for a silicon n.p.n. device with a minimum gain of 5:1 at a collector current of 5 amperes, and a collector-base rating of 500 volts. The power requirement was of the order of 5 watts.

(g) British markets

In Britain, unlike America, there was as yet no pressing need for ignition systems incorporating high-voltage transistors, and as a result, once the initial development work was completed, J.L.E. were quite happy to let Delco adapt the transistor to the precise needs of the automobile market, with special reference to American conditions.

However, a limited number of devices was manufactured, and since Lucas were engaged in building up outlets other than the automobile industry for their semiconductor devices, effort was put into finding markets for the high-voltage transistor. It will be appreciated that at first Lucas sales representatives were somewhat coolly received when they – widely considered as 'the automobile boys' – appeared offering new devices in this highly sophisticated technology, but as a result of their efforts, new markets were found in television cameras, including the Marconi Mark VII (see p. 231), in X-ray systems and in voltage regulators.[5]

In more recent years, the demand for high spark-rate ignition systems has developed in Britain, and the high-voltage transistor is now in quantity production as an essential part of the Lucas 'Opus' ignition system, having a spark rate of 700 sparks per second.

(h) Some key personnel

The team which developed the high-voltage transistor was small – only five or six in number. In the early stages of the work there was only one university graduate directly involved. Short histories of the key people are given below:

T. L. Hughes: no university training. Joined the company in 1959 as an electronics engineer. Previously he had worked with Ecko in Brighton, but had had no previous training in semiconductors. He was trained for the job by R. C. Lowe at the Lucas Group Research Centre. He is now a Senior Technical Assistant at the Group Research Centre.

R. C. Lowe: no university training. Joined the company in 1956 from Rank-Taylor-Hobson. He had a very extensive experience in thin film physics. Later became a manufacturing engineer in the Production Division.

M. Searle: Higher National Certificate in Electronics. He came to Lucas from the Atomic Energy Authority at Aldermaston where he carried out research on semiconductors. Currently, he is transistor project engineer in the Production Division.

D. J. Taylor: no university training. He joined the company in 1957 with an H.N.C. in Electronics. Previously he was with the General Electric Company working on airborne radar equipment. Now concerned with the marketing of the Lucas semiconductor products.

Dr G. C. Williams: has a Ph.D. degree from Swansea. At the time of development he was Head of Physics Research at the Group Research Centre. Later became General Manager of the Silicon Semiconductor Products Division of Joseph Lucas (Electrical) Ltd.[6]

The rapid development of the high-voltage transistor was due in no small measure to the skills of Lowe, described by Williams as an 'artist in fashioning silicon'. Such an individual is often important in early development work, and Williams makes the point that persons with these types of skills, provided they have technical expertise available, can often 'get on with the job' more effectively than those who are academically trained.

References

1. 'Lucas Group Research Centre', Lucas Publication No. 2788.
2. C. Freeman, 'Research and Development in Electronic Capital Goods', *National Institute Economic Review*, **34** 40 (1965).
3. C. S. Fuller, 'Diffusion of Donor and Acceptor Elements into Germanium', *Physical Review*, **86** 136 (1952).
4. C. S. Fuller and J. A. Ditzenberger, 'Diffusion of Lithium into Germanium and Silicon', *Physical Review*, **91** 193 (1953).

5. J. W. Sharpe, 'Electronic Ignition Systems', *Lucas Engineering Review*, **1**, no. 2 (Oct 1964) 2.
6. G. C. Williams, private communication, 1968.

26 LYTAG LTD: LIGHTWEIGHT AGGREGATE FROM PULVERISED FUEL ASH

Lytag Ltd gained a Queen's Award for 'innovation in building materials'. Lytag Ltd is a subsidiary of John Laing and Son Ltd, and it was the Laing research and development organisation that developed a new lightweight (i.e. low-density) concrete. Lytag Ltd is a subsidiary formed to manufacture and sell the new aggregate, 'Lytag', which is manufactured from the ash produced in modern power stations using pulverised fuel.

This innovation illustrates the difficulty of dividing innovations into new products and new processes. To the extent that the Laing group developed a new way of making lightweight aggregate, it is a process innovation, but Lytag Ltd was set up to manufacture and sell a new product, an aggregate which could be used by a customer to make lightweight concrete.

Other organisations played an important part in the success of this innovation. Government research at the Building Research Station (B.R.S.) provided both knowledge and people to be used by Laings. The Central Electricity Generating Board (C.E.G.B.) was responsible for financing some of the B.R.S. research and its publicity activities helped to sell 'Lytag'. Various organisations have organised symposia which have aided the diffusion of knowledge about 'Lytag'.

Two different ways in which one technology can influence another are illustrated by this innovation. A change in the technology of burning solid fuel by the C.E.G.B. led to a change in the waste product available for use in lightweight concrete manufacture. The importance of technology transfer is particularly clear in the development of a 'sintering method based on existing practice in the sintering of iron ore.

(a) The development of lightweight concrete

Lightweight (low-density) concrete can be prepared by trapping air inside the concrete. One way of doing this is to use a porous aggregate in the concrete mix. The Romans used pumice from volcanic regions as an aggregate in the preparation of lightweight concrete – for example, in the construction of the 143-ft diameter dome of the Pantheon.[1]

There are two main properties of lightweight concrete which have commercial attractions. One is its low density, which makes possible the construction of tall buildings with reduced loads on the foundations and structural members. The other is its effectiveness as a heat insulator. Following the Second World War, a shortage of steel for use in building applications led to an increased interest in concrete. The development of industrialised building systems further increased the use of concrete. This, together with an increased demand for good insulation to go with central heating, produced an interest in lightweight concrete, various types of which will now be described.

Clinker concrete. Furnace residues, known as clinker, have been used as an aggregate for lightweight concrete since the nineteenth century and the British Museum, which was finished in 1907, contains some clinker concrete. Because of its low price and availability in large quantities, clinker has been used extensively in England. However, the composition of clinker varies considerably. This variability was responsible for some building failures and led in 1929 to work being carried out at the B.R.S. Variability of clinker was also responsible for the origin of John Laing's research and development organisation, which started with the employment of a chemist to test clinker quality.

Expanded clay aggregates. The shortage of steel in the First World War provided an impetus to examinations of possible uses of concrete. One result was the production of a lightweight aggregate that was neither natural in origin (as is pumice) nor an industrial waste product. It had been known for a long time that when certain clays or shales are heated, the decomposition of minerals present in the clay produces gases which bloat the clay, forming an expanded cellular structure which can be retained on cooling. This knowledge was made use of in 1913 by the concrete ship section of the U.S. Emergency Fleet Corporation, who developed an artificial aggregate for use in lightweight concrete in ship construction.[3] This development led to the publication of results that might have helped to overcome a prejudice against lightweight concrete in load-bearing applications. In the words of the Editor of *Engineering News Record* in 1919:

> It has been the impression for years that the two properties, strength and weight, were more or less proportioned in concrete and that an aggregate which floats as does this new burnt clay could not possibly have a strength usable in structures. If these views are to be revised and the new aggregate has such strength as Government studies so far show it to have, its application is readily obvious to all structural engineers.[4]

In Britain, concrete barges and coasters were also built during the First World War using know-how from Denmark. Some 200 vessels were built and various techniques were used, including reinforced concrete and pre-cast slab production, but there appears to be no reference to the use of lightweight concrete. However, the American results were quoted in various British publications including *The Builder*,[5] which claimed that the lightweight concrete used in the American ship construction programme was costly in terms of price per lb but could, in certain structures, prove economical owing to the decrease in weight of concrete required.

The relationship between the chemical composition of clays and their ability to form an expanded aggregate has been investigated by many workers, but a direct empirical test is still more directly informative than a complex mineral analysis.

High-quality concrete is now produced in the United Kingdom using expanded clay aggregates such as 'Aglite' and 'Leca' which originated in America and Denmark respectively.

Aerated concrete. Aerated concrete is obtained by introducing air or other gas into a slurry of cement or lime and fine siliceous material such as sand. When the mixture sets hard the trapped gas bubbles produce a uniform cellular structure. Aerated concrete has properties that are different from lightweight concrete produced from lightweight aggregates and it has been suggested that the term 'cellular silicate' would be a more correct name.

The basic principles of aerated concrete manufacture have been known and understood since the beginning of the century,[1] but significant commercial production did not begin until 1929 when a factory was erected in Sweden to produce this material using a process based on the work of Axel Eriksson. In this country, attempts to produce aerated concrete commercially failed until in 1950 John Laings started the production of 'Thermalite' blocks with an adapted process and product designed to suit British conditions.

(b) *The role of the C.E.G.B.*

During the last twenty years there has been a gradual change in the method used to burn solid fuel in C.E.G.B. power-generating stations. New stations employ pulverised fuel which is mixed with hot air and blown into the combustion chambers. Most of the resultant ash, which has the consistency of face powder, stays in the air stream and is partly removed by mechanical means, and partly by electrostatic precipitators. The recovered ash is known as fly ash in the United States and pulverised fuel ash (P.F.A.) in this country.

The older firing process, using a chain-grate, produced clinker. The changeover has led to a shortage of furnace clinker, formerly the most important lightweight aggregate, and has presented the C.E.G.B. with an enormous waste disposal problem. To dispose of the P.F.A. they have set up a marketing organisation which investigates possible uses of P.F.A. and gives publicity to established uses, including Laing's 'Thermalite' and 'Lytag' which are both mentioned, for example, in a publicity film. The success of the C.E.G.B. in gaining outlets for P.F.A. is shown in the following table giving production and sales of P.F.A. in thousand tons:[6]

Year	Output	Sales	Sales as per cent of output
1956–7	4,300	150	3·5
1961–2	7,900	1,170	14·7
1966–7	8,475	3,500	42·0

In 1952 the British Electricity Authority (later to become the C.E.G.B.) commissioned the B.R.S. to investigate the use of P.F.A. in building applications. Following in the wake of extensive studies carried out in the United States, the Central Electricity Research Laboratories at Upper Boat carried out research into the replacement of cement by P.F.A. in concrete of normal density and the C.E.G.B. was able to specify the use of 20 per cent replacement of cement by P.F.A. in many of its contracted building programmes.[8] Research was also carried out at Glasgow University. John Laings were responsible for introducing aerated concrete 'Thermalite' using P.F.A. instead of sand, and an artificial aggregate 'Lytag' made by sintering P.F.A. By 1965–6, 938,000 tons of P.F.A. were being used in aerated concrete building blocks and 65,000 tons were used in the manufacture of lightweight aggregate; the replacement of cement by P.F.A. in concrete and mortar consumed only 50,000 tons. The major outlet for P.F.A. was 2·3 million tons as load-bearing fill in roads and building sites.[9]

The C.E.G.B. can claim to have been much more successful in finding markets for P.F.A. than the American power industry which is fragmented. The very term 'P.F.A.' is said to be an example of the C.E.G.B.'s superior marketing ability. It seems to be easier to interest people in testing P.F.A. than to interest them in 'fly ash', which is the American term.

(c) The Building Research Station

Prior to the Second World War, experimental studies of new types of lightweight concrete were carried out at the B.R.S. by several workers including Bessey. Foamed blast-furnace slag as an aggregate was investigated, as was aerated concrete and the structural use of lightweight concrete. This work was stopped by the onset of the war but in 1943 Bessey began work on aerated concrete again. Bessey, Dilnot and others including

Hobbs investigated various compositions and P.F.A./cement mixtures were used in part of the study. An important impetus to lightweight concrete research was then provided by the problem of pulverised fuel ash disposal.[7] During a visit to the United States, T. Whitaker of B.R.S. met Leftwich who is accredited with the first publication describing the use of P.F.A. for making an aggregate.[10] A successful attempt was made to manufacture a ball-shaped aggregate from P.F.A. but this work was shelved by B.R.S. after it had been reported to the C.E.G.B. C. Hobbs, later to be in charge of Laing's research and development, and other members of the future Laing team worked on this project.[19]

Before P.F.A. could be utilised in lightweight aggregate concrete, a satisfactory method of turning the ash into an aggregate had to be developed and potential users of the aggregate had to be persuaded that the properties of a concrete made with this aggregate would be satisfactory. In 1952 the Research Station restarted its attempts to produce an aggregate from P.F.A. and in 1954 it published a description of a system capable of producing cellular pellets by sintering P.F.A. (Sintering means heating to the point at which grains of powder fuse together where they touch, without actually melting.) The main part of this work was the development of a method of mixing the ash with water to produce 'green' pellets with sufficient strength to withstand handling to the sintering furnace.[11] Sintered P.F.A. is a hard, brick-like material with about 40 per cent of its volume consisting of voids.

In 1955 and 1956 W. Kinniburgh[12, 13] published the results of B.R.S. tests carried out on concrete prepared from the sintered P.F.A. pellets. The main outlet for sintered P.F.A. was seen as being a replacement for clinker in concrete block making and the test results passed the requirements of the appropriate British Standard for concrete blocks for insulation rather than load-bearing purposes.

The way seemed clear for commercial development and Kinniburgh could state in his 1955 paper[12] that 'several manufacturers are already doing development work'. However, he also stated that 'a large amount of development work on the pilot scale is still required before a decision can be made as to the right kind of machinery and furnace for full-scale production [of sintered P.F.A.]. Consideration is being given to travelling-band sinter machines, stationary sinter machines and shaft kilns. If these problems can be solved, then the building industry should be presented with a cheap and efficient new material.'

In fact, the first commercial attempt to produce P.F.A. aggregate by sintering failed and it was left to Hobbs to develop the successful technique that led to the production of 'Lytag'. The first commercial attempt was by Sinterlite Ltd. In agreement with the C.E.G.B. and the results of the Research Station, attempts were made around 1956 to set up plant capable of sintering the entire output of Battersea Power Station, some 70,000 tons per year of P.F.A.[8] There were development troubles and Sinterlite sold out to the Cementation Co. Ltd, who set up two vertical kilns designed by Dr F. P. Somogyi of Sinterlite Ltd. The resultant aggregate was christened 'Terlite' but, as could have been forecast had a decision been postponed until the field-scale results were available from B.R.S., the problems of scaling-up could not be solved and in 1960 the kilns were shut down.[10] The problems encountered in the use of vertical kilns are discussed in the next section.

(d) John Laing and Son Ltd

After the Second World War, Laing decided to reintroduce 'Easiform', a concrete housing system that had been used for a short while after the First World War. This system used cavity walls, the inner leaf being constructed of lightweight clinker concrete. Problems associated with the variability of clinker led Laing's engineer in charge of this project to consult the B.R.S. about quality control of clinker, which at that time was part of the Station's activities looked after by C. Hobbs.

Hearing that Laings were considering setting up their own materials testing laboratory, a B.R.S. worker, P. E. Starnes, approached Maurice Laing, the second son of J. W. Laing, and gained an interview. Starnes was engaged to supervise chemical testing and at his own request he was allowed to carry out some development of new materials. This development work was supported by Maurice Laing and led to Thermalite, which by 1947 had reached pilot-plant scale under the direction of

Starnes. It became necessary to recruit someone to take over the general running of the materials laboratory and Hobbs, who had left B.R.S. to work for the West Riding Rivers Board, was recruited by Starnes to be Chief Chemist in charge of the Laing Laboratory which had only four assistants. In 1949 Thermalite Ltd was incorporated and Starnes at the age of thirty was made General Manager having overall responsibility for both technical and commercial aspects of the project. By 1952 it was evident that Thermalite was going to be a commercial success but Starnes quarrelled with the Board of Laing's and resigned.[16]

Some Swedish know-how was used in Thermalite production and whilst in Sweden Maurice Laing became interested in expanded aggregates and bought an option on the spot for the United Kingdom production of a Swedish aggregate, 'Sillit'. The information about 'Sillit' was passed to Hobbs, who carried out a survey of suitable English shales for the Sillit process. Some suitable materials were found but they were all at a distance from potential markets. Hobbs, who had worked on P.F.A. at the B.R.S., wrote a report for Maurice Laing in which he pointed out that P.F.A. looked like a much better material for the manufacture of lightweight aggregate. Among its advantages was the fact that the bulk of P.F.A. came from large power stations in regions of high population density, i.e. near to potential markets.[17] Laing dropped the Sillit option and in 1952 Hobbs started a study of P.F.A. for the manufacture of lightweight aggregate.

The crucial problem seemed to be the method of sintering. This was the problem that was to beat Sinterlite Ltd but, fortunately, Hobbs had some knowledge of methods of sintering from his previous experience. The simplest type of kiln is the vertical shaft kiln and Hobbs had some knowledge of this type of kiln from visiting lime works where limestone is heated to give lime. Rotary kilns are used in the cement industry and during part of the war Hobbs had worked at a cement works in a Government-sponsored attempt to produce phosphate fertiliser by sintering phosphate rock. This attempt was unsuccessful but gave Hobbs a practical knowledge of sintering problems. Another area of technology involving relevant knowledge is the sintering of iron ore. Hobbs knew little about this but he was able to attend a symposium on the sintering of iron ore organised by the British Iron and Steel Research Association where he acquired details about the sinter strand method in which pellets are formed into a bed on a travelling grate which passes under an ignition hood.

By 1954, after two years' research at modest cost, Hobbs was sure that a lightweight aggregate could be manufactured from P.F.A. but the market for Thermalite was beginning to develop and much of the capital available in the Laing group was required for expansion of Thermalite capacity. The decision to go ahead was not made until 1959 and a factory was eventually set up adjacent to Northfleet Power Station in Kent. A separate company, Lytag Ltd, was set up to manufacture and market the new aggregate, 'Lytag'.[17]

The decision to go ahead with Lytag came after a market survey had been carried out by consultants. However, Hobbs is somewhat cynical about the real value of this market survey. He claims that the result was predictable from a knowledge of the total market for aggregates and the claimed cost and properties of the new material which only he could define. Predictably, therefore, it produced results identical with his own forecast and subject to the same errors, though by being presented by an outside organisation it clothed these forecasts with a greater air of respectability. Hobbs is convinced[17] that the personalities of the Laing directors were more important than a seemingly rational decision based on market research. Without Maurice Laing supporting the Lytag project, it is doubtful if it would have been started as other people wanted money for alternative investments. (Starnes makes similar comments about the interplay of personalities in regard to the Thermalite project.)

When the Northfleet Lytag plant was set up there were eighteen months of teething troubles. However, unlike the Sinterlite plant at Battersea, the Lytag plant became successful. Hobbs claims that the reason for Lytag's success was that he had chosen the most suitable sintering technique (sinter strand) and Somogyi had not. Another possible explanation is that Laings waited for five years from 1954 to 1959 to see if the Sinterlite plant would be successful. This might be taken as an example of the inadvisability of being first to try a new technique, but both Hobbs and Starnes claim that they knew Somogyi was wrong in 1954 and that Lytag would have been successful even if it had been the first attempt and not the second.

Hobbs's view of the relative merits of the shaft kiln and sinter strand techniques are given in a paper[14] published in 1959 at the request of *British Chemical Engineering*. This paper was based on work carried out in 1953–4. The origin of Hobbs's knowledge of the sinter strand technique is obvious from this paper; data are compared with corresponding data for the sintering of iron ore and he includes six references to the *Journal of the Iron and Steel Institute* and one reference to the German *Stahl und Eisen*. He states that the shaft kiln process should be slightly cheaper, provided that the kiln can be kept running at its maximum output, but he adds that there is some doubt as to whether a shaft kiln working on P.F.A. could run continuously at maximum output. Ash inevitably contains some unburnt fuel but the amount present can vary considerably and Hobbs claims that this variability would make it unlikely that a shaft kiln process could be kept going continuously. The sinter strand process, on the other hand, should be much more versatile and would accommodate changes in the composition of the ash.

Starnes describes the failure of Terlite as follows:[16]

The Terlite project was based upon an attempt to use a type of vertical shaft kiln, developed in Germany, for the manufacture of cement, for the purposes of sintering P.F.A. This possibility had been fully investigated by myself and others some years before and rejected. The primary reason for this was the impossibility of producing nodules of P.F.A. capable of withstanding the stresses occurring inside the kiln, and also the difficulties of maintaining control over the combustion of the ash. . . . It is interesting to note that the alleged feasibility of the process was 'demonstrated' by means of a model kiln only 3 or 4 ft high in which, of course, the conditions prevailing as regards mechanical stress were in no way comparable to those occurring in a 20–30-ft tall full-scale kiln.

It is probably true, therefore, that the delay in Laing's starting production was not due to a desire to wait and see what happened to Terlite and that Lytag would have succeeded even if Terlite had not existed.

The construction of the plant at Northfleet took place in 1959 under Hobbs's direction and he also controlled all aspects of manufacture, marketing and sales during this early period of operation. Skilled development work was required, particularly in areas such as large-scale P.F.A. handling and dust control which were not susceptible to pilot-plant study, before the operating costs were brought down to an acceptable level.

(e) Lytag Ltd

A separate company, Lytag Ltd, was set up to control the manufacture and sales of Lytag. One of the reasons for doing this was the fact that, in certain applications, Lytag would be competitive with Thermalite and it was felt that the Thermalite sales force might not sell Lytag to its best advantage. Hobbs remained in control of Lytag Ltd until the sales began to increase and the production costs to decrease. The operation was still not running at a profit when Hobbs handed over to a new manager, retaining his connection with Lytag by becoming Technical Director of Lytag Ltd. Hobbs returned to managing Laing's Research and Development Centre, which has now grown from its origin as a clinker-testing laboratory to a separate company within the Laing Group employing some 150 people and having specialist laboratories for chemistry, physics, soil mechanics, concrete and structural engineering.

The Lytag engineers brought down the production costs and an enthusiastic sales force gradually built up sales, to cope with which additional factories were built at Tilbury in Essex and Rugeley in Staffordshire. These factories both had twice the capacity of the Northfleet factory.

The build-up of Lytag sales was the most difficult and commercially testing part of the whole project and even now much remains to be done. The building and constructon. industry has several built-in factors tending to resist innovation. Factors discussed by Bowley[15] include the preponderance of small firms, the difficulty of training migrant labour in new techniques, the difficulty in getting new types of construction accepted until they have been used for several years and the delay in getting official specifications and regulations altered to permit innovation. Before new structures will be accepted by local authorities, they have to be shown to be safe. It has been suggested that the B.R.S. does not have sufficient resources to keep up with all the innovations produced since the Second World War, in terms of providing the data required for Codes of Practice and British Standard Specifications. Following the work of Lea[2] on clinker aggregates, the B.R.S. was able to produce B.S.877:1939 to cover foamed blast-furnace slag aggregate, but it was not until 1964 that a modern specification, B.S.3797:1964, was available for lightweight aggregates.

In addition to the factors common to most building innovations, the development of lightweight concrete in this country has been delayed until recently by a plentiful supply of cheap bricks and cheap coal for heating plus a population with Spartan tendencies producing a lack of interest in insulation. (The reverse of these factors helped to make the Scandinavian countries the first to use lightweight concrete.)

The development and use of lightweight concrete using Lytag is still continuing. Research by John Laing, Professor Evans at Leeds, the B.R.S. and others has played an important part, as has the diffusion of information by Lytag Ltd, the C.E.G.B., the Cement and Concrete Association and the Reinforced Concrete Association (now part of the Concrete Society). Various symposia have been held, including one on structural lightweight concrete organised by the Reinforced Concrete Association in 1962.

From the point of view of the Laing organisation, the Lytag venture has not yet (1969) become a financial success. The sinter strand method of manufacture borrowed from the sintering of iron ore is too expensive a process for the production of a low-price material under present market conditions. The situation could, however, be changed by any of a number of possibilities. The market conditions could change, research by Laing, the B.R.S. or the C.E.G.B. could produce an improved product or process, or research abroad such as the work of Ott in Austria could lead to new developments.[18] An important chapter in the development of Lytag remains to be written.

(f) Comments

The Lytag case illustrates the complexity of the process of

innovation and the gross oversimplification that is involved in regarding innovation as a linear process starting with invention and proceeding through development to manufacture, sales and profits. No one 'invented' Lytag; its manufacture arose out of a complex mixture of technical ideas, personal drives and changes in methods of building construction leading to predictions about the future. If the process leading to the manufacture for sale of Lytag is to be described in terms of a linear sequence, then what is the start of this process? Is it the Roman's use of pumice as a lightweight aggregate, the manufacture of an artificial lightweight aggregate during the First World War, the need of the C.E.G.B. to dipose of P.F.A. or the work carried out at B.R.S. which resulted in the supply of both ideas and people to the John Laing organisation? The Lytag innovation can also be considered to start with Maurice Laing's desire to venture into new areas of activity, but he would not have supported Lytag without a belief in a potential market and, from this point of view, the Lytag innovation starts with changes in society creating an increased demand for lightweight concrete. Lytag involves both a new process and a new product and another starting point is the development of the sinter strand by the iron and steel industry.

The Lytag case is also a good example of the different roles of people in the process of innovation. Technical ideas were transferred from B.R.S. to the Laing group via the movement of people. These ideas were used because Maurice Laing was personally motivated to innovate and had the power to support innovation. It is quite common for innovations to take much longer to be financially successful than was expected by enthusiastic supporters. The Lytag case is no exception and it is not surprising that the most commonly occurring factor in the Queen's Award winners is the presence of a person in a position of authority who supported the project (see Table 3, p. 69 above).

References

1. A. Short and W. Kinniburgh, *Lightweight Concrete* (C.R. Books Ltd, London, 1963).
2. F. M. Lea, *Investigations on Breeze and Clinker Aggregates*, Building Research Technical Paper No. 7 (H.M.S.O., London, 1929).

3. Anon., 'History and Properties of Lightweight Aggregates', *Engineering News Record*, **82** 803 (1919).

4. Editor, 'Lightweight Aggregate for Structural Concrete', *Engineering News Record*, **82** 3 (1919).

5. 'Lightweight Concrete', *The Builder*, **117** 244 (1919).

6. H. W. G. Dedman, Ash Marketing Officer, C.E.G.B., private communication, 1968.

7. G. E. Bessey, 'Utilisation of Pulverised Fuel Ash in the Building and Civil Engineering Industries', *Trans. Pulverised Fuel Conference of the Institute of Fuel*, June 1947.

8. H. W. G. Dedman and A. E. Hawkins, 'Utilisation of P.F.A.,' Paper 14, Second Conference on Pulverised Fuel, Institute of Fuel (1958).

9. H. W. G. Dedman, 'The Commercial Utilisation of P.F.A. from Power Stations of the C.E.G.B.', Fly Ash Utilisation Symposium (Pittsburgh, 1967).

10. R. F. Leftwich, 'New Lightweight Aggregate from Fly-ash and Slags', *Concrete* (Mill Section), **54** 14 (Jan 1946).

11. *The Use of Sintered P.F.A. as Lightweight Concrete Aggregate*, Building Research Station Note No. A34 (1954).

12. W. Kinniburgh, 'Sintered Pulverised Fuel Ash as a Lightweight Aggregate', *Architects' Journal*, 7 July 1955, p. 127.

13. W. Kinniburgh, 'Lightweight Aggregate from P.F.A.', *Concrete and Constructional Engineering*, **51** 571 (1956).

14. C. Hobbs, 'Building Materials from Pulverised Fuel Ash', *British Chemical Engineering*, **4** 212 (1959).

15. M. Bowley, *The British Building Industry* (Cambridge U.P., 1966).

16. P. E. Starnes, private communications, 1968 and 1969.

17. C. Hobbs, interview, 1968.

18. A. Wilson, C.E.G.B., private communication, 1969.

19. F. C. Harper, private communication, 1969.

27 MARTIN-BAKER AIRCRAFT CO.: AIRCRAFT EJECTOR SEATS

Continuous evolution of engineering design over more than two decades resulted in Queen's Awards to Martin-Baker (M.B.) in 1966, 1967 and 1968. Many improvements have been incorporated, yet refinement has been achieved without loss of essential simplicity. The firm, a private and highly centralised one, has a long tradition in innovations connected with aircraft, and the stimulus for starting the development of ejector seats in M.B. came through the association with a potential user, the R.A.F.

(a) Escape from aircraft

The idea of the parachute goes back at least as far as Leonardo da Vinci: 'If a man have a tent roof of calked linen 12 braccia broad and 12 braccia high, he will be able to let himself fall from any great height without danger to himself'[1] (a braccia was about a yard). Stewart[2] describes the discussion and argument which preceded the decision in 1919 to issue all personnel on military aircraft with parachutes.

A parachute by itself, however, does not solve the problem of abandoning and getting safely clear of an aircraft. On 13 August 1913, Pegoud 'did the first test of what might be held to be something resembling an ejection seat. He allowed himself to be heaved out of his aircraft by streaming his parachute from its container and waiting for the jerk. But Pegoud's work had little impact on the men in command of military aviators.' It was widely thought that the airstream 'directly it got into the canopy, would tear the canopy to pieces. Alternatively it was thought that it would carry the parachutist back and entangle him with the tail of the aircraft.'[2]

Mostly, the 'abandon aircraft' phase of an escape continued to consist of jumping over the side. However, the danger inherent in this inevitably increased with the operating speeds of aircraft.

Voiciechauskis's safety device. A British patent application filed in 1936 by the Lithuanian, Bolius Voiciechauskis,[3] describes a device 'to remove the occupant out of the cabin before the final catastrophe. My invention contrives to answer this purpose by an expedient of a rail (or rails) along which the seat with the passenger moves upwards or downwards so as to push him out through a comfortable opening in the cabin ceiling or floor respectively.'

For flight over water, the safety device would contain two parts which together form a box.

> The passenger's seat is attached to the lower part and the upper part closes on this so that the box closes round its occupant impermeably to water. The safety box is provided with an ejecting appliance consisting of resilient means which, when set in action, drive it along the above-mentioned rail and consequently eject it with its occupant through an aperture in the cabin floor. The upper part of the safety box is provided with a parachute which is expanded either by the

occupant of the box or automatically by a cord attached to the rail in the cabin. The purpose of opening the parachute may be attained by introducing a watch mechanism as well.

For overland flight, the sides of the box are omitted; the passenger wears his parachute on his back and separates from the seat after ejection.

Voiciechauskis's concept is notable for its originality, its completeness (for instance, provision is made for a distress radio transmitter) and for the similarities later design proposals bear to it. However, it does not appear to have been developed further. It may be described as part of mankind's heritage of creativity but not as part of its heritage of innovation, since it was not exploited for productive purposes.

Developments during and after the Second World War. With operating speeds up to the region of 300 to 400 m.p.h., it was recognised that some means of powered ejection of the pilot was becoming necessary. There was some work in Germany on ejector seats for high-speed planes such as the Heinkel 162 aircraft.[4] In a communication from the Swedish SAAB company it was stated that

> our company started developing ejector seats in 1940–41 and in January 1942 the first dummy ejections from the rear seat of a SAAB 17 light bomber were made with the aim of developing a seat for our SAAB 21A pusher fighter. The SAAB 21A flew for the first time in June 1943 fitted with a development of this earliest seat and was also produced in large quantity as one of the first aircraft in the world to have an ejector seat. In March 1947 our first jet fighter, the SAAB 21R, flew for the first time with a Model I seat and this aircraft was also produced in quantity. In 1948–49 the SAAB 18B and T18B series of twin-engine bombers and attack aircraft were modified to take two Model II for the pilot and the navigator. At the same time the bombardier's seat was deleted. Thus, the role of the aircraft changed from bomber to ground attacker. In September 1948 the SAAB 29 swept-wing jet fighter flew for the first time fitted with a lightweight Model II ejector seat. Nearly 700 aircraft of this type were produced between 1951–56 and the Model II

design was also sold to Britain where the Folland Aircraft Company Ltd has further developed this design into a fully automatic seat for the Folland Gnat lightweight fighter and trainer.[39] The first live ejection with the Model T seat was also its first use in an emergency, when the pilot of a J-21A saved his life with it on 29 July 1946 after colliding with another fighter.[5]

The Swedish Government issued a Royal Decree appropriating the M.B. patents in Sweden at no cost.[9] The Folland Gnat aircraft sold to the Royal Air Force were fitted with the Swedish seat and for this infringement of their patents M.B. were awarded royalty payments. Similarly a later SAAB seat, the Mark IV, was built for the SAAB 35 Draken aircraft. 'The Copenhagen Maritime and Mercantile court has ruled that British patent rights have been infringed in the contract under which the Swedish SAAB concern is supplying 46 Draken jets to the Danish Air Force.'[40] So, though it would appear that SAAB led M.B. in the initial development of the ejector seat, the strong patent position M.B. was later to develop gave it a commanding lead where normal commercial conditions existed.

(b) *The Martin-Baker Aircraft Co.*

The Martin-Baker Aircraft Co. was formed in 1934 to exploit a special system of construction for aircraft using steel tube.[6] This construction, evolved by Mr (now Sir) James Martin, was embodied in the company's first product, the MB-1, a light two-seat cabin monoplane aircraft.[7, 8] The original directors of the company were Francis Francis, Captain V. E. Baker and R. H. Parratt; Martin was Managing Director and Chief Engineer. Baker was killed during a forced landing whilst testing the MB-3 single-seat fighter aircraft which first flew on 31 August 1942. Following this accident the company produced a redesigned aircraft,[9] the MB-5, which first flew on 23 May 1944. Martin commented, 'If they'd ordered a squadron or two there might have been Martin-Baker aircraft flying today.'[38] After 1946 changes took place and Martin became the sole owner of all the shares.[36]

M.B. had been engaged in the development of many accessories for aircraft in addition to its prototype development work.

The M.B. patent blast tubes for machine guns which had been fitted on the MB-2 were subsequently used on Hawker Hurricanes and Typhoons, De Havilland Mosquitos and other aircraft. Many thousands were manufactured during the war years. The M.B. explosive cable cutter was developed between 1937 and 1939 and was standardised for British bombers just before the war. Each cutter weighed about 3 lb and was capable of cutting a three-ton balloon cable; many thousands of these were manufactured by the company. The company was also responsible for the design of a twelve-gun nose to be fitted to the R.A.F. Douglas Boston to convert it to the British Havoc night fighter. Other M.B. developments include a jettisonable hood for the Spitfire and other aircraft and a patented automatic oiling unit for contra-rotating airscrews.[8]

The link between Martin and the aero industry extends back to the year 1912, when he designed and built an airspeed warning unit which gave a 'too-low' signal at 35 m.p.h. and a 'too-high' signal at 40 m.p.h. In 1939 he moved from London to the company's present site at Denham, Bucks., and the labour force consisted of a man, a boy, and Martin himself. By 1968 over 1,000 people were employed on that site alone.[9]

(c) Evolution of ejector seat design

In January 1944 a test pilot attached to the Royal Aircraft Establishment, Farnborough, was killed whilst testing an early version of the Gloster Meteor jet aircraft. He managed to escape over the side of the cockpit but was injured in the process, lost consciousness and fell to the ground.[10] Shortly after this 'a Royal Air Force Technical Staff Officer in conversation with me [Martin] said that there was anxiety among pilots of Meteor aircraft in connection with baling out, and that some means of escape would have to be devised which would make it possible for pilots to abandon aircraft safely at high speeds as climbing out of the cockpit was out of the question'.[11] As a result of this, a 'swinging arm' scheme for the removal of a pilot from an aircraft was prepared. A model was shown to Air Chief Marshal Sir Wilfred Freeman and the Air Minister, Sir Stafford Cripps, on 11 October 1944 and they agreed to lend M.B. a Defiant aircraft for tests.

The swinging arm concept reflects the simplicity which characterises Martin's designs. The arm was to be hinged near the tail of the aircraft and motive power for lifting the pilot clear over the tail was to be provided first by a spring under his seat and then by air pressure on the exposed underside of the arm. Disadvantages of the scheme soon appeared, however, the main ones being its reliance on aircraft speed and the dangerous loads to which the pilot would be subjected in the event of the aircraft spinning.

The swinging arm soon gave way, therefore, to the telescopic ejector seat gun which used the gas from an explosive cartridge to extend the sliding tubes of the telescope and so propel the seat and occupant along guide rails out of the cockpit. By its very nature, this system had a major physiological disadvantage, for it required that the pilot be subjected to a violent jerk during the initial movement of the seat. To determine the mechanical stresses to which a pilot could safely be subjected, doctors of the R.A.F. Institute of Aviation Medicine had carried out a series of tests in the spring of 1944. They acted as 'guinea pigs' during tests on a rocket-powered trolley on the 2,000-ft rocket track at Farnborough. The trolley was propelled along the track and then decelerated at varying rates to determine the loadings they could safely endure. Martin commented:

> It must be pointed out that the maximum 'g' load on the sled during acceleration is only $3\frac{1}{2}$ 'g' so the information obtained from these tests was of no use to Martin-Baker. It was then that I designed what is known as our testing rig on which it was possible to accelerate a man up an inclined ramp with 'g's up to 20 or more and it was from hundreds of tests on this rig that I was able to design the type of cartridge to use which would not be likely to cause injury to the person's spine.[36]

Subsequently

> the Institute of Aviation Medicine was concerned with the problem of escape from aircraft and undertook studies to define the limits of forces that could be applied to man. They also provided advice to the firm as regards the integration of the functional clothing, such as 'g' suits and pressure suits with the escape system. They were also involved in the investigation of incidents that occurred in service, on the

interpretation of which a better seat was developed, and on one occasion the Institute provided a live subject for the testing of a new seat system. The technical development is however 100 per cent Martin-Baker. [37]

Data from the tests on the M.B. inclined ramp indicated the need to limit the rate of change of acceleration to 220 g/sec. and the mean acceleration to 16g. [12] So the concept of sequential firing of cartridges was introduced into the seat design. In this, the initial movement of the seat was generated by firing a single cartridge which ignited the auxiliary cartridges and boosted the seat velocity. [13] A later development of the seat aimed to make the position of the pilot adjustable with respect to the seat frame and permitted the introduction of a 42-in. stroke gun extending over the full height of the frame. [14] With this, the pilot left the aircraft with a velocity of 60 ft/sec and with an improved possibility of survival. Later still, with the introduction of the V-bomber force including aircraft such as the Handley Page Victor with its high tailplane, it became necessary further to increase the ejection velocity of the seat and a three-cylinder telescopic gun was produced. [15] This gave the pilot an ejection velocity of 85 ft/sec after a gun stroke of 72 in.

Reference was made earlier to the patent portfolio which M.B. have prepared on their ejection seat features and it is extremely rich in demonstrations of practical creative engineering. Over thirty-eight patents expressly related to ejector seat design have been awarded to the company, and over eighty patents have been awarded to Sir James Martin. The following developments illustrate quite strikingly the thoroughness with which the many aspects of the M.B. ejection seats have been developed:

(i) Leg, [16,17] head, [14] body [18] and arm [19] restraint devices to avoid injury during high-speed ejection.

(ii) A small parachute (drogue) to slow down and stabilise the seat on separation from the aircraft [20] and its later development into a duplex drogue system (two drogue parachutes operating one after the other) [15] to cater for high-speed ejection.

(iii) A sequential control unit to release and separate

the pilot automatically from the seat and deploy his parachute.[21]

  (iv) An anti-spin line attached to the seat and aircraft which corrects the spinning motion given to the seat by the ejection gun mounted behind the centre of gravity of the man–seat combination.[22]

  (v) A rocket motor unit to increase and sustain the seat velocity beyond the 85 ft/sec imparted by the telescopic ejection gun, giving the pilot a greater trajectory height and so a greater chance of survival in the event of a low-level or on-the-ground ejection.[23]

  (vi) An arrangement of auxiliary rocket nozzles[24] mounted horizontally under the seat frame to provide more precise spin control of the seat than was possible with the static-line described above.

  (vii) A manually operated adjustment to vary the position of the rocket motor with respect to the ejector seat to ensure a spin-free flight.[25, 26] These developments related the thrust line of the rocket motors to the centre of gravity of the pilot–seat combination.

  (viii) A pivoted seat pan connected to the rocket pack via a simple bell-crank and spring mechanism for spin-free flight.[27]

  (ix) An underwater escape system for use in conjunction with the aircraft ejector seat. This comprises a compressed-air operated ejection gun,[28] a canopy penetrator,[29] a pressure-sensitive valve to initiate the escape sequence,[30] and a drogue gun disabling system.[31, 32, 33]

  (x) A command system of ejection for multi-seat aircraft[34,35] to permit the pilot to eject a crew member without waiting for his command to be obeyed.

(d) Comments

Centralised control. The most striking feature of the M.B. ejector seat development is the simplicity of the solutions adopted to answer the problems of ejection from high-speed aircraft. Without their power to simplify, the designers of the company might well have produced an automatic ejector seat which required accelerometers to sense, computers to control and servo-mechanisms to operate it. This feature of simplicity reflects the constantly applied design philosophy of one of the company's founders and its Managing Director and Chief Designer – Sir James Martin. His work in the field of ejector seat design has won him many honours including the C.B.E., the Wakefield Gold Medal (1951), the Laura Taber-Barbour Air Safety Award (by the Flight Safety Foundation Inc. of the U.S.A. (1958 – first award to a non-American), the Cumberbach Trophy (1959) and the Royal Aero Club Gold Medal (1964).

Martin retains a high degree of control over all the activities of the firm both with respect to its policies and its technological developments. In his own words:

> as far as the success of the company is concerned, I think it is fair to say that it is entirely due to the fact that I have designed one of the best ejection seats in the world and as I am the person responsible for every detail, I spend long hours on the problems. This is the sole reason why we are successful. I get here [to work] at about 7.30 in the morning and leave at about 8.30 at night and I am here until 4.30 on Saturdays. So really, if the purpose of this report is to be of information to any Government Department, then this is the key thing.[36]

Each drawing and modification is personally approved by Martin before being sent to the production departments. Similarly, the production philosophy of ensuring that every component is finished to a high quality is to be attributed to him. Many production engineers would question the necessity for the quality and expense of the finish which each component of the seat receives, but there is no doubt as to its effectiveness in inspiring user confidence.

There are also other indications of the centralisation of control in this privately owned company. 'Martin-Baker is not a trade union shop and few, if any, of the members belong to any trade union.'[36] A high degree of selection is exercised by Martin with respect to those whom the company employs.

Development policy. It is the stated policy of M.B. to maintain complete freedom in deciding its lines of development. This policy is implemented in a practical way by its refusal to accept any development contracts for its ejector seats. The procedure is that a potential customer delivers the appropriate section of the aircraft fuselage to the company's headquarters at Denham. There, whatever modifications or developments necessary to match the ejector seat to the aircraft are carried out and, when the combination has demonstrated its effectiveness, it is delivered to the customer on a sale-or-return basis. The company retains the right to use any development made for a particular application in whichever way it desires and so this policy of private funding gives it great flexibility in its development. The determination to retain internal control of its operations is an aspect of company policy which can partly be traced to earlier experiences such as the non-adoption of the MB-5 aircraft and the fact that

a considerable amount of design work had been done [on a fully automatic ejector seat] when, on 19 June 1947, Martin-Baker Aircraft were requested not to undertake any further work connected with parachutes or their equipment on behalf of the Ministry of Supply. The company regarded this decision as unfortunate, to put it mildly, since a considerable amount of experience of high-speed operation of drogues and parachutes had by that time been acquired – this experience being directly relevant to the automatic seat problem.[10]

Another interesting aspect of the company's policy is that it firmly declines to reveal the extent of its operations.

Self-sufficiency. An examination of the list of principal suppliers to the company[10] shows how well developed is its policy of maintaining maximum independence. Except for extremely specialist items such as parachutes and pressure-responsive elements, it relies on outside firms only for components such as tubes and forgings for which there is a clear economic advantage in subcontracting and alternative suppliers are available. This self-sufficiency even extends as far as the company producing its own upholstery and packing cases for its ejector seats, and it is reflected in the facilities which it uses, having its own airfield at Denham for testing.

The measure of success. The success of the M.B. ejector seats should be measured first of all by the number of lives which have been saved by the use of the seats, secondly by the number of customers who have used them and thirdly by the extent to which they have become accepted as design standards for the ejector seat industry by both users and manufacturers. By 1969 over 2,000 successful escapes from damaged aircraft had been made using M.B. ejector seats, and in 1967 this total grew at the rate of over one successful escape per day. The success rate of the company's seats in service with the U.S. Air Force in 1967 was 96·5 per cent and with the U.S. Navy it was 95 per cent. M.B. seats are in service with the air forces of thirty-five different countries and they have been fitted to over forty different aircraft types.

Testing. It is appropriate to mention the fact that in the development of ejection seats much courage has been required not only by Martin as the Chief Designer of the M.B. ejection seat but also on the part of test 'pilots' such as Bernard Lynch, Squadron Leader Fifield and W. T. Hay. Describing his condition after testing one of the rocket-assisted seats under experimental delayed ignition conditions, Hay wrote:

> The evidence of the rocket's thrust was there, quite horribly, to be seen. My legs and arms were puffed and swollen into mottled purple; my belly, once rippled and rigid, was swinging down almost to my knees like some obscene and pendulous sack of frogspawn, the muscles in and all around it ripped, slack, and useless like ruptured elastic. My wife stood at my bedside, shocked and embarrassed – on my behalf – into near-tearful silence. Laughter, for both of us, was the only possible antidote. 'Jenny', I whispered, 'call the doctor, but for God's sake no reporters. I think I'm pregnant.'[41]

References

1. I. B. Hart, *The Mechanical Investigations of Leonardo da Vinci* (Chapman & Hall, London, 1925).
2. O. Stewart, *Aviation: The Creative Ideas* (Faber, London, 1966) p. 218.
3. British Patent 474,862 to Bolius Voiciechauskis, filed 1936, published 1937.

An abridgement of this patent appeared in vol. 42 of the *Journal of the Royal Aeronautical Society* (1938) on p. 262, only six pages after an abridgement of one of the many patents granted to James Martin.

4. R. Smelt, 'A Critical Review of German Research on High-Speed Airflow', *Journal of the Royal Aeronautical Society*, **50** 933 (1946).
5. 'Ejection Seat Development in Sweden', *The Aeroplane*, 692 (1953).
6. *Jane's 'All the World's Aircraft'* (B.P.C. Publishing, London, 1936) p. 68c.
7. *Jane's*, 1947, p. 50c.
8. Ibid., p. 60c.
9. Sir James Martin, interview, May 1968.
10. C. G. Keil, 'The Development of Martin-Baker Ejection Seats', *Aircraft Engineering*, **37** (1965).
11. J. Martin, 'Developing the Ejection Seat', *Flight Safety* (a journal of the Royal Aeronautical Society), **1**, no. 4 (1968).
12. British Patent 583,258 to James Martin and M.B. Aircraft Company, applied for 2 May 1945, published 12 Dec 1946.
13. British Patent 583,257 to above, applied for 28 Feb 1945, published 12 Dec 1946.
14. British Patent 640,520 to above, applied for 10 July 1947, published 19 July 1950.
15. J. Martin, 'Ejection from High Speed Aircraft', *Journal of the Royal Aeronautical Society*, **60** 569 (1956).
16. British Patent 607,527 to above, applied for 16 May 1951, published 10 Sep 1952.
17. British Patent 679,063 to above, applied for 16 May 1951, published 10 Sep 1952.
18. British Patent 848,208 to above, applied for 21 Dec 1956, published 14 Sep 1960.
19. British Patent 831,472 to above, applied for 21 Dec 1957, published 30 Mar 1960.
20. British Patent 616,238 to above, applied for 31 Aug 1946, published 18 Jan 1949.
21. British Patent 711,234 to above, applied for 7 Sep 1951, published 30 June 1954.
22. British Patent 843,269 to above, applied for 28 Dec 1956, published 4 Aug 1960.
23. British Patent 920,545 to above, applied for 7 Nov 1960, published 6 Mar 1963.
24. British Patent 941,683 to above, applied for 20 Jan 1961, published 13 Nov 1963.
25. British Patent 959,879 to above, applied for 30 Mar 1961, published 3 June 1964.
26. British Patent 959,898 to above, applied for 11 May 1961, published 3 June 1964.
27. British Patent 1,014,278 to above, applied for 3 Dec 1963, published 22 Dec 1965.
28. British Patent 991,251 to above, applied for 16 Feb 1962, published 5 May 1965.
29. British Patent 991,252 to above, applied for 16 Feb 1962, published 5 May 1965.
30. British Patent 991,253 to above, applied for 16 Feb 1962, published 5 May 1965.
31. British Patent 991,254 to above, applied for 18 May 1962, published 5 May 1965.
32. British Patent 991,255 to above, applied for 18 May 1962, published 5 May 1965.
33. British Patent 991,257 to above, applied for 18 May 1962, published 5 May 1965.

34. British Patent 1,046,588 to James Martin applied for 11 Sep 1964, published 26 Oct 1966.
35. British Patent 1,046,589 to above, applied for 11 Sep 1964, published 26 Oct 1966.
36. Sir James Martin, letter to W. G. Evans, 25 Feb 1969.
37. G. H. Dhenin, Director of Health and Research, Ministry of Defence, letter to W. G. Evans, 21 Feb 1969.
38. 'In the Hot Seat Business', interview with Sir James Martin, *Sunday Times*, 1 June 1969.
39. U. Delbro, SAAB Aktiebolag, letter to W. G. Evans, 18 Mar 1969.
40. 'SAAB Broke U.K. Patent say Danes', *Financial Times*, 3 Apr 1970.
41. Doddy Hay, *The Man in the Hot Seat* (Collins, London, 1969).

28 METALS RESEARCH: IMAGE-ANALYSING COMPUTER

It often happens that a highly successful innovation emerges in an area different from that in which the firm was previously engaged. The Metals Research Ltd (M.R.) innovation – the image-analysing computer known as 'Quantimet' (Q.T.M.) – was a new departure from its other interests in metallurgical research and the manufacture of metal crystals.

(a) The Quantimet image-analysing computer[1]

The possibility of automatic analysis of microscopic or photographic images has developed rapidly during the last fifteen years. Among the first experimental instruments were those of Lagercrantz[2] in 1952 and Roberts and Young[3] in 1956. Both of these early instruments were applied to biological problems of cell counting and discrimination, but the methods and techniques were soon extended and applied to the measurement of particle dimensions. The technology advanced rapidly with the development of the 'flying spot' scanner at Imperial College by Roberts and Young, the 'wide track' scanner at the National Coal Board Research Laboratories and the 'photograph scanner' at the Mullard Research Laboratories. About the same time, the 'cytoanalyser' for pre-screening of cervical smears, the sanguinometer for counting blood cells and the 'nuclear track scanner' were developed in America.

The Q.T.M. is among the first instruments to have been

designed specifically for metallurgical applications as opposed to biomedical or particle analysis. The metallurgical problem is very similar to other image-analysis applications in that the metallurgist is interested in the assessment of non-metallic material (inclusions) that look similar to powder particles when viewed through a microscope as well as the investigation of metal grains that exhibit a boundary structure very similar to the cell matrix found in many biological problems. The metallurgical problem differs in one important respect – the particles are almost invariably either circular or, if the material has been rolled or drawn, lie parallel to some common 'rolling direction'. The problems of measuring such parameters as particle size distribution is much simplified in these situations. The use of modern television techniques has made it possible to incorporate large television displays which remove most of the need for operator skill and have increased speeds to allow many hundreds of features to be analysed for parameters like size distribution in one second.

The Q.T.M. consists of three main component parts: an optical system, a television system and an analogue computer. The microscope, fitted with a beam-splitting prism, projects an image of the metal surface simultaneously into a conventional binocular eyepiece for direct viewing and directly on to the tube of a television camera. The output from the camera is displayed on a monitor to give a television microscope picture for focusing and selecting the appropriate area to be investigated. The camera output is also fed into an electronic detector which responds to the changes in output voltage as the scanning spot in the camera tube passes over the features of the specimen (e.g. grain boundaries, inclusion pores, cells, bacteria, powder particles, etc.). The 'discriminator' in the detector can be set to respond to areas darker or lighter than a selected threshold and the signals obtained from these areas are then fed into the computer which almost instantaneously computes the various geometrical and statistical parameters which describe the field of view. The features which have been detected are automatically marked on the television display so that the operator has an instantaneous visual check on the correct setting and operation of the instrument.

The Q.T.M. has very wide application in both scientific

N

research and production or quality control laboratories. Typical applications are: non-metallic inclusions in metals, powder technology, biomedical research, particle track photographs in nuclear physics, fibre measurements in the cotton, wool and glass fibre industries and porosity measurements in ceramic materials. The instrument has been bought by the world's leading steel makers and by car and aircraft manufacturers.

(b) The company[4]

Metals Research is an independent private company which was founded in 1957 by Dr M. Cole for the purposes of carrying out sponsored research in metallurgy and physics and manufacturing metal single crystals. The philosophy of the firm was to try to fill a gap in the market; Cole wanted to embark on a product line with a high technological content which was too sophisticated technologically for most small firms and had too small a market potential for the large firm. The idea originally was that M.R. would be a research and marketing organisation only. It was felt that new ideas could be developed by M.R. and then subcontracted to an engineering firm for production. Subsequently, M.R. would market the finished product. Cole himself seemed particularly well suited to engage in this kind of activity. He had obtained a Ph.D. at Cambridge in metallurgy, after which he spent some time researching in materials science. He felt that he knew very well what type of equipment and, in particular, what types of metal crystals would be required by scientific laboratories. The subcontracting of production did not work very well in practice, primarily because of the low priority which the engineering firms gave to the M.R. work. Because the orders were small 'custom jobs', the firms would only do 'our work during times of economic slump but during a boom the bigger orders from the larger firms tended to have priority'. It was subsequently decided that M.R. would have to do its own manufacturing.

(c) The development of Quantimet

In 1960 the firm set up an instrument division to manufacture some of the instruments which it had developed for its own research and crystal technology projects mainly concerned with the growing and cutting of metal crystals. During the previous three years the firm had established many contracts in the metallurgical industry. It was also well known that they were interested in diversifying into new products. According to Cole:

> Metals Research was never short of its own ideas for new products, but as it became known as an enterprising manufacturer of sophisticated equipment, scientists started to come to [M.R.] Cambridge with their inventions. Ideas for new scientific products started flowing in from sources in Government, university and industry. Most of them, though promising, needed too much capital or special marketing facilities, and had to be rejected.[5]

One of them, however, turned out to be very important, although it was not adopted directly. The original suggestion for Q.T.M. came from Griffin, a scientist at the National Engineering Laboratory at East Kilbride. Basically, the idea was to use an electronic scanning technique on a photograph from which one could determine the concentrations of black and white areas. Griffin, who owned some Metals Research equipment, enquired whether M.R. would be interested in manufacturing his machine. In order to test the marketability of the device, M.R. did a routine market survey in which they circularised some 100 likely buyers to discover if they would purchase such a device if it were offered at £500. The response was discouraging, but some steel companies they had approached said that if an alternative device were offered, they would buy several even if they cost thousands of pounds. The suggestion made by the steel companies was that the device should look through a microscope 'directly at the metal' without needing an intermediate photographic step. After further market research, it was decided to drop the original idea and develop the instrument as suggested by the steel companies. A specification was obtained from them. It was soon realised that the instrument had a much wider applicability than had originally been thought. None the less, it was decided to develop the instrument for the steel industry in the first instance because it could be sold immediately.

At the beginning, M.R. development work appears to have been personally organised by Cole, but in December 1963 Dr C. Fisher, who had just completed his doctorate at Cambridge in plasma physics, joined the company to set up a more formal development department. The first task was to develop with all speed the new instrument later known as Quantimet. In fact, Fisher's worth was well known to M.R. as he had previously designed an extremely successful spark erosion machine during a summer vacation job. The company had kept close contact with him throughout his academic career and three years after he joined M.R. he became Technical Director. The first model of Quantimet was completed and sold in 1964.

(d) Comments

Reasons for success. According to Cole, 'the instrument was developed in record time and proved an immediate success'. This success seems to be attributable to three major factors: manpower, technology and marketing. Fisher is regarded as an important figure in the development of Quantimet. He is an exceptionally able engineer/physicist who seems to combine understanding of physical phenomena with technical skill. In the opinion of some at M.R., if the technical direction of the project had been under someone of a lower calibre, at least in the early days, Quantimet would not have been produced.

Of equal importance to the success of the innovation was the availability of the techniques of television. After they had marketed Quantimet, M.R. discovered that others had tried to produce a similar instrument. In particular, Radio Corporation of America (R.C.A.) had, four or five years previously, attempted to make an image-analysing computer. One reason why they failed appears to be that television technology was not sufficiently developed. Similarly, the Rank Organisation because of the lack of appropriate television techniques, decided to back the 'flying spot technique'. Although the Rank instrument was technically successful, it was too expensive and consequently did not sell very well. Still, one should not draw the conclusion that Fisher and his team simply borrowed from television technology the systems they required. It would be more accurate to say that television technology had advanced enough to make it feasible for M.R. to make the further advances required. Another area where many advances were made at this time was computer technology and M.R. was able to cash in on these too. There were a large number of innovations in computer techniques, particularly in logic circuitry, which ultimately resulted in faster information processing.

Metals Research also has a thoroughly professional marketing organisation which attempts to ensure that the company's resources are applied to its products and markets in the most effective way. It should be recalled that a significantly new design for Quantimet emerged after considerable market research on prospective customers.

Finance and growth. During its first ten years of existence the company grew from three to 225 employees, about 70 per cent of whom are engaged in development and production. The major part of the growth that has taken place in the company since 1960 has come from the manufacture and sale of sophisticated electronic instruments rather than from the production of metal single crystals and the consulting work in materials science for which the firm was originally established. The important thing, while the company was growing fast, was to maintain control over all functions of the business to 'ensure a smooth passage between the Scylla of finance and the Charybdis of "boffincy"'. It was Dr M. Cole's brother, Mr D. Cole, an arts graduate, who built up the commercial side and laid the foundations for an international marketing organisation which has since proved its worth: 60 to 70 per cent of the company's sales are overseas. Finance for the Quantimet itself was obtained from the profits of previously successful developments. Dr Cole pointed out that

> the profits from the very successful Servomet Spark Machine provided the money to invest in the development of the Quantimet in the way that crystals research profits helped to develop Servomet. The Servomet, therefore, leap-frogged crystals and the Quantimet the Servomet, each product having a much larger market than the one which paid for its initial development.[6]

The firm is partly financed by the Industrial Corporation

Finance Company (I.C.F.C.) and between 1960 and 1967 increased its annual turnover from £30,000 to about £500,000. It has acquired a 17-acre parkland site in the village of Melbourn, near Cambridge, where it has built a modern factory.

References

1. C. Fisher and M. Cole, 'The Metals Research Image Analysing Computer', *The Microscope*, **16**, no. 2 (Apr–July 1968).
2. C. Lagercrantz, *Acta Physiologica Scandinavica*, suppl. 93, **26** 1 (1952).
3. F. Roberts and J. E. Young, *Proc. Inst. Electr. Eng.*, **99**, IIIA, 747 (1956).
4. Interview with Dr M. Cole, Managing Director, Metals Research Ltd.
5. M. Cole, private communication, 1968.
6. M. Cole and C. Beadle, private communication, 1969.

29 OXFORD INSTRUMENT CO: HIGH MAGNETIC FIELDS AND LOW TEMPERATURES

The complex of 'spin-off' companies around the Massachusetts Institute of Technology (M.I.T.) at Boston is by now a well-known example of how small firms can challenge the giants in areas of advanced technology. Considerable study of the 'Boston Ring Road' or 'Route 128' phenomenon, as it is sometimes called, has been carried out by Professor E. Roberts[1] and others at M.I.T.'s Sloan School of Management to document the social origins, managerial structure and economic importance of this group of companies.

In Britain, the question is increasingly asked, 'Why are there no such firms on this side of the Atlantic?' The answer is that there are some small 'spin-off' firms in existence but, as yet, there are not enough of them to attract widespread attention. One such company is Metals Research (p. 370). Another is the Oxford Instrument Company (O.I.) which was formed in the late 1950s by Mr Martin Wood, a member of the staff of Oxford University's Clarendon Laboratory.

(a) *High magnetic fields and low temperatures*[2,3]

High magnetic fields. Superconductivity is a phenomenon observed in many metals and alloys below about 20°K (i.e. below −253°C.) in which the electrical resistance of the material disappears. With zero resistance, there is no energy loss when an electric current is created and that current will continue indefinitely unless the temperature of the superconductor is raised above some critical value.

Superconductivity was observed, apparently unwittingly, by low-temperature physicists such as Dewar as early as 1898, but they ignored it, attributing any anomalies to faults in their instrumentation. Kamerlingh Onnes in 1911 was among the first physicists to look seriously into the behaviour of the resistance of metals at low temperatures. He determined experimentally that the resistance of a superconductor was only some 10^{-11} times that at room temperature. One of the first applications which occurred to Onnes was to use a superconducting coil to produce a very strong magnetic field having an intensity of about 100,000 gauss. The conventional method of producing such a field was to force a large current through a solenoid of normal conductor which, because of its electrical resistance, consumed a great deal of power. A superconducting solenoid, it appeared, would be cheaper to make as well as to operate.

Unfortunately the early superconducting materials lost their special properties in a magnetic field of a few hundred gauss. This magnetic field intensity is far below that required even in a modest electrical machine. The breakdown of superconducting properties occurs above the 'magnetic threshold', which for a given substance at a given temperature is the maximum magnetic field at which the material remains superconducting. The magnetic threshold is a physical property of a material and varies with temperature; consequently, it may be exceeded by increasing either the magnetic field or the temperature. There were three main reasons for the low magnetic thresholds observed in the early experiments. Firstly, as H. London demonstrated in 1934, the threshold value was low because the wires were *too thick*. He pointed out that the diameter of the wire should be of the same order of magnitude as the depth of penetration of the magnetic field into the wire, which he showed to be about 10^{-5} cm. Secondly, it was shown that some *alloys* have a higher magnetic threshold than the pure metals which Onnes had used. The third reason for the breakdown of super-conductivity was, as Onnes suspected, that there were *faults* in the wire, where heat was developed which could not be withdrawn and which locally warmed the wire above the vanishing point of resistance.

The property of switching to the normal state at quite low current densities and in the presence of modest magnetic fields turned out to be a major drawback to the development of high-field superconducting magnets. However, these problems were subsequently solved by improved engineering techniques in manufacturing the wires and by research into the magnetic thresholds of alloys. According to Wood, 'Not for another fifty years, until 1960, . . . were alloys found which remained super-conductive in much higher fields. A compound of niobium and tin was found to remain superconductive in fields approaching 200 kilogauss.'[4]

Low temperatures. Helium-3 (a rare light isotope) has a lower boiling point than any other liquid. Below $0 \cdot 35°$K, however, the vapour pressure is so small that little cooling can be effected by evaporation. Until the late 1950s this had been the lowest temperature that could be maintained continuously.

In 1951 H. London,[5] a physicist at the Atomic Energy Research Establishment (A.E.R.E.) at Harwell, suggested that a process of diluting liquid helium-3 with liquid helium-4 (the more common isotope) would produce a cooling effect in much the same way in which a gas cools on expansion. In 1956 it became known that below $0 \cdot 9°$K two distinct layers form in a mixture of helium-3 and helium-4, with the helium-3 lying above the helium 4. It was predicted that as molecules of the upper helium-3 crossed the boundary into the lower liquid helium-4, a cooling effect would occur.

Refrigerators based on this effect were first operated independently by Hall[6] and Neganov,[7] reaching temperatures of $0 \cdot 055°$K and $0 \cdot 025°$K respectively. Recently, in a more sophisticated apparatus, Wheatley[8] succeeded in reaching $0 \cdot 0045°$K and indications are that still lower temperatures are possible.

(b) *The Oxford Instrument innovations*[9]

The company. The origins of O.I. lie in the pioneering work on high magnetic fields done at the Clarendon Laboratory of Oxford University. In 1959 Martin Wood, who was working with Dr N. Kurti, F.R.S., Head of the High Field Laboratory, on the design of high-power magnets, was asked to make a magnet for an ex-Clarendon scientist, Professor D. H. Parkinson of the Royal Radar Establishment at Malvern. In the preceding year Wood had helped in the design of several pieces of equipment involving magnetic fields for other university and Government laboratories and had given advice on a number of related problems, including such devices as the magnetic powder clutch system installed in Hillman cars under the name 'Easidrive'. The 60 kilogauss magnet for Parkinson was the first complete manufacturing project to be tackled.

The company was formed in 1959 and Wood appears to have received a good deal of encouragement both from Kurti and the Head of the Department, Professor B. Bleaney, F.R.S., C.B.E. The initial idea was to provide a means whereby other laboratories could benefit by obtaining devices discovered and developed at the Clarendon Laboratory for its own needs. According to Wood, these ideas rarely get the publicity they deserve because the organisation in which they germinate has no interest in or mechanism for promoting them.[10]

The company set up first in a spare room in Wood's house, and later in a shed at the bottom of the garden where an old lathe was installed together with equipment for manufacturing fibreglass casing and for winding magnets. The first magnet was completed with the help of Mr J. Milligan, who had previously worked in the low-temperature workshop of the Clarendon; it has been used for producing magnetic fields of up to 85 kilogauss for experiments in optical spectrometry, magneto-resistance and superconductivity.

The company was subsequently asked to make another magnet, similar to the one requested by Parkinson, for the polarised neutron project on the Dido reactor at A.E.R.E., as well as some coils for a mirror machine for experiments on controlled thermonuclear fusion. The shed was only just large enough to cope with these two magnets so Wood decided to look for a better site:

> . . . so in 1961, by a process of cycling round the back streets of north Oxford, and peering over walls, we discovered a secluded and disused stables and slaughterhouse for which we were able to negotiate a five-year lease. It had not been used since the last bullock was slaughtered in 1939, except by woodworm, bats and spiders, but when the straw and cobwebs had been swept away and the mangers and stalls, derelict delivery bicycles and slaughtering equipment removed, it was converted into a pleasant enough workshop.[10]

Up to that time the company had been concerned with winding high-power solenoids for specialist applications using traditional techniques. The market for this type of product was limited by the number of laboratories in the world with adequate power plants and cooling systems; the number must have been less than ten. It appeared to Wood that if he restricted himself to traditional techniques for producing high magnetic fields, his company would always remain small and therefore a new type of product was needed.

Superconducting magnets. In order to understand the origin of O.I. interest in superconductors, it is necessary to digress for a moment and follow a chain of developments that had taken place in the technology of superconduction. In 1955 G. B. Yntema at the University of Illinois put a small superconductive wire on an iron core and succeeded in generating a field of 8 kilogauss. In 1960 a field of 4·3 kilogauss was produced by a coil with no iron yoke at all by S. A. Autler at the Lincoln Laboratories in Massachusetts. Then J. E. Kunzler in the Bell Telephone Laboratories produced a coil of molybdenum and rhenium which provided a field of 15 kilogauss. According to Wood, 'the race was on':

> Everything was now set for rapid advances. Magnetic fields were required urgently by research groups in many branches of the physical sciences, the theoretical side of the behaviour of superconductors was making quite rapid advances, metallurgical expertise was capable of developing the special

skills required and the cryogenic industry was sufficiently advanced to provide for all the needs of low temperature work. Already by 1963, there were three companies offering superconductive magnets and now the number of firms has doubled and the maximum field strength available is well over 100 kilogauss.[4]

In the autumn of 1961 Wood attended a conference at M.I.T. on high magnetic fields. It was while he was there that researchers from the Bell Telephone Laboratories announced that a very high current could be passed through a niobium–tin alloy, even in the presence of a high magnetic field. The implication of this was that high fields would, in future, become available as a tool for research in many laboratories by the use of superconducting magnets. Wood appears to have decided there and then that his company would move into this new technology, and on his return to England he set about making his first superconducting magnet.

Niobium–zirconium emerged as the first reliable superconductor and O.I., which had become a limited company in October 1961, ordered its first niobium–zirconium wire early in 1963 from the Wah Chang Corporation. The superconducting magnet had to be wound with great care because of the 'astronomical cost' of the wire. It was tested in liquid helium at the Clarendon Laboratory in April 1962 and produced a field of over 40 kilogauss.

This magnet has been used many times for various experiments and at the Royal Society for demonstrations of how energy can be stored without loss in superconducting devices and released at any time it is required.

Encouraged by the success of this first magnet, the company undertook to make a number of other superconducting magnets and started developing a transistorised power supply to provide a stabilised current as well as a cryostat for housing the magnets which operate in liquid helium. Subsequently, the company entered the field of cryogenic engineering. It continues to maintain close informal contacts with the staff of the Clarendon Laboratory and has developed similar contacts with most other universities in this country. In July 1966 O.I. produced the first 100-kilogauss superconducting magnet in

Europe. These magnets cost about £10,000 and can be operated by the research workers themselves. By contrast, the cost of producing the same field by earlier methods, taking into consideration power and cooling plant, was of the order of £100,000 and required a good deal of space as well as engineering maintenance staff.

O.I. have sold magnets to laboratories in nearly every British university, to the National Physical Laboratory, the Royal Radar Establishment and the A.E.R.E., as well as many private companies. Abroad, O.I. magnets have been sold, for instance, to the National Standards Laboratory in Australia, the German Plasma Physics Institute and many other national research institutes.

The Harwell refrigerator. By the early 1960s London's earlier ideas about the feasibility of a helium-3 dilution refrigerator seemed feasible.[11] After sponsoring some academic work under Professor H. E. Hall at Manchester University, the Atomic Energy Authority (A.E.A.) investigated the possibility of placing a development contract with one of a number of commercial firms, including Arthur D. Little Inc. of Cambridge, Massachusetts, and Philips of Eindhoven in Holland. Finally, the A.E.A. awarded O.I. the development contract and also asked them to study the commercial feasibility of the idea. The main problem was to adapt the London dilution principle to continuous operation. In March 1966 the first prototype was made to work and it was shown at the Physical Society Exhibition. World-wide interest encouraged O.I. to go into production. The first production model was tested in July 1966.

The refrigerator is manufactured under licence from the U.K.A.E.A. to whom O.I. pay royalties. Informal relations with Harwell were maintained and both London and Hall were consulted on points of detail. The total market for the refrigerator is estimated at several hundreds. The sale price is between £8,000 and £10,000, but varies somewhat with particular customer requirements.

(c) Personnel

The company's first two directors were Martin Wood and his wife, Mrs K. A. Wood. Mrs Wood studied chemistry, biology and English at Cambridge and has mainly been responsible for the publicity side of the enterprise. Martin Wood, who had worked for some time as a coal-miner in South Wales, took an engineering degree at Cambridge. After his university career he returned to coal-mines, but left again because of the contraction of the industry. A short time later he joined the technical staff of the Clarendon Laboratory. He is now Chairman and Sales Director of the company.

Early in 1963 Mr Rackstraw, a skilled designer and engineer, joined Mr Milligan in the former slaughterhouse as the first full-time employee. He has an H.N.C. in mechanical engineering and is now Production Manager. Later in the same year F. D. Thornton, a research physicist from the University of London, joined the company. Since then the staff has increased to about ninety.

(d) Comments

Finance. It appears that O.I. was financed with Wood's personal resources from 1959 to 1967. Any money in the form of profits was ploughed back into the firm to allow for the rapid growth necessary in the early stages of the firm. According to Mrs Wood, the company was grossly over-extended at the time of the Queen's Award and was expanding more quickly than profits. Various companies were interested in investing in O.I., but none of the negotiations came to fruition. Finally an agreement was struck with Technical Development Capital, who granted O.I. the long-term capital necessary for future expansion.

Technology transfer. Although the development of the niobium–zirconium superconducting wire was a product of both basic and applied research, the essential step as far as the innovation was concerned was the transfer of this developing technology to the coil-winding technology. Coil winding, especially for specialist application, is still something of an art as well as a science, and without the wide experience of Wood in this field a superconducting winding might have proved too difficult to make.

Management techniques. In the early days of the company none of the directors had any experience of management. It seems that they learned the necessary management techniques as they went along. However, as the company grew, it became more and more important to exchange amateur skills for professional management techniques. It appears that some of the major problems now faced by this young company are concerned more with management and finance than with science and technology.

'Spin-off' firms around Oxford. Martin Wood has vigorously resisted the common fate of the small, science-based firms on the Boston Ring Road – swallowing up by one of the giants. Instead, Wood seems determined to interest others in starting up science-based enterprises around Oxford University. So far he has met with only limited success, but these are early days as yet.

References

1. E. Roberts *et al.*, 'Technology Transfer and Entrepreneurial Success', Twentieth National Conference on Administration of Research, Florida, 27 Oct 1966.
2. The author would like to thank Mr Paul Drath for his assistance in preparing this section.
3. K. Mendelssohn, *The Quest for Absolute Zero* (Weidenfeld & Nicolson, London, 1966).
4. M. Wood, 'The Technology of Superconduction', *New Scientist*, 29 Dec 1966, 5 Jan 1967.
5. H. London *Proceedings of the International Conference on Low Temperature Physics* (Clarendon Laboratory Oxford, 1951).
6. H. E. Hall *et al.*, *Cryogenics*, **6** 80 (1966).
7. B. Neganov *et al.*, *J.E.T.P.* (*Soviet Technical Physics*), **50** 1445 (1966).
8. J. Wheatley (to be published).
9. Interview with Mrs K. A. Wood, Director, Oxford Instrument Company, 1968.
10. Oxford Instrument Company, private communication, 1968.
11. H. London *et al.*, *Physical Review*, **128** 1992 (1962).

30 PLASTICISERS LTD: SYNTHETIC MATERIAL FOR CORDAGE

This case study demonstrates how a small family firm can be a successful innovator and it also illustrates how patents can be an incentive to innovation in that Plasticisers had to examine unconventional methods of fibre-forming because existing patents prevented them from using normal techniques. The 'synthetic material' that gained Plasticisers a Queen's Award is a fibrous form of polypropylene produced in an unconventional manner from a film of polypropylene without the use of a spinneret, which has been the traditional way of converting synthetic polymers into fibres. The process developed by Plasticiers has been adopted by British Ropes, providing an example of innovation resulting from collaboration between two organisations of differing size.

(a) Fibrillation

Fibrillation is a rather loose term which has been applied to a variety of processes for the subdivision of various materials into a fibrous state. In antiquity, leaves and stems of plants were converted into fibrous masses by processes such as beating. The shredding of coarse fibres into finer fibres or fibrils is also known as fibrillation. Fibrillatable materials, whether natural or synthetic, are characterised by having different tensile strengths in longitudinal and transverse directions so that mechanical process such as beating, cutting or even twisting can cause the material to split in one direction with the production of a fibrous mass. Different types of products can be produced by fibrillation depending on the degree to which the process is carried. A carefully controlled amount of splitting can result in a net-like structure if the fibrous mass is opened out. At the other extreme, sheets of material can be split into separate fibres which can then be chopped into staple prior to spinning into a yarn.

One of the variables that affects the properties of polymeric materials is the degree of orientation of the polymer molecules. If the polymer chains are entangled in a completely random manner, then transverse and longitudinal tensile strengths are the same. If, however, the polymer chains are pulled out to lie mainly in a longitudinal direction, then the longitudinal tensile strength of the material is higher than its transverse strength and fibrillation becomes possible. The possibility of fibrillating synthetic materials was realised in the 1930s and the basic patent in this field (to I.G. Farbenindustrie in 1936) states: 'artificial threads are obtained by splitting, e.g. by twisting, brushing or rubbing films, bands, tubes or threads of an organic thermoplastic material which are strongly orientated parallel to their length'.[1]

The vast majority of synthetic fibres have been produced by extrusion through a spinneret, a process reputed to have been invented by Louis Schwabe[2] in 1842. The commercial use of fibrillation did not become possible until the advent of a new class of polymers, the crystalline polyolefins. A film of highly oriented polyolefin has its polymer molecules arranged longitudinally and has an almost negligible transverse strength which makes fibrillation comparatively easy.

The crystalline polyolefins, high-density polyethylene and polypropylene, became available commercially in the 1950s following the work of Ziegler and Natta who shared a 1963 Nobel Prize for their discoveries. Various systems for the production of high-density polyethylene were devised throughout the world and there is no patent-based monopoly, but the situation with polypropylene is different. Natta worked in collaboration with the Italian firm of Montecatini who established a firm hold on patented methods for the production of polypropylene. Both plastic and fibre production were included in the Montecatini patents. In this country Shell and I.C.I. both obtained licences to produce polypropylene plastic but I.C.I. obtained the sole rights for polypropylene fibre.

Plasticisers Ltd wished to produce polypropylene fibre by conventional means but were prevented from doing so by the patent situation. As a result, Plasticisers were forced to look at non-conventional means of producing polypropylene and came up with the idea of using fibrillation, a process outside the scope of the Montecatini patents. Various groups of workers have examined the fibrillation of polypropylene but Plasticisers were the first to obtain a commercial process. They did this without reference to anyone else's work, but when they began to apply for patents, they found that there were already in existence a large number of fibrillation patents going back to the 1936 I.G. Farben patent referred to above. Plasticisers managed to obtain patents of their own but have not always managed to enforce them as their validity is still being challenged.

The complex patent situation is one reason why I.C.I. have not paid much attention to fibrillation. Because fibrillation is a fairly simple process and because there is no firm patent position, anyone can do it, which means that competition forces prices down, preventing large organisations like I.C.I. from recouping R. & D. and other overheads. In fact, I.C.I. have marked a polypropylene tape, 'Nufil', which is highly crystalline, oriented and capable of being fibrillated by anyone with the necessary machinery, but in 1965 Plasticisers Ltd were selling a fibrillated polypropylene tape at the same price as I.C.I.'s unfibrillated Nufil.

(b) Plasticisers Ltd[3,4,5]

Plasticisers Ltd is a family business owned by the Slack family which has a history of weaving in Yorkshire stretching back more than a hundred years. In the 1930s the Slack family was mainly concerned with importing bristles of various kinds, processing them and supplying them as fillings to brush manufacturers throughout the country. The main raw material for high-quality brushes was pig's bristle imported from China and India. A shortage of this material in the late 1930s, intensified by the Second World War, meant that the Slack family was keenly interested in alternative fibres for brushes, an interest that was to have three important effects: the invention of 'flagging', the formation of Plasticisers Ltd and enthusiasm for synthetic fibres when the became available.

Flagging was invented in 1939 by the father of the three Slack brothers who, at present, control Plasticisers Ltd. Pig's bristle is a fibrillatable natural fibre with a tip that splits into about twenty fibrils. In an attempt to make horse hair an acceptable replacement, flagging, a mechanical method of fibrillating horse hair, was developed. It was partly as a consequence of this that twenty years later Philip Slack could think of fibrillation as a method of obtaining polypropylene fibres when prevented from using more traditional techniques.

One of the problems with horse hair as a brush fibre is its

brittleness, and in 1943 the Slack family set up Plasticisers Ltd at Drighlington, near Bradford, with the original intention of plasticising horse hair to make it less brittle. As synthetic fibres became available, Plasticisers Ltd was used as a kind of research and development organisation to examine the potentialities of synthetic fibres firstly for brushes and later for other applications. Plasticisers Ltd did not start trading until 1957 and was financed by the other Slack companies, E. and G. Slack, Slack Bros, Drighlington Hair Blenders and Slack Sales.

Three Slack brothers are concerned with Plasticisers Ltd. Owen Slack, the eldest brother, and David, the second brother, have been directors since 1943. Owen remains Chairman, but now gives most of his attention to the brush side of the family companies. The present Managing Director is Philip T. Slack, the youngest brother, who has been described as brilliant.[3] Both David and Philip are highly inventive individuals who seem to be happiest when solving an interesting technical problem. None of the three brothers has any academic qualifications but a fourth brother gained a Ph.D. for research concerned with animal hairs. He worked with Plasticisers for a time, was bought out and became Director of Fishery Research for the New Zealand Government.

After the war, when nylon became available, the Slack brothers read the available literature and visited machinery manufacturers. They were offered plant for making nylon brush fibres but only on condition that they bought six at £80,000 each. They declined. Machinery for extruding nylon monofilament was available but it did not contain any stretching device. It seems likely that the Slack brothers were prevented from being early innovators in the use of nylon by lack of appropriate machinery and since then they have preferred to develop their own machinery.

The French Government backed polystyrene exports and the price fell, encouraging the Slacks to experiment with chopped polystyrene monofilament, but polystyrene was not satisfactory for brushes.

In the mid-1950s high-density polyethylene became available from the United States. It was not suitable for brush fillings but Philip Slack tried winding and knitting the polyethylene monofilament, with the result that in 1957 Plasticisers Ltd sold its first products, polythene pot scourers. In the same year both Courtaulds and Plasticisers announced the availability of polythene monofilament for fishing twine. They both had one extruder each and polythene slowly became acceptable in trawl and sports netting.

The brush trade was an impoverished industry with many small firms competing by price cutting. There seemed no signs that the British brush industry could follow the American example of actively marketing their products and persuading the housewife to buy more brushes, so the Slacks felt they should continue to examine new applications for synthetic fibres.

In 1958, when polypropylene became available, Plasticisers tried to obtain a licence from Montecatini to enable them to manufacture monofilament but they were refused. In 1959 they took patent advice, were assured that the Montecatini patents were sound and so they sued for a licence on the grounds of non-exploitation. It was not until 1965 that the case became due for hearing and the day before the case Plasticisers were offered a limited licence to enable them to manufacture polypropylene monofilament of one particular diameter. This they accepted and then sued for a licence for other grades, again on the grounds of non-exploitation. Plasticisers now produce polypropylene monofilament but they claim they came too late into the market to make much of a success.

One of the Plasticisers' directors has stated that the main reason that they got involved in fibrillation was I.C.I.'s 'bloody-mindedness'. I.C.I. had the sole licence for the United Kingdom production of polypropylene fibre, and the delays in the legal process whilst Plasticisers tried to obtain a licence made the Slack brothers search for a way of producing poly-propylene fibre that was outside the scope of the Montecatini patents. With the previous experience of 'flagging', fibrillation seemed a possible way of avoiding the Montecatini patents. The first experiments were carried out with polythene and involved attempts to make a film and then shred it into fibres. Various methods of stretching and cutting were attempted. When Plasticisers were able to obtain polypropylene for their experiments, they found that it was easier to fibrillate than polyethylene and a simple stretching and twisting process was

developed, resulting in the experimental production of fibrillated polypropylene twine in 1960.

By 1962 Plasticisers were in a position to commence filing
patent applications. Three applications were published by 1967,
the most important of these being 1,040,663. This patent concerns fibrillation by twisting, and although there are many such
patents dating back to 1936, there is a chance that it will
succeed in spite of the application being opposed. The main
difference between Plasticisers' patent and earlier ones is that
the process of fibrillation is stopped at the stage when the fibrils
are still interconnected. Earlier patents were concerned with
the production of staple which could then be spun into a yarn.
Plasticisers' process is much simpler. A film is produced and
stretched to the point at which it tends to split into fibres.
Fibrillation is then produced by twisting and a twine is produced
which appears to have been formed from separate fibres although
in fact they are still joined together in places. By using a wide
film, twine can be made that is sufficiently thick not to need
plying (e.g. three small-diameter twines do not have to be plied
to make a larger-diameter twine); thus the process is essentially
simple and cheap.

Plasticisers attempted to sell machinery for the conversion of
polypropylene granules into fibrillated products as well as the
products themselves. Since their experience after the war the
Slacks have never been very keen on other people's machinery,
and David Slack in particular enjoys developing machinery.
Diversification of effort between selling both machinery and
products has caused complications, but against this, being both
manufacturers and users of machinery puts the Slacks in a good
position to appreciate the problems involved.

Through their manufacture of polyethylene monofilament,
Plasticisers had established contacts with the cordage trade and
they began to interest people in their simple twisting process for
the production of polypropylene cordage. A major problem
was that, for the process to be economic, Plasticisers had to be
able to supply the machinery at a moderate price which they
could only do if they were to produce the machines on a
production line as opposed to the 'one-off' method. Plasticisers
lacked the resources to go into large-scale production of
machinery without an assured market, but the problem was

resolved in January 1964 when British Ropes Ltd signed a contract with Plasticisers.

(c) Plasticisers and British Ropes Ltd

As a 1968 winner of the Queen's Award for Technological Innovation, British Ropes Ltd falls outside the scope of the present study which deals with 1966 and 1967 winners. However, the present position of Plasticisers Ltd has been influenced considerably by the agreement reached between the two companies in 1964 and it is therefore relevant to describe some of the developments of fibrillation that took place within British Ropes.

The British Ropes group of companies is claimed to be the world's largest manufacturer of wire, wire rope and fibre rope. It was established in 1924 from a number of separate companies, some of which were over 200 years old. In the ten-year period 1956–65 the group's turnover increased from £18 million to £39 million. Within the Fibres and Plastics Division of British Ropes there is a strong tradition of development work carried out at manufacturing sites rather than in separate establishments. For example, in 1946, when synthetics were being developed for peace-time use, British Ropes recruited A. W. Smith, a physicist, to assist their development work. He joined the staff of the Leith factory rather than a research establishment and subsequently became an authority on ropes made from man-made fibres.[6]

The first person in British Ropes Ltd to become enthusiastic about fibrillating polypropylene was the late R. Cairns, who was Production Controller in the North-east. It had been believed by Cairns and others that a gap existed between natural fibres such as manila and sisal and the very high-priced synthetics, nylon and terylene. Although sisal was £60 per ton in 1947 and £70 per ton in 1968, there were occasional violent price fluctuations[7] and the price of sisal had been as high as £240 per ton at one point between 1947 and 1968. In 1964 the price of nylon staple was of the order of £700 per ton; the price of polypropylene granules was about £200 per ton and falling. A cheap process for the conversion of polypropylene granules into twine was therefore attractive.

British Ropes carried out work with polyolefin monofilaments and Cairns heard that Plasticisers were carrying out interesting work with polypropylene when he went to see them to look at machinery for producing monofilament.[3] One thing that impressed Cairns was the lower cost of extruding a film and stretching it compared with producing fine filaments by a spinneret. He brought a British Ropes team to Plasticisers, and after they agreed that fibrillation looked promising, terms for an arrangement were discussed.[3,7] British Ropes agreed to pay royalties to Plasticisers and in January 1964 a contract was signed. Large samples of oriented film were sent from Plasticisers to British Ropes' Willington Quay factory in County Durham where experiments were carried out with existing sisal equipment. In April 1964 the first extruder purchased from Plasticisers was installed at Willington Quay and attempts were made to produce baler twine. A member of British Ropes' work-study team was sent to work for eight months at Plasticisers to gain know-how.

The attempts to produce baler twine were looked after by normal factory staff who were 'guided and encouraged' by R. Cairns until his death in 1965. A considerable amount of empirical development work was required to obtain a satisfactory and reliable production method. In the words of one of the workers on the extrusion process:[6]

I rather enjoyed the exploration of the mysteries of the material, trying to decipher what was wrong, eventually putting things right after eight or nine tries and then not being sure exactly what I'd done to put it right. We thought each extruder was identical but we soon learnt that you had to treat each one as an individual.

Drastic modifications to existing twisting equipment had to be made and British Ropes have filed patent applications based on improved machines and their discovery that improved yarns could be made by using 'over-tension' and spinning drag to cold draw the polymer. The first sale of 'Formula S' baler twine (the 'S' stood for secret) was made in January 1965 and by the autumn of that year the manufacturing process was firmly established and was transferred to a new production line built at another factory in the North-east. The first market for the baler twine was the agricultural market, which is noted for resistance to new ideas. Extensive trials were carried out and in 1966 'Red Star' baler twine was launched nationally. A market has also been established in industrial packing twine. Some newspapers, the Post Office and others now use 'Red Star' in their tying machines.

Another British Ropes factory, at Charlton in London, manufactures marine ropes. The physical testing laboratory at Charlton was involved in the early work and the possibility of using fibrillated polypropylene in ropes was established. Extruded film is now transported from the North-east to Charlton where it is 'spun' into twine which is then spliced to make marine ropes.

Sales of twine and ropes are increasing and in 1967 British Ropes calculated that £500,000 worth of sisal, which has to be imported, was replaced by a polypropylene produced in Britain.[7] British Ropes regard this as just a start.

British Ropes have acknowledged their debt to Plasticisers who in turn admit that the market for rope and twine would have taken three to five years longer to penetrate without the resources available to British Ropes. However, it is claimed that the contract with British Ropes has delayed other lines of advance. The agreement between the two firms prevented Plasticisers from supplying machinery to other companies interested in manufacturing twine or rope. As a result, it is claimed that the chance to win a world market for British-made machinery has been lost.

British Ropes' publicity states that they are looking at new markets for polypropylene and they have set up a research and development department to investigate new applications. However, none of these other applications met with the rapid success of the twine development and Plasticisers' hope of expanding machinery sales to British Ropes was not realised. Plasticisers have therefore switched their aim from machinery sales to product sales. The inventive Slack brothers have developed a second generation of products which are fibrillated to a greater extent than the coarse baler twine and are nearer to traditional textile materials in appearance. In 1969 a new £400,000 factory was built for the manufacture of these new products.

(d) Comments

(i) *Science and innovation.* The fibrillation process developed by
Plasticisers depends for its simplicity on the unique properties
of polypropylene, a synthetic material made available by the
Nobel Prize-winning work of Ziegler and Natta. Clearly, this
is an example of economic benefit from scientific research. It is
not, however, an example of benefit arising unexpectedly from
'curiosity-oriented' research. Ziegler was the Director of the
Max Planck Institute for Coal Research at Mülheim where his
workers discovered new catalyst systems for the production of
high-density polythene. This discovery has been discussed by
Allen[8] who points out that, whilst Ziegler was undoubtedly a
scientist of high repute, he was also very much aware of the
commercial potentialities of his work. Ziegler's name appeared
on nineteen patents before he became Director of the Max
Planck Institute, and the work on new catalysts for olefin
polymerisation was clearly of commercial interest as is shown
by the fact that American oil companies were carrying out
similar work at the same time.

Ziegler's discovery of new catalyst systems for polyethylene
was made in 1954 and the first published details in the form of
Belgian patents appeared in 1955. Natta at the Milan Poly-
technic was able to use Ziegler's new catalysts in 1954 and by
September 1954 he was in a position to inform American
scientists at an international symposium that he had prepared
new polymers, including a new form of polypropylene.[9]
Natta's patents filed in 1954 and published in 1955 were assigned to the Italian firm, Montecatini Società Generale, who
were producing commercial quantities of the new polypropylene
by 1957. In 1959, referring to his discovery, Natta wrote,[9]
'It is perhaps the first time in the history of macromolecular
chemistry that a scientific discovery has been followed so
rapidly by such a vast amount of research in scientific and
industrial laboratories'.

One reason for the short time-lag between the work of
Ziegler and Natta and the manufacture by Montecatini of a
new polymer could be that Ziegler and Natta's work had
obvious commercial implications even before it was successful.
A considerable amount of industrial research aimed at new

methods of polymerising ethylene and propylene was already being carried out (Fontana, for example, produced a high molecular-weight polypropylene in 1952) so that there was considerable industrial interest in the new catalysts and new polymers produced by them.

Before highly oriented and easily fibrillatable polypropylene became commercially available for Plasticisers' experiments, various problems had to be solved by scientists working in industry. For example, the poor heat and light stability of polypropylene required the development of stabilising chemicals which could be added to the polymer. At one stage further away from 'scientific' discovery, there were processing problems, i.e. problems in the conversion of the new polymer to saleable articles, and it was in this area that the Slack brothers with no academic qualifications were able to make their contribution, enabling this country to profit from discoveries made abroad. Thus in this case, the route from scientific discovery to innovation is a complex one involving different groups of workers with different objectives that happened to overlap.

(ii) *New products and new processes*. Innovations are sometimes divided into two categories, new product and new process. This study illustrates how the division is, in many instances, a meaningless one. Ziegler discovered a new process for making polymers but the new process led to new products. Similarly, Plasticisers developed a new process, fibrillation, but the results of fibrillation can be described as new products. Part of Plasticisers' success lay in the simplicity of their new process. Other workers attempting to use fibrillation developed more complex processes. Workers at the Shirley Institute (the Cotton, Silk and Man-Made Fibres Research Association), for example, developed the 'Polystress' process for the conversion of polypropylene film into a fibrous network with a texture similar to conventional continuous filament yarn. From their previous experience of resistance to innovation, the workers at the Shirley Institute thought that fibrillated products would have to resemble existing products if they were to succeed commercially. They succeeded in this aim with the Polystress process but the machinery required turned out to be somewhat complex with the result that no industrial concern thought it worth while to develop. Plasticisers eventually bought the Shirley patents for Polystress just in case the process turned out to be useful.

The simple process developed by Plasticisers resulted in a product quite unlike anything else in appearance, being composed of much coarser fibrils than the products of other workers' developments. Despite its unusual appearance, British Ropes was able to sell 'Formula S' baler twine in large amounts, mainly because of its low cost which reflects the simplicity of Plasticisers' process.

The failure to take sufficient account of the fact that the cost of the process is a vital property of a new product has prevented another group of workers from achieving commercial success. Starting in 1953, a Danish research team led by Ole-Bendt Rasmussen has carried out extensive studies on the splitting behaviour of polyolefin films. Several patents have been filed and they are notable for the ingenious ways which have been devised for obtaining fibres and non-woven fabrics. However, the novel processes are expensive and, in the words of a textile writer[19] in 1964, 'Rasmussen made his first invention in the field ten years ago . . . ten years is a long time and it is still only an experimental project but there have been very big difficulties both technical and economic, especially the latter'. Some of the Rasmussen patent rights have now been bought by Phillips Petroleum Company and it is possible that eventually Rasmussen's work will be utilised.

(iii) *Innovation by a small firm*. This case study illustrates one advantage of a small family firm in that the short-term satisfaction of shareholders can be replaced as a management objective by the conviction that a new idea will eventually prove financially successful. Such firms, however, are often hampered by a shortage of capital. The Government-financed National Research and Development Corporation (N.R.D.C.), which was set up to help in this type of situation, was found by Plasticisers to be of little help. The length of time required for a decision to be reached and the share in any profits required turned Plasticisers against N.R.D.C. The agreement with British Ropes enabled Plasticisers to achieve sales of machinery with a corresponding loss of control over the situation which is one of the penalties that small firms often have to suffer in return for support. There are now in existence certain finance companies specialising in venture capital for new technology, and Plasticisers have been able to obtain financial assistance with the minimum of interference in their control. It will be interesting to see if their second generation of fibrillated products is a success.

References

1. British Patent 479,202 to Johnson and I. G. Farben (1936).
2. Ford and Reynolds, *Shirley Institute Bulletin,* **38** 24 (1965).
3. D. Martin, a director of Plasticisers Ltd, interview, 1968.
4. Plasticisers Ltd, Press release on occasion of Queen's Award, Apr 1967.
5. P. Slack, private communication, 1969.
6. British Ropes. newspaper, *Strands,* Aug 1968.
7. Rees, British Ropes Ltd, private communication, 1968.
8. J. A. Allen, *Studies in Innovation in the Steel and Chemical Industries* (Manchester U.P., 1967).
9. Natta, foreword to *Polymer Reviews,* **2** (1959).
10. *Text. Inst. and Ind.,* **2** 258 (1964).

31 RENOLD GROUP (J. HOLROYD): ROTOR MILLING MACHINES

The award to the Renold Group in 1967 was for the development by its wholly owned subsidiary, John Holroyd Ltd, of a method for speeding the production and improving the quality of rotors for compressors and similar equipment. This case study shows how an interesting triangular relationship was established between Holroyds as manufacturers of rotor milling machines, Svenska Rotor Maskiner Aktiebolag (S.R.M.) as holders of the patents covering the design of the rotors, and various users of the rotors such as Howdens and Holmans in the United Kingdom as well as many overseas licensees.

(a) Rotary compressors

Many hundreds of designs of pumps and compressors have been made and published in the technical literature in response to the needs of industrial and domestic users. The capacities of the

pumps and compressors may vary over a range of several orders
of magnitude but they may be divided into three general
classes: reciprocating units where the compressing or pumping
element moves linearly; centrifugal units where the rotary
movement of the pumping element imparts energy to the
pumped fluid by giving it a motion relative to the element; and
rotary units with a combination of the characteristics of the
previous two classes. Obviously these are very broad categories
and many units have been built with a unique combination of
features that can only be adequately described by reference to
their details. Suffice it to say that from the 1930s to the present
day the rotary class of compressors and pumps with some of the
characteristics of centrifugal and reciprocating units came into
increasing prominence. They were needed not only to supplement the conventional designs of pumps and compressors where
these were proving inadequate for the service conditions involved, but also to provide new answers to problems such as
noise level acceptability, mechanical reliability and freedom
from contamination that were being posed by the emergence
of new industries such as those concerned with air-driven tools,
spraying equipment and supercharged diesel engines.

Two problems were evident to the pump/compressor industry.
The first was to design new pumps/compressors with characteristics matched to the needs of the user. The second was to
develop the methods whereby these designs on paper could be
converted into usable hardware. The first of these was effectively
an exercise in geometrical ingenuity – seeing how enclosed
forms rotating in space could be made to pump or compress
fluids with the maximum possible economy of effort. The
second required no less ingenuity, but of a different kind, for
here the problem was one of accurately reproducing in metal
many thousands of times the designs as they appeared on paper.
As might be expected, the simplest designs of rotary unit were
not as efficient or as suitable as many applications required,
but they were relatively easily produced and so they became
established in the engineering industry. Prominent among these
were gear pumps[1] which, because they combined the continuous-
discharge characteristic of centrifugal pumps with the positive-
displacement characteristic of reciprocating pumps, became
increasingly important in industry. They included pumps using

conventional gear wheels of either the spur or herringbone type to transport the fluid and they also included a family of lobed pumps. These resemble conventional gear pumps in that rotating 'teeth' mesh and in so doing transport the fluid. But they differ in that the 'teeth' on the rotors differ from each other. The most common of this type of pump is the Roots design[2] which typically has two or three 'teeth' per rotor, the lobed rotor resembling half-cylinders placed axially on a central cylinder, and the channelled rotor having corresponding segments removed from a central cylinder to accept the lobes. This design was simple to produce but had a rather poor efficiency[3] and so screw pumps with helically shaped lobes were developed.

In 1935 a patent was granted to Aktiebolaget Milo, a Swedish joint-stock company of Stockholm, for the design of a 'rotary engine' such that it had 'for its object to form the teeth or threads of the rotors in such a manner that the leakage area connecting the high-pressure side with the low-pressure side and formed by the clearances [between the rotors and between the rotors and stationary casing] is kept as small as possible and that the rotors may be as short as possible'.[4]

So a new form of compressor had been designed and it was clearly interesting in theory. The problem was to produce it in practice, for the design called for a helical lobe and channel to be machined on male and female rotors and with the sides of the lobes and channels to be accurately machined such that clearances of the order of one-thousandth of an inch could be maintained between the rotors (which could be over 30 in. in diameter) as they rotated at speeds of over 2,000 r.p.m. Further, the shapes of the channel or lobe did not correspond to standard geometrical figures but rather had to be generated to ensure that a minimum leakage would take place between the high-pressure and low-pressure zones of the compressor. The complete patent was accepted and published in Britain in April 1937. But little more was heard of the design in the technical Press for two reasons: the difficulty of manufacturing rotors accurately so that the necessary clearances could be maintained, and the problem of ensuring that the design of the exit and entry ports would comply with the fluid flow in the pumps so as to minimise energy losses.

In the 1940s further attempts were made to popularise the screw-type pump. In America a new design with improved rotor forms and ports was patented and demonstrated.[5] In Sweden, where air compressors were particularly important because of the amount of rock drilling done in the country, intensive effort was being devoted to the screw pump development. In 1943 the Elliott-Lysholm compressor was developed[6] and by 1944 thirty of these units had been built.[3] In 1946 Aktiebolaget Ljungströms Angturbin published a follow-on patent[7] to the earlier patent[4] of 1935 to which they had purchased the rights. But the performance of screw compressors and their prospects were clearly being limited by the difficulties of producing the rotors to sufficient accuracy.

Aktiebolaget Ljungströms (later known as Svenska Rotor Maskiner AB, S.R.M.), the Swedish firm which held the principal patent rights on rotary screw compressor design, granted licences in the late 1940s to a range of firms throughout the world to produce their compressors – if they could. One such licence was granted to the firm James Howden and Co. Ltd of Glasgow, who used shaping machines with profile control on the cutting tool to generate the rotors. This method of production was extremeley expensive in machining time and in its developed form it took over 200 hours of machining to produce each rotor. Because of the expense involved in this operation, together with the low production rates achieved, the rotary screw compressors appeared to be limited in application to aircraft, etc. – applications where the smooth delivery, large capacity and lightweight construction of the compressors would compensate for their high cost.

(b) The contribution of John Holroyd and Co. Ltd[8,9]

In 1950 two directors of the firm John Holroyd and Co. Ltd, Mr F. H. Stott and Dr H. Walker, went to Stockholm to assist their Swedish agents to mount a display of the firm's worm gear speed-reduction gearboxes. It was an annual event for Stott and Walker to take a week's golfing holiday together and that year they had decided to combine their holiday with the opportunity of fostering their sales in Sweden.

Holroyds had long been established in the worm gear

production business. The firm had been set up in 1871 by John
Holroyd to manufacture sewing and knitting machines. The
details of the company's development have been outlined in a
recent book.[10] By 1888 the firm was engaged in the production
of a wide range of machine tools including a screw cutting
machine, later developed into a thread milling machine by one
of the firm's three Managing Directors. The success of this
machine led to the development and patenting in 1897 of a
long-thread milling machine by Harry Liebert and it was soon
followed by other designs which led the way to the development
'of machines specifically designed for worm gear milling. These
were supplemented in 1904 by others for cutting wormwheel
teeth; from 1906 gear cutting appears as a distinct production
category. Then came machines for grinding worm threads after
hardening'.[19] As the firm developed, three main product lines
became evident: machine tools; worm and other gears; and
casting non-ferrous metals, mainly bronze.

In 1936 the board of John Holroyd and Co. Ltd (by then a
public company) consisted of E. O. Liebert (Chairman and
Managing Director), C. Meek and G. W. Taylor (Joint General
Managers) and F. T. Stott (Works Director and Chief Engineer).
'This board remained unchanged for twenty-eight years,
although the status of its members altered, as all four became
Managing Directors. Also on the staff was Mr Harry Walker,
an engineer in charge of gear design, who was to become, as
Dr Walker, well known and highly regarded in the gear world
as one of the foremost authorities on worm gears'.[10] Stott was
'an able engineer who joined the company as a boy in 1914 and
who returned after the First World War, during which he had
a distinguished career in the Royal Flying Corps'.[10] By the end
of the 1940s Holroyds had built up a reputation for the design
and manufacture of special machine tools, and in 1945 had over
200 of these in Fords of Dagenham alone, in addition to their
business as one of the two principal manufacturers of worm
gears in Britain. It also had licensees for the production of its
gear wheels, among them the American firm Delroyd, utilising
the specialist skills which Holroyds had developed over many
years in the field of gear cutting and particularly with respect
to the cutting of worm gears. To this expertise Walker and Stott
had contributed significantly, and a patent[11] of 1946 shows a

novel method for precision cutting or grinding of worm gears.

So, when Stott and Walker went to Stockholm in 1950, they took, in addition to their golf clubs, a great deal of experience in the production of helical-shaped forms on rotors – just the kind of expertise that for so long had been lacking in the efforts to produce rotary screw compressors. Stott had been asked by the firm's Swedish agents only a few weeks before the trade fair if the firm could assist in producing the rotors for screw compressors, but he had rejected the request on the basis that the problem (on paper) looked far too complex. Indeed the problem was complex, but on their way back from the golf course one day, one of the agents pointed out the firm Aktiebolaget Ljungströms (now S.R.M.) and they called on the designer responsible for the rotary screw compressor project, H. Nilson. He had a wooden model of one of the compressors on his desk, and after hearing of their efforts to produce the rotors using shaping machines, Stott offered to try to produce a pair of rotors using milling techniques based on their existing machines for producing worm gears. Stott asked for and obtained short sections of the wooden rotors to experiment with. Back in Milnrow (Rochdale) they mounted the rotor segments in one of their standard worm gear milling machines, replaced the normal steel cutting blade with a soft lead metal blade, and by driving the cutting head (together with the lead blade) over the rotor segment obtained an impression of the shape the cutting blade would have to have if it were to cut the required rotor shape. In fact it was a process reminiscent of a dentist taking an impression for a set of false teeth. Using the shape of the lead former the first set of cutters were made in tool steel and these were in turn used to cut the first set of rotors.

Nilson came to Rochdale, and he and Stott, together with the toolroom foreman, Mr Rankine, modified the tools for clearance. The next batch of rotors was clearly superior to their first efforts. But it was evident that if they wished to obtain really high-quality products they would have no option but to design a machine specifically for milling rotors. The machine used for the initial proving trials had been simply a rather robust unit designed for removing the major portion of the metal to be machined from worm gears prior to their receiving a final machining and grinding on more precise (but more delicate) machines. Nilson specified a machine size capable of generating 630-mm diameter rotors, but Stott would only guarantee 350-mm rotors not to need hand finishing. In fact the first set produced 'were 420 mm diameter and the rotors came out "spot on"'.[8]

The new milling machine used the accumulated experience of Holroyds in the design and construction of worm gear milling machines. Unless the lobes and channels of the rotors could be located precisely on the perimeter of the rotors, and unless they had precisely matched helical positions on the rotor, not only would leakage of the compressed fluid occur but also the rotors would not even rotate freely. Any errors in lead (giving the rotor lobes or channels their helical form) or divide (giving the lobes or channels their positions on the perimeter of each rotor) would have to be compensated for by increased clearance between the rotors. This could not be tolerated if a high-efficiency unit was to be built. The worm gear milling machines designed by Holroyds had also to be extremely accurate in 'lead' and 'divide' operations. Any 'backlash' or slackness in the drive mechanism would be unacceptable in manufacturing accurate worm gears and so the company had readily available the necessary expertise to design and build rotor milling machines of the required accuracy.

Once the problems of 'lead' and 'divide' accuracy in the rotor milling operation had been resolved by building special-purpose machines, the next problem which was evident was the need to make the cutters for producing the rotors more accurate. The cutting tools, whether for male or female rotors, would only generate the correct shape of rotor if the tool shape was correct within the available clearance limits and the tool was mounted so that it cut at the correct distance from the rotor axis. Obviously the tool would wear in service and so it was necessary to develop a highly reliable method of generating new cutting tools and of renewing old tools that had become worn. The method that Stott had used to prove the feasibility of producing rotors could not be relied upon in a production system although it had given satisfactory results in the trial stage.

So it was necessary to develop a new method of cutting the

o

tool blades accurately and economically. This problem was solved by Stott, Rankine and Mr Ternent (a special assistant to Stott) when they cut a thin section of hardened steel to the exact profile of the required rotor cross-section. They then mounted a bronze blade in the milling machine and cut it to shape by passing it over the stationary hardened steel section. This bronze blade then became the master profile against which all subsequent blade forms (of that type and size of rotor) were to be checked. Each master blade is used to generate one template in steel and this in turn is used to generate all further cutting blades. To copy the master profile it was necessary to design and build a special tool grinding machine. This consists of a stylus which copies the shape of the master profile, a grinding wheel to machine the tool blades to the correct shape, a grinding wheel trimming device to ensure that the wheel was of exactly the same shape as the stylus, and a shaft and indexing mechanism to ensure that each blade on the cutting head would be held in precisely the correct position whilst it was being sharpened. This machine

is designed for the accurate reproduction of any desired profile shape on rotary milling cutters and for the sharpening of the cutting edges to give a constant angle of cutting relief at all points on the profile. The cutter so produced is much superior to the older type of form-relieved cutter due to the constant angle of relief at all parts of the cutting profiles. A form-relieved cutter suffers from the defects that the cutting relief varies according to the shape of the profile, and under certain conditions suffers from the defects that the cutting relief on some parts of the profile may be as small as to cause 'rubbing back' of the cutting edges and the necessity for frequently sharpening with the removal of excessive metal and a short life for the cutter.[12]

A further development which was of significance in ensuring that the desired accuracy could be obtained in machining the rotors concerned the number of blades fitted to the cutters of the milling machine. In an American milling machine textbook[13] published in 1951 it was stated that

face mills are made in light duty and heavy duty types. The light duty face mill has a lighter body and larger number of teeth, being used principally for finishing cuts. The heavy duty face mill has a stronger, heavier body and fewer blades and is employed for heavy roughing cuts. The general practice is to obtain the desired quality of finish and accuracy by one cut whenever practicable. This eliminates the additional time involved when the operation is performed in two cuts known as roughing and finishing cuts. The combining of two cuts in one is sometimes applied to face mills having blades tipped with sintered carbide material. The roughing and finishing blades are mounted on the same body, with a limited number of finishing blades set to a smaller diameter and extending slightly farther from the face than the roughing blades. . . . The object of the inserted tooth construction is economy in first cost as well as in maintenance, [making it] possible to quickly replace the teeth or adjust them to compensate for wear.[13]

This latter advantage had been evident for some time and it is interesting to note that the Ingersoll Machine Co. of the United States were using adjustable cutters in the 1890s[14] and that these were in common use in the 1920s.[15] But Holroyds reversed the trend in number of blades. Their

finishing cutters, however, are entirely special in that the number of blades provided is considerably less than would be incorporated in an orthodox cutter. Moreover, the arrangement of the blade is peculiar to the process. It is contended by Holroyd that with any large multi-blade milling cutter, even though it has been carefully sharpened, there are minute differences in the heights of the blades. Therefore, while each blade removes approximately the same amount of metal from the work during the feed applied at each revolution, the highest blade performs a final shaving action and provides the finished size.[16]

So Holroyds adopted a finishing cutter which had few blades rather than many as was the normal convention, and to one of these blades it gave the final 'shaving' duty. Generally this final 'shaving' blade is mounted mid-way between two of the other blades which are mounted equally spaced on the periphery of the finishing cutter. The master template is used for checking the profile of this final 'shaving' blade and once it has become sufficiently worn it is relegated to a semi-finishing position on the cutter and is replaced by a freshly ground and set blade.

After the design and construction of the first specialist rotor milling machine and its 'partner', the cutter profile grinding machine, and after the initial production methods had been established, Stott took a set of the rotors to Sweden in 1955 and demonstrated them to the assembled S.R.M. licensees from the United Kingdom, the United States, Japan, Germany, Sweden, etc. The process was received enthusiastically by all the licensees except for one firm who first tried to build a rotor milling machine according to the principles outlined by Stott in his demonstration but later scrapped their efforts and purchased a Holroyd machine instead. Holroyds then started production of a series of machines all built to the same basic design but with different rotor capacities. These culminated in a machine capable of producing 32-in. diameter rotors as used for blowers in steelworks and wind tunnels. These rotors can be produced accurately enough to ensure that no hand finishing of the rotor forms is necessary and by 1962 the first of these 32-in. rotor milling machines was delivered to an S.R.M. licensee in Japan.

In 1964 the company announced its first fully automatic rotor milling machine at the London Machine Tool Exhibition. This eliminated the necessity for hand setting of the position of the rotor relative to the cutter for each machining stage and so increased both the productivity and accuracy of the machining process. In 1966 the firm opened a new £250,000 rotor production facility to enable them to increase their output of finished rotors which by that time were being used not only for pumps and compressors but also for high-capacity in-line flow meters.

Another feature of the production process developed by Holroyds is the use of a fly tool unit for finishing the female rotors. This consists of a single-point tool mounted on a wheel which is spun at high speed with its axis parallel to the rotor axis and at a precisely set distance from it. This therefore cuts the circular channel form in the female rotor characteristic of the S.R.M. rotor design. Since this simple operation can be performed with extreme accuracy (owing to the use of a high cutting speed, very fine feed rate and accurate milling machine), it has the desirable effect of leaving the tolerance required of the machining operation to produce the lobes of the male rotor much less stringent – a welcome development in particular for the 32-in. diameter rotors. Further developments by S.R.M. have resulted in the design of a two-stage compressor with two rotors machined on the same shaft.[17] These too are being produced using Holroyd rotor milling machines.

In 1964 Renold Chains Ltd acquired all the share capital of John Holroyd and Co. Ltd and in 1967 the expanded group became Renold Ltd. In 1967 F. T. Stott retired from Holroyds after fifty-three years' service, twenty-eight of them as a member of the company's board.

(c) Comments

The innovation made by John Holroyd and Co. was to devise a production technique for a design which already existed but which could not be used commercially on a large scale because of the complexity of the problem of producing the helical screw rotors. S.R.M. had licensed the design to companies in the United States, the United Kingdom, Japan, Germany and Sweden, but it nevertheless remained virtually unexploited.

Holroyds became involved through a contact made by Stott, a Managing Director, and Walker, the Technical Director (Gears), while on a combined business trip and golfing holiday in Sweden in 1950. Although they had earlier rejected the problem as too complex, they found after inspecting a model of the screw rotors that it could be tackled by relatively straightforward extension of the considerable experience the firm already possessed in the milling of worm gears provided they could devise a method for producing the cutters of the milling machine to the required shape. They did this by an ingenious method and the first rotors produced were extremely encouraging. The method was further developed to give greater accuracy in the cutters, and in the process a new cutter grinding machine was developed. The company's experience in building worm gear milling machines with virtually no backlash (i.e. undesired movement between the machine's parts) played an important part in ensuring that the rotors would be sufficiently accurate so as not to require further machining. The design of the cutters too was altered, reversing a previously established practice, to produce more accurate forms on the machined rotors. In 1955 Holroyds demonstrated its machines to the S.R.M. licensees and during the next ten years all rotors built to S.R.M. designs were produced on Holroyd machines. Under the terms of their agreement with S.R.M., Holroyds sold the completed rotors they made to S.R.M. licensees whilst the licensees were awaiting the delivery of rotor milling machines from Holroyds. This had advantages both for Holroyd and for S.R.M. Holroyd got profitable production experience while they were still developing their milling machines; S.R.M.'s licensees got the necessary opportunity to build up the market for compressors, etc., produced to S.R.M. designs without having to invest in the expensive milling machine facility until the market would justify it.

References

1. F. A. Kristal and F. A. Annett, *Pumps* (McGraw-Hill, London, 1953) p. 137.
2. P. H. Schweitzer, *Scavenging of Two-stroke Cycle Diesel Engines* (Macmillan, New York, 1949) p. 112.
3. E. W. F. Feller, *Air Compressors* (McGraw-Hill, New York, 1944) p. 308.
4. Aktiebolaget Milo, 'Improvements in Rotary Engines', British Patent No. 464,476. Applied for (U.K.) 16 Oct 1935, published 16 Apr 1937.
5. J. E. Whitfield, 'Improvements in and relating to Rotary Screw Pumps, Motors, and Like Fluid Devices', British Patent No. 558,740. Applied for (U.K.) 22 Apr 1942, published 19 Jan 1944.
6. W. A. Wilson and J. Crocker, 'Fundamentals of the Elliott-Lysholm Compressor', ASME Paper 45:A:45, *Mech. Eng.*, **68** 514 (1946).
7. G. K. W. Boestad and Aktiebolaget Ljungströms Angturbin, 'Improvements in or relating to Rotary compressors or engines', British Patent No. 611,258, applied for 25 April 1946, published 27 Oct 1948.
8. F. T. Stott, H. Walker, G. W. Taylor, (previous directors of John Holroyd and Co. Ltd) and C. J. Meek (Managing Director, John Holroyd and Co. Ltd), interviewed by W. G. Evans, 23 May 1969.
9. John Holroyd and Co. Ltd, 'Statement in Support of Application for the Queen's Award to Industry for Technological Innovation'. (1967).
10. B. H. Tripp, *Renold Ltd 1956–1967* (Allen & Unwin, London, 1969).
11. H. Walker and John Holroyd and Co. Ltd, 'Improvements in and relating to the Cutting or Grinding of Worms, Spiral Gears and Helical Gears', British Patent No. 574,988. Applied for 10 Feb 1944, published 29 Jan 1946.
12. John Holroyd and Co. Ltd, 'Profile Grinding and Sharpening Machine for Form Milling Cutters', Publication No. 13T (1969).
13. *A Treatise on Milling and Milling Machines*, 3rd ed. (Cincinnati Milling Machine Co., 1951) p. 65.
14. 'Adjustable Cutters Produced by Ingersoll Machine Co. of U.S.A.', *The Engineer*, 3 Feb 1893.

15. T. R. Shaw, *The Mechanisms of Machine Tools* (Oxford Technical Publications, 1923) p. 66.
16. J. J. Marklew, 'Producing Helical Rotors for Compressors', *Machinery and Production Engineering*, 13 Dec 1967.
17. Svenska Rotor Maskiner Aktiebolag, 'Improvements in Rotary Devices having Intermeshing Screw Rotors', British Patent No. 851,262. Applied for (U.K.) 27 Feb 1959, published 12 Oct 1960.

32 SANDERS AND FORSTER: STRUCTURAL STEELWORK

In the eleven years prior to their receipt of a Queen's Award for Innovation in 1966, Sanders and Forster made a series of improvements in the design and production of steel-framed buildings so that they could produce a wide range of industrial buildings at a lower cost despite the increases in costs of steel and labour during that time. Sanders and Forster, a medium-sized company with some 500 employees, is a member of the Chamberlain Group.

This case provides an example of innovation which was made possible by research carried out at a university being developed commercially because of specific market factors. It utilised new materials and new techniques from outside the firm.

(a) Steel-framed buildings and J. F. Baker[1,2,3,]

Most of the early development of steel-framed buildings took place in America, especially in Chicago where a large fire in 1871 cleared the way for multi-storey redevelopment of the town centre.

In 1929 the majority of the companies in the British structural steel fabrication industry formed the British Constructional Steelwork Association, which had the objectives of a trade association. The newly formed association gave financial aid to the setting up of the Steel Structures Research Committee with the aims of simplifying and standardising the procedures for using structural steelwork. Research on steel structures was started at the Building Research Station, where one of the people concerned was J. F. Baker. In 1933 Baker left the Building Research Station to become Professor of Engineering at Bristol University where he continued his research for the Steel Structures Research Committee.

During the 1930s attempts were made to provide a rational basis for the design of structural steelwork and experiments were carried out on full-scale buildings during their construction. These experiments confirmed laboratory findings that the beam-to-column connection in steel structures was not a simple free joint and that bending moments could be transferred to adjoining parts of the structure.

Baker developed a theory in which the load on each individual component is taken up by the structure as a whole. The degree of rigidity of the joints has a considerable effect on the distribution of the load and the calculations involved were complex. With a team of workers, first at Bristol University and then at Cambridge, Baker worked at the theoretical and experimental development of the plastic design method based on the concept that the design criterion for a steel structure should not be a safe stress at a particular point but the collapse load of the whole structure.

Baker's work at Cambridge from 1947 onwards was supported by the British Welding Research Association, as welded joints were an essential part of the plastic design method, being almost rigid and therefore allowing the transfer of stresses throughout the structure. This transfer of stresses meant that less steel was needed than if each part of the structure was considered separately. Welding techniques were improved during the 1939–45 war and the construction of welded hollow-steel frames would have been possible after the war, but the shortage of steel and the reluctance of the industry to use on-site welding meant that only a small number of such constructions took place. In 1955 it was claimed[2] that fourteen buildings designed by Baker's plastic design method had been erected.

(b) Sanders and Forster Ltd[4, 5, 6, 7]

Sanders and Forster Ltd, founded in 1888, was acquired by the Chamberlain Group Ltd which was founded by Mr A. G.

Chamberlain. In 1952 A. G. Chamberlain had retired from the Group board and his brother W. E. Chamberlain, then aged fifty, was Chairman of the Chamberlain Group and also Managing Director of subsidiary companies, including Sanders and Forster. W. E. Chamberlain realised that there was no one to take over his unique position of control. He advertised for three new directors and as a result M. H. Briggs joined the board of Sanders and Forster. Briggs, a graduate engineer, had varied experience in structural work and some patents to his credit. He was interested in the possibilities of industrialised building based on the use of standardised components and designs, though at that time (1952) the majority of steel structures were 'one off' constructions.

Briggs's interest in standard buildings was shared by W. E. Chamberlain. They had both experimented in this area in the 1930s when Briggs obtained a patent for a type of 'prefab' and the Chamberlain Group had put up almost standard factories for letting. Briggs looked at the possibilities of a standard building system based on aluminium instead of steel, which was in short supply; the economics of such systems, however, looked unfavourable. The Sanders and Forster design team were asked to produce designs for a standard building system based on steel but no satisfactory scheme emerged.

In 1953 Briggs read about Baker's plastic design method in the *Journal of the Institute of Structural Engineers*, which contained the Presidential Address of Lt.-Col. R. F. Galbraith describing a steel-framed storage building designed for the War Office in accordance with Baker's theory. Briggs was impressed by the saving of steel obtained by Baker's design but the large amount of site welding was considered by Briggs to be expensive and somewhat impracticable. Briggs then adapted Baker's design for use in an industrialised system with no site welding. He did this by simplifying some of the structural components, reducing the amount of welding required and replacing on-site welding by shop-welded assemblies which could be bolted to the other components on site.

The new design was first used for a timber warehouse which had already been designed and priced as an orthodox steel-framed building. The construction of the warehouse to the new design proved to be satisfactory and a 30 per cent saving on the original price of the steel frame was obtained.

An unforeseen advantage of the new system was that it enabled Sanders and Forster to purchase more steel than had previously been possible. Although the licensing system for steel had ended, the steel rolling mills were operating an unofficial quota system based on previous purchases of steel. This created difficulties for Sanders and Forster, who were attempting to expand their usage of steel. The steel mills were, however, prepared to supply any quantity of steel if they were paid for carrying out simple fabrication themselves. The simple stanchions and purlins designed by Briggs could be supplied by the steel mills in a fully fabricated state. By allowing the steel mills to do this, Sanders and Forster were left with additional capacity to fabricate the more complicated rafters and they were able to obtain all the steel that they required.

In 1954 Sanders and Forster introduced their range of standard steel buildings based on Briggs's modification of Baker's design. An advertising campaign stressed the price reduction possible with the new design and orders were obtained for factories, warehouses and other kinds of single-storey buildings.

Having started with a new method of design, Sanders and Forster then proceeded to develop a new method of production geared to the mass production of components for a system instead of the 'one off' method. The first step in this direction was a mechanised transfer plant for the semi-automatic sawing and drilling of structural sections. This system allows steel sections to be moved from machine to machine without man-handling or the use of an overhead crane; it uses four men where about twenty would be required using traditional methods. The Sanders and Forster transfer plant, claimed in its publicity to be 'the second plant of its type in the world', was an improvement on the first such plant, which had been built in 1954 by the British firm Boulton and Paul Ltd. Sanders and Forster were assisted in their development through being a member of the Chamberlain Group in that capital was available and another subsidiary, Chamberlain Industries Ltd, was able to manufacture the conveyors required for the transfer plant.

Sanders and Forster continued to improve the efficiency of

their production methods. From the arrival of sections in Sanders and Forster's own rail sidings to the construction of buildings by teams of their own workers, the whole process now resembles a factory production line with a high degree of mechanisation. The efficiency of this process is assisted in the design stage by the use of a computer which provides a minimum cost solution taking into account both the cost of material and labour.

As mentioned in the case of Concrete Ltd (p. 191), industrialised building involves the substitution of temporary labour by capital invested in factory production units, with the attendant uncertainty about continuing demand for the products of the factory. The problem of providing an adequate market has been tackled in this country by Government encouragement to local authorities to make use of industrialised systems. In the late 1950s a consortium of local authorities developed the CLASP system which used a steel framework supplied by Brockhouse Steel Structures Ltd as the basis for schools, old people's homes, etc. In 1960 the War Office decided to carry out most of their own building using the CLASP system and invited tenders for the design and supply of the steel framework. Sanders and Forster redesigned the CLASP frame and made use of a special type of lightweight beam known as an open web joist which Briggs had seen being used in North America. Briggs had purchased special machinery from the United States to manufacture the lightweight beams which substantially reduced the cost of a steel frame. Sanders and Forster gained the War Office contract for the CLASP system and followed this by securing contracts for the second consortium of local authorities, SCOLA, set up in 1962 and the third consortium, SEAC, formed in 1964. They have also produced a special system for the Co-operative Wholesale Society.

Industrialised building must have a continuing market to be economically successful. Sanders and Forster have guarded against the possible effects of a 'stop–go' economic policy by building up a considerable export market to some sixty countries including Canada, Sweden and West Germany. The low price of the Sanders and Forster systems which has enabled them to compete in the world market is partly a result of their ability to use new ideas. Sanders and Forster have no separate research and development department; instead they have successfully utilised other people's advances. In addition to Baker's theory, the transfer plant, the open web joist and the use of a computer for structural design, all of which have been referred to above, Sanders and Forster have pioneered the British manufacture of the Litzka lightweight beam invented by the German engineer, F. Litzka. They have also combined the Wheelabrator shot blasting system with the spray coating of a zinc-rich epoxy primer to give a new way of attacking the corrosion problem.

In the development of new ideas, outside help has been utilised when required. The computer programme for the lowest-cost design was developed by A. G. Crowe and Partners in co-operation with Ferranti Ltd, and the use of Litzka beams involved another firm of consulting engineers, W. S. Atkins and Partners, who developed a computer program for the preparation of the detailed design tables (German tables could not be used as German sections differ from the British).

From their initial use of Baker's theory in 1954 to their Queen's Award in 1966, Sanders and Forster had supplied over 2,000 industrial and other buildings totalling over 20 million sq. ft. of floor space, all based on the plastic design theory.

(c) Comments

This case provides one of the few examples in our sample of innovations where an idea from a university was developed and 'applied' by industry. The comparative rarity of such examples might mean that university research has little to do with the needs of modern industry or, alternatively, it might mean that industry lacks awareness of university developments. Neither of these simple formulations does full justice to the present case. Baker's early work did not lead to a practical design method. The plastic design theory[3,8] used by Sanders and Forster was published in stages starting in 1949, so there was a delay of four years before it came to their notice. Although Baker's plastic design reduced the amount of steel required in a structure, the structural steel industry was reluctant to use the new method of design because of the high degree of welding considered to be essential for the provision of the rigid joints on which the design rests. Sanders and Forster were able to use Baker's design by making it part of an industrialised system with welding carried out in a factory where it could be subjected to a degree of control difficult to obtain on a traditional building site. To see Sanders and Forster's success entirely in terms of the application of a new theory is therefore an oversimplification; it can also be seen in terms of the development of industrialised building systems based on factories producing standard components.

Sanders and Forster were attempting to produce a standard building system before Briggs heard of Baker's theory, which was one of several new concepts which Briggs was able to adopt. It is at least possible that Sanders and Forster could have developed an industrialised system without Baker's theory. In saying this, it is not intended to deny that an important and substantial contribution was made by Baker's theory to Sanders and Forster's success. The intention is rather to show that a linear model (see p. 72) of industry applying the results of Baker's work at Bristol and Cambridge Universities is an oversimplification of a complex process which involved the coming together of several separate strands of development. One such strand was the development of a market capable of placing large contracts for an industrialised building system. The post-war rise in the birth rate faced local authorities with the problem of providing more schools. In many places new schools were required rapidly and the ability of industrialised building to reduce construction time helped to overcome reluctance to depart from traditional methods and led to the setting up of the consortia which provided Sanders and Forster with a substantial market.

The lack of a clearly identifiable research and development team illustrates one of the difficulties in answering questions about optimum size of research teams. Sanders and Forster have a design team which works on both routine assignments and the development of new systems. This minimises organisational resistance to the use of outside assistance.

The Chamberlain Group remains in the control of the Chamberlain family, with L. F. Chamberlain, a son of the founder as Chairman. The decision of W. E. Chamberlain to recruit three new Managing Directors for the subsidiary companies is an interesting solution to the problem of management succession in a firm which had been run by one man. Briggs's success with Sanders and Forster seems to be contrary to the requirements of some kinds of management theory which would have suggested the need for a management development plan to ensure the provision of successors to W. E. Chamberlain from within the company.

It is not suggested that one should generalise too much from the Sanders and Forster case, but their own formula for success seems to have been: 'Don't do research and development; don't bother with management development; make use of good people and good ideas from any source.'

References

1. M. Bowley, *The British Building Industry* (Cambridge U.P., 1966).
2. W. F. Cassie and D. W. Cooper, 'Comparisons in Modern Structural Steelwork', *R.I.B.A. Journal*, Apr 1955, p. 231.
3. J. F. Baker, 'A Review of Recent Investigations into the Behaviour of Steel in the Plastic Range', *Institute of Civil Engineers*, Paper 5702 (1949).
4. 'Chamberlain's Structural Company Gains Queen's Award', news release from Chamberlain Public Relations (1966).
5. H. E. Andrews of Sanders and Forster, private communication, 1967.
6. M. H. Briggs, private communications and interview, 1970.
7. 'S. & F. Steel Framed Buildings', trade brochure (1967).
8. J. F. Baker, 'The Design of Steel Frames', *The Structural Engineer*, **27** 135 (1950).

33 SHORT BROTHERS AND HARLAND: SEACAT GUIDED WEAPON SYSTEM

The 'Seacat' missile forms part of a complex integrated guided weapon system and, as such, it is composed of many different subsystems. Short Brothers and Harland (S.B. & H.) at Belfast, Northern Ireland, won a Queen's Award for Technological Innovation in 1967 and subsequent Awards for Export Achievement in the three succeeding years for the Seacat guided weapon system which was developed and sold by the Precision Engineering Division (P.E.D.). S.B. & H. were responsible for the design, development and production of the missile part of the system and of control, guidance and test equipment for use on ship and shore. Other contractors developed specialised subsystems on direct contracts from the Ministry of Supply (as it then was) and the Admiralty. Thus, the warhead fuse was developed by E.M.I. Ltd, the rocket motor by Imperial Metals Industries Ltd, the warhead by the Armament Research and Development Establishment, the launcher by Rose Brothers Ltd and the transmitter by Pye Ltd. The successful integration of these subsystems was carried out by S.B. & H. A land-based version of the Seacat known as Tigercat has also been developed by S.B. & H.

(a) The Seacat missile system

The weapon system consists of *Seacat missiles*, a *launcher* which also carries a *transmitter aerial*, a human operator (the *aimer*), an aiming station (known as a *director*), an *encoder* and a *transmitter*. As soon as the target is detected by the ship's main radar system its position co-ordinates are fed automatically to instruments in front of a control officer standing outside the Seacat director bin. The control officer when directing traverses on to the target bearing indicated on his fire control dials, and the aimer picks up the target in his binoculars. After the missile has been fired, its control surfaces are locked in a neutral position for a short period until it appears within the wide-angle field of vision of the aimer's binoculars along with the target. The aimer then guides the missile manually towards the target by means of joystick control. The joystick is a thumb-operated control incorporated in the right handgrip of the director. The aimer's movements of the joystick to correct the missile's course are converted to electronic signals and fed to the signal shaping unit which encodes the signals prior to transmission to the missile. After being received and decoded by the missile receiving system, the decoded signals are amplified and used to control an electro-hydraulic servomechanism to alter the settings of the two pairs of flight control surfaces in conformity with the aimer's commands.

The Seacat control system is a *radio* command link system with guidance commands transmitted to the missile from a remote source. As such it may be compared with a *wire guide* system used in the comparably-sized Nord Aviation SS-11 and SS-12 series missiles. In these, guidance commands are transmitted to the missile along a wire that is trailed behind the missile as it flies towards it target. The wire guide system was also used in the German X-7 during the Second World War[1] and offers the advantage of a high degree of command signal security but at the expense of mechanical reliability (the wire is prone to snapping if it touches an obstruction or hits the water during flight). The Germans had also developed the HE293, HE294 and HE298 air-to-sea remote controlled 'guided bomb' which, with the exception of the signal shaping unit, had several features similar to the Seacat.

The choice of a radio command link system for the Seacat was the result of a combination of factors. These included (i) the availability of a suitable transmitter and receiving unit that had been developed previously by the Royal Aircraft Establishment (R.A.E.) at Farnborough as part of a general missile development programme; (ii) the difficulties in trials of the 'Malkara' anti-tank missile with failure in its wire guidance system; and (iii) the inherent simplicity of the radio command link system if suitable signal security could be guaranteed. The command link system is inherently the least complex system and therefore most in line with the Seacat's emphasis on simplicity; most of the guidance components are located at the launch site, which means that the missile part of the Seacat system carries the bare minimum of electronics, thereby increasing its reliability and decreasing greatly the cost of the material expended in each shot. The visual link avoids the tendency towards complexity associated with continuous monitoring by radar of both missile and target and it is very much cheaper. The system has a limitation in that behaviour of the missile is controlled from the launching site with consequent degradation in control accuracy as the range increases.

Seacat was designed to provide ships with their own effective defence against low-level aircraft which had succeeded in penetrating a naval unit's outer defence fighters and long-range guided weapons. The vast majority of countries still rely mainly on conventional bombs, torpedoes, air-to-surface rockets and gunfire. 'Analysis of trials with modern high-speed aircraft has shown that, for the majority of such weapons, releases occur within a few thousand yards.'[2] There is little advantage in striving for a medium-range capability as long as it is possible to 'kill' the target efficiently at ranges greater than the attacking aircraft's customary striking distance.

(b) The Short Brothers and Harland innovation[3]

S.B. & H. began in 1898 when the brothers Eustace and Oswald Short set up a small factory under Battersea Bridge for the manufacture of balloons. A few years later, joined by a third brother, Horace, they began to make powered aircraft. They became the first manufacturers of aircraft in Britain when in 1909 they set up production of Wright biplanes under contract from the Wright Brothers. In 1919 the partnership was converted into a limited liability company – Short Brothers (Rochester and Bedford) Ltd – which continued to make substantial contributions to the aircraft industry until it was reorganised as a holding company in 1947.

The evolution of S.B. & H. falls into three distinct phases: 1898–1936, the evolution and operation of Short Brothers (Rochester and Bedford) Ltd; 1936–47, the foundation and consolidation of Short & Harland Ltd as a daughter aircraft and manufacturing unit; 1947 to date, the consolidation of S.B. & H.'s design and technical organisation in Northern Ireland, the exploitation of the undertaking as a balanced design and manufacturing unit and the initiation of a degree of diversification.

The critical years in the company's development occurred in the third phase during the early 1950s when the first Short's aircraft indigenous to Northern Ireland were designed, when the flying boat – Short's traditional product – failed to retain its market and when a guided weapon activity was started in the newly formed Precision Engineering Division (P.E.D.). The P.E.D. was set up in 1952, largely under the direction of Sir Matthew Slattery, about three miles from the aircraft factory, as a form of diversification to undertake the design, development and manufacture of (i) precision engineering products generally; (ii) the special equipment and instrumentation associated with sophisticated aircraft control systems; and (iii) guided weapons and their associated equipment.

From one point of view, diversification into guided weapons

technology was a risky venture because S.B. & H. had no experience in guided weapons systems at all. On the other hand there had been considerable enthusiasm about guided weapons and rocketry throughout both industry and government. This seems to have emerged from the discovery after the Second World War from the German archives of just how much the Germans had accomplished in these fields, and many people in England (as well as in America and Russia) were anxious to pick up where the Germans had left off. About the same time, the Royal Navy was coming to recognise that its standard 40-mm anti-aircraft gun used against close-range air attack would not be able to cope with the targets of the future. It was proposed to replace the gun with a missile but the type of missile required was laid down very stringently because there was a very large number of 40-mm guns – they had been fitted as standard equipment to nearly all the vessels of the Royal Navy and the Royal Merchant Navy – and a missile to replace it had to be cheap, reliable and easy to use. The Navy made it abundantly clear that it was not interested in a scaled-down version of the Seaslug or Bloodhound missiles because it was felt that these systems were too expensive to make and operate and that they required too many highly trained technicians on board ship.

Thus, in 1952, a requirement was issued by the Royal Navy for a small, simple and cheap missile to replace the 40-mm gun. An essential part in the fulfilment of this specification was the choice of a sufficiently accurate guidance system. There are three recognised systems which may be used to allow a missile to seek and home on the target.[4] One way this can be done is by illuminating the target with a radar beam transmitted by the missile (active homing) or by a source outside the missile (semi-active homing) and by fitting a 'seeker' in the missile which enables it to pick up and 'look' towards the target which acts as the source of the reflected radar energy. This system requires that much complicated and expensive equipment be carried within the missile.

Another method is to lock a narrow radar beam to the target and cause the missile to measure its own position in relation to the centre line of the beam and fly as close as possible to this line. This method is normally called beam riding. On each of these systems the missile, once launched, guides itself towards the target.

An alternative system which permits simpler missile equipment can be provided by continuous measurement by radar from the ground or ship of the positions of the target and the missile. This information is fed into a computer which transmits to the missile command information in the form of steering instructions designed to ensure interception of the target. This system is called 'radar command guidance'. The S.B. & H. innovation largely centred around the choice of a variant of radar command and guidance and the design of the appropriate control system.

Among the various ideas considered for the control of the missile was the concept of visual command control, and this system was ultimately selected for further study because of its inherent simplicity. The use of a guidance system of this type had been studied by the Royal Aircraft Establishment (R.A.E.), Farnborough, in 1951 for air-to-air missiles and in 1952 for anti-tank missiles and it had been concluded that a small single weapon using visual command guidance could give very high accuracy when used against targets with small crossing rates (i.e. low angular rates of change of the target sight line). Before any decision could be reached on the use of visual command control for the Navy requirement where the missile would be launched and controlled from a ship and the targets might have high crossing rates, it was deemed necessary to undertake further study.[5] The design and testing of the visual command guidance principle required the parallel development of a control system to manœuvre the missile accurately towards its target. The control system in Seacat converts information from the aimer into the appropriate electronic information which, when transmitted to the missile, provides signals to guide the missile. The heart of the Seacat control system is contained in the 'shaping unit'. This was a development that was necessary if the simplicity requirement was to be met. The basic problem is that, in guiding the missile towards its target, the aimer gives it a lateral acceleration. In order to deduce the missile position from this, two integrations are necessary. But it is common knowledge that any closed-loop control system in which there are two integrations is inherently unstable. The usual approach to remove this difficulty is to introduce gyroscopes in the missile to damp the instructions from the aimer so as to keep the missile on the appropriate course. S.B. & H. developed an electronic device to replace these gyroscopes and which could be located in the ground equipment. This design introduced a saving of some £400 per missile.[6]

S.B. & H. obtained a Ministry of Supply contract and launched a considerable campaign to recruit the required manpower to Northern Ireland. The development took place in three stages:

(i) *Laboratory methods*. Simulator methods were developed in association with the staff of R.A.E. which permitted the design team to determine the feasible operational limits of the missile and human controller. (The simulator is a computer and optical complex which attempts to perform the major functions eventually to be required by the missile and human operator.)

(ii) *Experiments in the field*. This was done with real aircraft but using a dot light in the field of vision of the binoculars to simulate the action of the missile. R. J. A. Paul, Head of Electronics for the Seacat project, has recalled that 'we used the fast-flying aircraft at Farnborough Air Show as targets for our simulated missile, which was housed in a temporary erection at the end of the runway'.[7]

(iii) *Missile development*. When it was apparent that the concept of visual command guidance was acceptable, R.A.E. handed over development to S.B. & H.

Early in 1956 it seemed that a visual command system had a reasonable chance of meeting the requirements of the Royal Navy, with which S.B. & H. had since 1952 been in close contact. A separate design study of a weapon using this form of control was started by the firm as a private venture (S.B. & H. Project SXA7) and, with the active encouragement of Earl Mountbatten of Burma (then First Sea Lord), proposals for this weapon were issued in June 1956. In July 1956, when a panel was set up by the Royal Navy to study a complete weapon system for naval use, they were able to base their considerations on the SXA7 proposal and a most favourable report was issued by them in October 1956.

Work on the detailed design of the SXA7 missile continued and was stepped up as more and more evidence in support of the SXA7 emerged. By September 1957, when the SXA7 proposal was reviewed jointly by the Ministry of Supply and Admiralty representatives, S.B. & H. were able to report that orders had been placed with E.M.I. for the design of a fuse, with I.C.I., Ardeer, for the charge of the blast-start gyro, with Bristol Aerojet Ltd for the motor casing and with I.C.I., Summerfield, for the motor charges. Contact with the Armaments Research and Development Establishment on the warhead and with the United States for the thermal battery had also been established so that work was proceeding on all aspects of the weapon.

From September 1957 until February 1958 discussions with Ministry of Supply and Admiralty Departments continued on a semi-official basis, culminating in June 1958 in a contract for S.B. & H. covering the design and development of a ship-launched guided weapon and ancillary equipment.

The first Seacat, as the SXA7 missile was later called, was fired at Aberporth in March 1959 and in December 1960 a Seacat was launched from the trial ship H.M.S. *Decoy*. The first operational ship to fire a Seacat was H.M.S. *Barrosa* in July 1962. Just four years after the development contract had been placed, the Seacat guided weapons entered service with the Royal Navy. The first foreign ship to fire a Seacat, the Royal Swedish Navy ship *Södermanland*, fired proving rounds as early as June 1962 and carried out its first operational firing in October 1962, only three months after the Royal Navy.

(c) Comments

In general, the trend of developments in present-day guided weapons is towards ever more highly sophisticated, complex and automated systems. This inevitably leads to high cost. Seacat marks a distinct reversal of this trend. From the outset the aim was to produce the simplest possible weapon system capable of fulfilling the assigned task. There is some evidence to suggest that the task of restraining scientists and engineers from indulging in the 'technically sweet' for its own sake was a difficult one and in some quarters the simplicity of the visual guidance concept was ridiculed.

But the utilisation of a simple concept does not imply that the technology was trivial. There is a good deal of sophisticated technology in the Seacat system, originating both within (see references 8, 9, 10, 11, 12, 13) and outside the firm. For example, S.B. & H. worked in close collaboration with R.A.E. Farnborough in the following areas: the original theory of visual guidance (which was studied by a team of scientists and medical doctors); the design of the control signal transmitter and receiver unit; servo and valve technology. With respect to the development of the precision servos H. G. Conway, who had worked on the Seacat project, commented as follows on the closeness of the collaboration in his Presidential Address to the Institute of Mechanical Engineers:

> The servo-work for missile applications required the learning of many new tricks. And here I must pay tribute to what was learnt from the Government Research Establishments. It was the Admiralty Gunnery Establishment (A.G.E.) at Teddington where all the significant early work on electrically operated precision hydraulic servo devices was done. These had to be evolved for electro-hydraulic servos, fed from electronic amplifiers and operating steering or driven mechanisms in weapons. The German V.2 and, to a lesser extent, the V.1, had pointed the way. The A.G.E. work, later exploited for aerial weapons by the R.A.E., was freely available to us and formed the starting point.[14]

When Conway joined the Seacat project he brought with him Mr R. S. Ransom. Prior to joining S.B. & H. as Managing Director, Conway had been Chief Engineer of the British Messier Company which specialised in the development and production of aircraft undercarriages – an area closely associated with the development of hydraulic control systems. Ransom, a design draughtsman, worked with Conway in British Messier and was described by him 'as one of the most creative people I have worked with'. Although Ransom holds no formal technical qualifications, he is named as inventor or co-inventor in eight out of the eighteen patents granted to the Seacat team.

Advances in warhead, propellant and battery technologies

were also utilised. For example, the servo amplifiers in Seacat are energised by a special type of thermally activated battery with a long storage life. This battery, which was developed in the United States can supply full power to the missile within one second of its firing; it was a fortunate discovery for the Seacat project because some of the engineers regarded the lack of a suitable energy source as a serious bottleneck to further development.

On the other hand, because of the political or commercial implications, a firm cannot always rely on obtaining information from outside sources and consequently it may be thrown back on its own resources. Such a restriction was encountered with the missiles' fast starting gyroscopes. These instruments are required to go from rest to 40,000 r.p.m. in under one second because they supply the guidance information for the missile. S.B. & H. were aware that such devices were available in France, and Conway went to negotiate a licence with Nord Aviation. While he was there he saw an assembled unit, but the licence fee requested was regarded as unacceptably high. S.B. & H. subsequently decided to develop their own unit in the knowledge that such a device had been made. In this, they were completely successful; their miniature gyroscope, based on Hero's turbine, with a cordite charge fixed inside the rotor, has since been sold to a number of foreign countries.

The interviews have clearly indicated that the small size of the Seacat design team, the enthusiasm of its leader and members, the relative isolation of the new and expanding Precision Engineering Division, and the discipline with which the project objective was pursued were important contributory factors in the successful development of the Seacat missile. Dr R. J. A. Paul, who was the Head of the Electronics Department from 1952 to 1956, has recalled that

In the initial work for the 'Seacat' my small team comprised Mr G. Barnes (Engineer), Mr G. Brown (Engineer) and Mrs Ruth Telford (Mathematician). During this early work we worked in close collaboration with colleagues at the Royal Aeronautical Establishment, Farnborough, in the feasibility studies including simulator work, both at Farnborough and Shorts.[7]

The project team exhibits a multiplicity of educational backgrounds. In the P.E.D. itself in 1967, there were approximately 49 engineers with university degrees, 3 with diplomas, 56 with H.N.C.s, 48 with O.N.C.s, 27 with City and Guilds training, 12 with G.C.E. or equivalent, 32 with no formal educational qualification and 38 secretarial and administrative staff. While not all of these worked on Seacat itself, the proportions were said to be indicative of the skill mix that worked on the missile.

Another factor in the rapid success of Seacat is the development of simulation techniques as a method of training. In order to meet the Royal Navy specification that the missile should be easy to use, the R.A.E. and Shorts decided to use computer techniques to simulate the mechanical operation of the missile and included the human operator as an integral part of the circuit. This made a visual presentation necessary and it seemed logical, therefore, to extend the use of the simulation to training. Trainees firing Seacat by simulation techniques often score direct 'hits' at first attempts. On transferring to 'live' firings, they appear to find very little difficulty in scoring a high number of 'hits'. These simulation techniques have proved to be a useful adjunct in stimulating Seacat missile sales.

References

1. E. Burgess, 'German Guide and Rocket Missiles' (Part 1), *The Engineer* Oct 1947, p. 308; see also Part II, Oct 1947, p. 332.
2. 'Seacat – the guided missile for the Defense of Ships', *Flight*, 5 4 (1963).
3. Interviews with P. F. Foreman, R. M. Armour and J. Foy, Short Brothers and Harland, Belfast, Northern Ireland, and H. G. Conway, previously Joint Managing Director, S. B. & H., subsequently Managing Director, Bristol Engine Division, Rolls-Royce Ltd, 1968 and 1969.
4. G. W. H. Gardner, 'Guided Missiles', *The Engineer*, Nov 1954, p. 728.
5. R. M. Armour, private communication, Dec 1968.
6. Statement of Claim by Short Brothers and Harland Ltd, in an application for the Queen's Award to Industry on behalf of the company's Precision Engineering Division in respect of the Seacat guided weapons system. Private communication, 1968.
7. R. J. A. Paul, private communication, July 1969.
8. Blast-start gyro – U.K. Patent No. 842,775 (1958).
9. Oil supply, wing servos – U.K. Patent No. 845,142 (1958).
10. Wing locking mechanism – U.K. Patent No. 895,143 (1958).
11. Cooling electronic pack – U.K. Patent No. 895,144 (1958).
12. Servo-pilot valve – U.K. Patent No. 839,153 (1958).
13. Container – U.K. Patent No. 843,037 (1958).

14. H. G. Conway, 'Engineering in the Air', *Chartered Mechanical Engineer*, Nov 1967, p. 470.

34 SMITHS INDUSTRIES: AIRCRAFT AUTOMATIC LANDING EQUIPMENT

Smiths Industries received the Queen's Award to Industry for technological innovation in 1966 for the part played by their aviation division in promoting the development of a blind automatic landing capability in the fleet of Trident aircraft supplied to British European Airways (B.E.A.) by Hawker Siddeley Aviation, to whom Smiths were contractually responsible for the development and manufacture of the flight control system. The initial £30 million contract between B.E.A. and Hawker Siddeley (then De Havillands), signed in August 1959, was for the delivery of twenty-four aircraft. In August 1965 a further contract was signed for the delivery of fifteen Trident 2E's – of similar capacity but slightly longer range than the Trident 1's. In 1968 a final contract was signed for the delivery of twenty-six Trident 3B's. Smiths Industries were contracted only to De Havillands, and they in turn subcontracted to Sperry about 25 per cent in value of the total system. In 1970 B.E.A. and Hawker Siddeley Aviation received Queen's Awards for technological achievement with respect to the same development.

The time scale of the innovation stretches from the post-war setting-up of the Blind Landing Experimental Unit of the Royal Aircraft Establishment to the point in time, hopefully (but by no means surely, for the integrity of the ground equipment has yet to be established) in the 1970s, when a B.E.A. Trident performs its first scheduled automatic landing with the total system comprising both ground-based and air-borne equipment cleared for operation by the British safety authorities in conditions of virtually zero visibility. This event would mark the attainment, on at least one airport runway, of the ultimate objective of the innovation, which is an all-weather landing capability for the Trident fleet.

This brief account of the innovation[1] concentrates on the ten-year period 1957–67, from the point at which Smiths Industries decided to initiate development of a new-generation civil autopilot potentially safe enough, when used with ground equipment of a similar level of integrity, to permit automatic landing in scheduled service in zero visibility, to the point at which a B.E.A. Trident made its first automatic landing in scheduled service in clear visibility. This was in May 1967. In one week of scheduled operations the same aircraft made twenty-seven automatic landings at fifteen different airfields in nine European countries, carrying a total of 1,400 passengers.

Perhaps the most striking feature of this innovation is the intimate relationship between government and industry, and the consequent vulnerability of a medium-sized equipment manufacturer to the outcome of semi-political decisions which it may not always hope to influence, let alone predict: such decisions may involve the procurement of aircraft for which its equipment is designed, the procurement of ground equipment with which its own equipment must be compatible, and the elaboration of procedures for demonstrating compliance with an official safety requirement, which might affect both the scale of the development effort required from each of the participants in the innovation and the market prospects of a particular generation of equipment.

In this type of environment the financial implications of a single innovation, taken in isolation from the rest of the firm's activities, are unlikely to provide an adequate basis for assessing the quality of decision-making from which the initial commitment derived, nor of its subsequent managerial and technological performance.

(a) *The requirements*

Government involvement in the innovation derives from three main factors governing the introduction of a new technology such as automatic landing into the civil aviation environment. First is the requirement for joint clearance of the design and operation of the airborne equipment by the British safety authorities – technical clearance in the form of a certificate of airworthiness issued only on the recommendation of the Air Registration Board, and operational clearance in the form of amendments to an aircraft's flight manual authorised by the Directorate-General of Safety and Operations at the Board of Trade. The second factor is government procurement and airfield installation of the necessary ground equipment, of a standard acceptable both to the safety authorities and to the airlines using the equipment. The third factor is the supply of pilots trained by the airlines and licensed by the Board of Trade for operating the system in progressively more difficult conditions.

These conditions have been categorised by the International Civil Aviation Organisation according to the height at which the pilot must decide whether or not he can see enough of the runway either to take over and land the aircraft himself after an automatic approach or to allow the aircraft to be landed by the automatic system. The limiting factor in both cases was the risk of something going wrong with the automatics. In 1959 a committee on landing aids composed of representatives of all interested parties in the aviation industry and chaired by the Chief Technical Officer of the Air Registration Board had decided that the use of an automatic landing system 'in the worst permitted circumstances' should not introduce a fatal accident risk greater than one in ten million in the last thirty seconds of flight. It was later decided that demonstrating compliance with this safety factor would probably require about three thousand fully recorded automatic landings in scheduled service in conditions of visibility such that the pilot, in the event of the worst conceivable kind of system malfunction, would have adequate visual reference to recognise the occurrence of a malfunction and enough time either to initiate an overshoot or to take over and land the aircraft himself. Only when this proving period was satisfactorily completed could an aircraft be cleared for category 3 operations, which meant the ability to operate blind to the surface of the runway. This particular category was divided into three subcategories according to whether external visual reference was available to permit manual control of the ground run (category 3A, requiring a runway visual range of at least 200 metres), visual taxiing (category 3B, requiring a runway visual range of at least 50 metres), or non-existent (category 3C). The runway visual range was the distance measured along the runway of the farthest visible light, and for categories 1 and 2 the relationship between this and decision height was based on the principle that the probability of having adequate slant visual reference at a given decision height should be as high as possible. Thus, a category 1 capability meant the ability to operate down to a decision height of 200 ft when the runway visual range was at least 800 metres, and a category 2 capability meant the ability to operate to decision heights between 200 ft and 100 ft with runway visual ranges between 800 and 400 metres. Pending the completion of this proving period, it was stipulated in 1965 that an automatic system should not be relied upon for landing an aircraft when decision height was lower than 150 ft. Attainment of this mid-category 2 capability was to depend, moreover, on evidence as to the satisfactory operation of the system in category 1 conditions.[2]

How, though, did one set about designing any system to a one-in-ten-million safety factor? The conventional method of ensuring safety under automatic control was to limit the authority of the autopilot over the control surfaces of the aircraft by means of some kind of cut-out device, so that in the event of a fault in the automatics resulting in the most dangerous type of malfunction, which drove a control surface hard in one direction, the system would cut-out without imposing too difficult a corrective manœuvre on the pilot. Clearly, this method was inapplicable for an automatic system required to land an aircraft in marginal visibility conditions, in which no height loss in the event of a fault could be tolerated. Moreover, control of sophisticated aircraft in the terminal phase of flight meant extending the authority of the automatics, not limiting it. The answer lay in the design technique known as redundancy

– in other words, rendering the probability of a malfunction conditional on the occurrence of not one, but at least two faults in the automatic system of the critical period.[3]

An autopilot could be provided with redundancy in the form of two or more absolutely separate subsystems, each of which was capable of controlling the aircraft on its own. The principle behind Smiths Industries multiplex concept was that these independent subsystems should work in parallel. By multiplying the control channels in each axis, the system could be arranged so that no fault or sequence of faults in a single subsystem could affect overall performance. In a duplex system, the two sub-channels in each axis would monitor each other. So long as their outputs were similar, the system would continue to operate. If a fault occurred in either of the subchannels, the discrepancy between their outputs would cause the whole system to cut-out, but without causing the aircraft to deviate materially from its flight path as defined by the radio guidance inputs to the flight computers. It would, therefore, 'fail steady', and the aircraft would be in a safe condition for the pilot to take over. In a triplex system, three subchannels in each axis would monitor each other. A fault in one subchannel would be recognised by the two others, which would operate to cut it out, leaving the system at duplex level. A triplex system could, therefore, be described as 'fail-operative' in that it could survive a single fault and would fail steady on the occurrence of a second fault. The key issue was the probability of two faults occurring in more than one subsystem in the last thirty seconds of the landing manœuvre. If, as Smiths Industries claim, this probability could be shown to be less than one in ten million, a flight control system designed on triplex principles (in conjunction with ground equipment of a similar level of integrity) promised a geuinely blind category 3 landing capability to any airline ready to bear the cost of buying, proving, maintaining and carrying the weight penalty of the extra airborne equipment. An airline's readiness to do this would depend on balancing this total cost against the projected increase in the cost of disruptions to the passenger-carrying and maintenance sche-dules of its front-line fleet, as the projected units of this fleet became larger and more expensive. A Trident 2E, for example, was to cost B.E.A. over five times as much as the price they had had to pay for the Viscount – £2,650,000 as against £470,000.

(b) The collaborative background

Smiths Industries' decision to risk their own money on the development of a fail-operative civil autopilot with an automatic landing capability may be traced to an internal report issued in June 1956 by the operational advisor to their Aviation Division, A. M. A. Majendie, on requirements for automatic control in the second generation of jet aircraft then on the drawing-board. These aircraft, subsequently designated 'Trident' and 'VC-10', were scheduled to come into service over the decade 1962–72. The background to the report was ten years' successful collaboration between Smiths Industries Aviation Division and the Blind Landing Experimental Unit (B.L.E.U.) of the Royal Aircraft Establishment (R.A.E.) on the development of a 'simplex' or non-redundant single-channel automatic landing capability for military purposes. The outcome of this development was the promulgation of an operational requirement for a simplex automatic landing capability in the V-bombers, to which the Smiths Industries Mark 10 autopilot was to be fitted, and in the consequent transfer of important technical information from the government research establishment to the private firm to enable a Mark 10B automatic landing version of this autopilot to be built and tested.

B.L.E.U. was set up in September 1946 as a result of military concern about weather interference with bombing operations. This had led to the formulation of an Air Staff operational requirement for a low visibility approach capability in military aircraft, with blind landing as the ultimate target. The Unit then consisted of a team of twelve scientists – six from the R.A.E. and six from the Telecommunications Research Establishment (now the Radar Research Establishment) at Malvern. It was based at the R.A.F. airfield at Martlesham Heath, Suffolk, close to a large disused runway at Woodbridge which had been used during the war for the landing of stray or damaged aircraft. In 1957 it moved to Bedford, where it is now.

The existing technique for low visibility approaches was to feed in signals from the ground-based instrument landing system (I.L.S.) radio approach aid to a cross-pointer indicator in the cockpit. The I.L.S. system consisted of two transmitters (localiser and glidepath), one beside and one at the end of the runway, which provided limited angular deviation signals from the appropriate horizontal and vertical approach planes. The intersection of these planes defined the optimal approach path. A pilot could usually manage to guide his aircraft along this path until he reached decision height. At this point, he had to decide whether he had adequate visual reference to land the aircraft, and if not, to overshoot.

Unfortunately, the reading of the cross-pointer indicator depended only on the location of the aircraft in relation to the radio beams. To manœuvre his aircraft down the defined approach path, the pilot also had to know the heading and the attitude of his aircraft – in other words, to co-ordinate the reading of this indicator with the readings of a number of other instruments. It was the extreme difficulty of this exercise, and the fact that there seemed then to be no technical solution to the problem of co-ordinating these readings on a single display instrument of the type that was later to become known as a 'flight director', which seemed to render any extension of this technique unfeasible for the landing manœuvre and which focused the Unit's attention on the idea, which had been around for some time, of coupling the radio guidance signals directly into the autopilot.[4]

It was soon apparent that for automatic operation to touch-down more accurate and reliable radio guidance would be required. This was resolved through the twin development of magnetic leader cables to provide azimuth directional guidance in place of the localiser from just beyond the runway threshold, and of a high-precision radio altimeter to measure the height of the aircraft in the crucial seconds before touchdown.[5]

The leader cable system consisted of two cables, one on each side of the runway, lying 250 ft from the centre-line and extending from about 5000 ft beyond the threshold in the undershoot area to as far down the runway as guidance was required. The cables were fed from alternators supplying current of a different frequency to each cable. The principle of operation was that at any point between the two cables the difference in the magnetic fields due to the current in them was a function of the horizontal distance of this point from the runway centre-line. A simple airborne receiver could discriminate between the two frequencies and compare the signal strengths.

The problem of extending vertical guidance to touchdown was solved, not by substituting an alternative form of ground radio guidance, but by dispensing with the need for ground equipment and deriving guidance from a radio altimeter situated in the aircraft, specially developed by B.L.E.U. in conjunction with industry. The 3° slope of the glide path was too abrupt for safe and comfortable landing, and it seemed that the most convenient way of checking an aircraft's rate of descent and flaring it on to the runway was to arrange for the vertical component of its flight path to follow an exponential curve starting from the glide path.[6] This could be done by reducing the aircraft's rate of descent in proportion to its height above the runway, provided its height could be accurately and continuously measured. The principle of radio altimetry was that by continuously altering the frequency of a radio beam directed at the ground, the echo time – and thence the height of the aircraft, to an accuracy of within plus or minus 2 ft – could be simply determined by comparing the frequencies of the returning and outgoing waves.

Smiths Industries' relationship with R.A.E. had originated in the late 1920s when they secured entry into the field of automatic flight control by obtaining the manufacturing rights for the R.A.E.-designed Mark 1 autopilot, whose technology was derived from the perception by F. W. Meredith of the effect of tilt on the performance of a gyroscope. Production started in 1931, and over the next fifteen years Marks 2 to 8 were developed from Mark 1 along closely related principles and were widely used by the Royal Air Force during the war. Meanwhile, in 1937, F. W. Meredith, the principal author of some of these designs, had joined Smiths Industries direct from R.A.E. where he was Head of the Physics and Instruments Section. In 1945 he initiated development of a new-generation three-axis all-electric autopilot which was to become known as SEP-1[7] or, in its military version, as Mark 9.[8]

In 1948 B.L.E.U. took delivery of one of the first models of the new Smiths Industries autopilot. With special coupling units designed and built by the Unit to enable radio guidance to be fed into it, it was fitted to a Devon for automatic flare investigation and to a Viking for leader cable trials. By the end of the year experimental landings were being made under partial automatic control in both aircraft, and in the autumn of 1949 a first public demonstration was staged. All the elements of a full automatic landing capability were on display – I.L.S. coupling, radio altimetry, magnetic leader cables, automatic throttle control – though the various airborne components had yet to be combined in a single aircraft.

Then, for several years, the programme hung fire. There were technical difficulties with localiser beam reflections on the approach. But more significantly, development work at R.A.E. Farnborough on runway lighting systems was leading the Air Staff to modify their appreciation of the urgency of an automatic solution. With improved high-intensity runway lighting patterns, manual landings in marginal visibility following an accurate automatic approach might be less difficult than expected.

In 1954 there was a revival of military interest. Approach speeds were drastically increasing and it looked as if future aircraft might be difficult enough to land manually even in clear visibility. Anxiety had specifically arisen over the performance of the projected all-weather thin-winged version of the fighter aircraft designated 'Javelin'. By the end of the year an operational requirement for an automatic landing capability in this aircraft had been formally accepted by the Ministry of Supply.

From 1949 onwards, close collaboration with the Unit in resolving some of the early difficulties of landing phase autopilotry had enabled Smiths Industries to construct a prototype autopilot with built-in automatic approach facilities. First known as 'Type D', it was to be developed into SEP-2 for civil transport aircraft and into military version Mark 10. In 1955 the Unit was provided with a Varsity aircraft fitted with 'Type D'. and it began to build its own analogue computer. A fully automatic solution to the low visibility landing problem had been re-established as its principal target. By the time the thin-winged version of Javelin was cancelled, a requirement had been promulgated for an automatic landing capability in

P

the V-bombers. Smiths Industries' Mark 10 autopilot had already been selected for these aircraft as a general-purpose autopilot. Early in 1956 B.L.E.U. representatives visited Smiths Industries Aviation Division at Cheltenham to discuss the development and manufacture of an automatic landing version, Mark 10B. There was an initial transfer of information to Smiths Industries to enable them to consider the matter. In July Smiths Industries returned the visit and confirmed their willingness to proceed. Three months later B.L.E.U. formally requested the Directorate of Air Navigation at the Ministry of Supply to place a contract with Smiths Industries for the development of the Mark 10 autopilot, together with the necessary coupling units for use in V-class aircraft, to a full automatic landing capability. In November there was a second major transfer of B.L.E.U. information to Smiths Industries, this time by correspondence, including the control equations for automatic landing and on the subject of automatic throttle control.

By October 1958, in spite of delays through the loss of a Canberra aircraft and its installed equipment and through the move from Martlesham to Bedford, where a new runway had to be prepared and equipped with full I.L.S. and leader cable facilities, the Unit was ready to stage a second major public demonstration. This time a Canberra and two Varisities were on display, and the demonstration was witnessed by military and civil authorities from all over the world.[9]

(c) Genesis of a decision

At about the same time, Smiths Industries' Aviation Division in Cheltenham were carrying out flight trials in a specially purchased Dakota aircraft of an experimental version of their new generation SEP-5 civil autopilot incorporating a triplex elevator channel. If successful, the multiplex concept might permit the transfer of an automatic landing technology, developed by B.L.E.U. in close association with Smiths Industries, from a military to a civil environment in which the safety requirement would be more exacting by possibly two orders of magnitude. For Smiths Industries, the development of this autopilot represented a bid to retain their commanding position in the United Kingdom civil autopilot market. The origins of the development must be traced to their recruitment of A. M. A. Majendie in 1954 as operational advisor to the Aviation Division, and to the recruitment of K. Fearnside a year earlier to become Head of Theoretical Studies in the Research and Guided Weapons Department under F. W. Meredith.[10]

Smiths Industries' interest in recruiting Majendie was that Majendie, as flight captain of the B.O.A.C. Comet 1 fleet, had direct operating experience of the performance of their SEP-2 autopilot, with its implications for the design of a successor. Thirty-six years old, his career in aviation had already been spectacular. After leaving Cambridge in 1940, where he read moral sciences in preference to an earlier inclination to study physics, he became a flying instructor with Coastal Command flying boats on North Atlantic patrols. At the end of the war he joined B.O.A.C. as an airline captain on flying boat routes to the Middle and Far East and to South Africa, and in 1949 he was appointed to special duties in connection with the proving and introduction of Comet aircraft. Over the next three years, with the original Comet Unit, he helped pioneer the basic techniques of flying commercial pure-jet aircraft at what then appeared to be extraordinary speeds of over 500 m.p.h. In May 1952, as flight captain of the Comet fleet, he had the distinction of commanding the world's first scheduled jet passenger service in a Comet 1 on the first stage of its flight from London to Johannesburg.

Majendie's interest in joining Smiths Industries derived from his conviction – itself a product of his combined interests in visual psychology and the theory of automatic control systems – that automation of the landing process in civil aviation was inevitable. Through the work of R.A.E. and the partnership between Smiths Industries and B.L.E.U., Britain was well placed to pioneer this development. Once he had joined Smiths Industries, in 1954, this conviction soon brought him into conflict with F. W. Meredith who opposed the idea of providing automatic trim facilities for the SEP-2 autopilot on the grounds it took authority from the pilot. Majendie argued that the logic of automatic flight control necessarily implied taking one kind of authority from the pilot, whose role he foresaw transformed from that of system operator to system manager.[11]

The SEP-2 autopilot was then in service with Comets, Viscounts and Britannias. It had grown into a very complex piece of equipment, and Research Department opinion was that the same level of performance could probably be achieved with a simpler design. Early in 1956, before leaving Smiths Industries, F. W. Meredith commissioned a performance survey of the autopilot, which was to take into account anticipated design requirements for the next generation of civil aircraft and the state of the art in the United States where Sperry and Bendix were known to be making progress in the field of I.L.S. coupled approaches. Issued in April 1956, the survey suggested that in view of the possible emergence of a requirement for automatic control to touchdown during the lifetime of a successor to SEP-2, '. . . all aspects of the manœuvre should be investigated and allowance made for its inclusion at a late stage in the design'. It was a cautious, defensive recommendation.

Majendie, meanwhile, was preparing his own report. Circulated to the Divisional Board in June 1956, it vigorously presented the case for initiating immediate development of a new-generation autopilot incorporating the principles of multiplex redundancy to meet what he judged to be the two principal airline requirements for the next generation of civil aircraft – uninterrupted automatic control in the worst conditions of turbulence and a blind automatic landing capability to certification standards of safety. Majendie stressed the 'immense' commercial advantage to be secured by the first autopilot manufacturer able to offer a fully certificated automatic landing capability for civil aircraft, and Smiths Industries' unique opportunity to secure this advantage by exploiting the technological expertise derived from their association with B.L.E.U. in joint development of a similar, though merely simplex, capability for the V-bombers.

The following month, on the strength of this report, the Divisional Board decided to commission a twelve-month multiplex feasibility study from the Research Department, to be directed by Kenneth Fearnside.[12]

Fearnside, like Majendie, had joined the R.A.F. after

graduating from Cambridge, where he took first-class honours in the natural science tripos. After four years with a radar and signals unit in the Middle East he became a radar area commander in France and spent the year following the end of the war supervising the installation of the Rhône–Carcassonne Gee Chain which provided navigational and positional radio guidance to aircraft. In 1946 he went to Harwell, at the invitation of Cockcroft, his former physics supervisor at Cambridge, and in 1950 he joined Isotope Developments Ltd as technical director after spending his last year at Harwell investigating the industrial uses of isotopes. In 1953, attracted by the prospect of working on guided weapons and at the prompting of F. W. Meredith, he joined Smiths Industries to become Head of Theoretical Studies in the Research and Guided Weapons Department of the Aviation Division. Over the next three years he built up a research team with a strong capability in systems analysis, and it was from this assemblage of skills that he was to manage the launching of the multiplex project, carrying out two reorganisations of divisional research and engineering activities in 1957 and 1960 to accommodate its growth. With Majendie's promotion to the managing directorship of the Aviation Division in 1963, Fearnside was to remain the one senior executive continuously associated with the project over the twelve years from its inception in 1956 to the time he left Smiths Industries to join Plessey Components Group as Technical Director in July 1968. Over the preceding five years the final technical decision within Smiths Industries on every point affecting system performance was his alone. This, together with organisational skills derived from his war-time experience, he was subsequently to consider his most valuable contribution to the technical success of the multiplex project.

In July 1957 Fearnside reported to the Divisional Board that a fail-operative multiplex autopilot could probably be built within the weight limits set by Majendie's assessment of what would be acceptable to the airlines, that several of the requirements for a civil autopilot with an automatic landing capability were already being investigated on government contract for various military projects, and that owing to the cancellation of a number of other government contracts the manpower would be available within the Division to tackle the multiplex project.

The one area of uncertainty was the applicability of the B.L.E.U.-developed leader cable system for terminal azimuth guidance in the civil environment. Opinion in government circles was that leader cables offered the most satisfactory solution. Opinion in industry was that their installation at civil airports would be difficult, in some cases impossible, and that the I.L.S. localiser beam could probably be developed to an adequate degree of precision and reliability for automatic azimuth guidance to touchdown, if the necessary effort was made. This one area of uncertainty would not affect the design of the new autopilot except in detail. But it certainly affected the customer appeal of the automatic landing facility which it would take a great deal of money and effort to design into it. For this reason, Fearnside's recommended development programme did not extend beyond the end of 1961, by which time the matter would presumably have been clarified.

His recommendations, submitted in July, were not instantly accepted. In August 1957, nearly fifteen months after his original report, Majendie circulated a memorandum confirming and updating his appraisal of the operational requirements and pressing for a Divisional Board decision on further development. Preliminary studies for a supersonic airliner indicated that the most difficult operational aspect would be the approach and landing manœuvre, and that on technical and economic grounds an automatic landing capability should be an integral part of the design formula. If Smiths Industries could introduce a new generation SEP-5 autopilot, capable of automatic landing, by the mid-1960s, they would be singularly well placed to secure this contract. In the relative absence of comparable work in the United States which could be applicable to routine civil operations, here was an opportunity for world leadership.

Towards the end of September 1957 the relatively cautious decision was taken to develop and produce a new-generation autopilot as a successor to SEP-2 in three distinct stages. The first stage, aimed at main route aircraft coming into service during 1962–9 such as the Trident and the VC-10, was to have simplex rudder and aileron channels and a triplex elevator channel with three gyro units and up to three radio altimeters to provide a fail-operative automatic flare capability; at airports where the I.L.S. localiser beam was good enough for control to touchdown in azimuth, it would also have a full emergency automatic landing capability. The second stage was to be simplex in all three channels, to replace SEP-2 once the first stage was in production. The third stage was tentatively envisaged as quadruplex in all three channels, with four radio altimeters, to provide a scheduled automatic landing capability for main route aircraft coming into service towards the end of the 1960s when suitable ground equipment for terminal azimuth guidance could definitely be expected to be available.

By January 1958 a preliminary specification for the new autopilot had been fully elaborated and two models of an experimental version with triplex elevator and simplex rudder and aileron channels were under construction for fitment to the Dakota aircraft, newly acquired for this purpose by the divisional flying unit, in preparation for the flight trials that were to begin that autumn, at about the same time as the second public demonstration to be staged by B.L.E.U.

Meanwhile, in April 1958, Smiths Industries submitted proposals to B.E.A. and B.O.A.C. for the fitment of a fully triplex fail-operative version of the SEP-5 autopilot, with an automatic landing capability, to their respectively projected fleets of Tridents and VC-10's.

(d) Marketing and development

The prospect of offering their new autopilot to two domestic aircraft – the Vickers (now B.A.C.) VC-10 and the De Havilland (now Hawker Siddeley) Trident, then known as the DH-121 – had given considerable impetus to Smiths Industries early design work. The development decision of September 1957 had been based on the tacit assumption that they were unlikely to be threatened by serious domestic competition for these two flight control contracts. For the past decade they had enjoyed a virtual monopoly in the British autopilot market for civil transport aircraft. But they underestimated the competition. The VC-10 contract went to Elliott Flight Automation, for the fitment of an anglicised Bendix PB-20 autopilot, which had certain operational advantages for B.O.A.C. in that it would be virtually the same equipment as that to be installed on the Boeing 707's they had on order.

In July 1958 the project and development branch of B.E.A.'s Engineering Department issued to interested firms their formal specification for an integrated flight control system in the Trident. Written into it, in qualified terms, was the potential requirement for an automatic landing capability, should adequate ground aids for terminal azimuth guidance become available. But an automatic flare capability – that is, automatic control to touchdown in the pitch plane only, using a radio altimeter – was to be provided from the outset. Since the purpose of this was to enable Tridents to achieve decision heights of 150 ft, in conditions where visual reference might be adequate for manual control of the aircraft in azimuth but not in pitch below 200 ft, the requirement was for a duplex fail-steady elevator channel. The specification stipulated that the system should be so designed that it could be upgraded to triplex level in all three control channels without major retrospective modifications. But until the situation over ground aids was clarified, and the safety implications of triplex automatic landing more fully evidenced, a duplex pitch channel was as far as they were ready to commit themselves.[13]

Having failed to secure the VC-10 contract, and having failed to extract from B.E.A. a more than tentative commitment to a triplex fail-operative capability in their new SEP-5 autopilot around which the Trident flight control system was to be built, Smiths Industries' incentive to maintain the dynamic of their multiplex development effort, funded until then from their own resources, was critically strengthened in 1959 by the promulgation of a Ministry of Defence operational requirement for a blind automatic landing capability in the Belfast strategic freighter aircraft, to be supplied by Short Brothers and Harland for R.A.F. Transport Command. For troop-carrying aircraft it would have been hard to justify a less exacting safety requirement than that considered to be attainable and satisfactory for civil operations. Since B.L.E.U. had already reported favourably on Smiths Industries' triplex proposals for the Trident, it was decided to entrust to Smiths Industries the development of a triplex system for the Belfast. Terminal azimuth guidance would be provided by the B.L.E.U.-developed leader cables then being installed at a number of military airfields for the use of the V-bombers.

The Belfast decision provided the framework for government technical support to Smiths Industries which was of the utmost relevance to their long-term commitment to the development of a triplex system for the B.E.A. Trident. A Varsity aircraft had already been made available to the Division's flying unit by the Ministry of Aviation for flight control development work on a simplex automatic landing capability for the V-bombers. A second Varsity was now made available for multiplex flight trials, and a Comet 2E to B.L.E.U. for performance evaluation of the projected triplex system in high-altitude operating regimes.[14]

The fixed-price development contract for the Belfast flight control system – the largest government contract Smiths Industries had ever received – was finalised in July 1960. The same year, the Research and Development Board of the Ministry of Aviation, which controlled expenditure on the Department's major R. & D. projects (both military and civil), set up an automatic landing management board to co-ordinate programmes for the V-bombers, the Belfast, Trident and VC-10, and to provide a forum for discussion between representatives of the interested technical and operational directorates of the Ministry and the Board of Trade. At some of these meetings, representatives of both state airlines were invited to express their views.[15]

One of the key issues was, of course, safety, and to this purpose the Air Registration Board (A.R.B.), exercising its responsibility for the formulation of British civil airworthiness requirements under powers delegated from the Board of Trade by the Civil Aviation Act of 1936, had set up in March 1959 a committee on landing and take-off aids. It was chaired by the Chief Technical Officer of the A.R.B. and consisted of representatives of all interested parties – the Ministry of Transport and Civil Aviation (as it then was), the Ministry of Supply, R.A.E. and B.L.E.U., Smiths, Elliotts, Vickers, De Havillands, B.E.A., B.O.A.C., Standard Telephones and Cables (ground radio equipment and radio altimeter manufacturers), and the British Airline Pilots Association. Over the next two and a half years the committee held seven meetings, which culminated in the issue of provisional draft airworthiness requirements for the automatic landing manœuvre in November 1961, in which the one-in-ten-million safety factor, anticipated by Majendie in 1956, was formally incorporated.[16]

Majendie and Fearnside attended the first five of these meetings. It was evident that Smiths Industries' prospects of getting B.E.A. to upgrade their procurement committment to a triplex fail-operative capability depended on the degree of confidence which could be made to prevail in the industry with respect to the development of adequate ground aids for terminal azimuth guidance. The general opinion in the industry was that leader cables were unsuitable for civil operations. Government opinion, based on B.L.E.U.'s expert assessment of the matter, was that there was no feasible alternative. Majendie and Fearnside questioned the grounds on which B.L.E.U. dismissed the possibility of developing the I.L.S. localiser beam to an adequate level of precision and stability. Their lobbying reached a climax at the fifth meeting of the committee in March 1961 when they presented a paper whose principal contention was that the available evidence simply did not suffice to justify B.L.E.U.'s misgivings. Given the potential advantages of the I.L.S. localiser over leader cables in the civil environment, '. . . a programme of extensive flight trials on a well-sited modern localiser system [should] be initiated to supply more practical evidence'.

Then, towards the end of 1961, on the advice of its automatic landing management board, at several of whose meetings representatives of both state airlines had persisted with the arguments developed by Smiths Industries in the A.R.B. landing and take-off aids committee, the Research and Development Board of the Ministry of Aviation decided to commission a two-year feasibility study on the development of the I.L.S. localiser beam for terminal azimuth guidance at civil airfields. It was to be conducted jointly by B.L.E.U. and R.A.E. Radio Department.[17]

The first Tridents were due to be delivered to B.E.A. in March 1964. Flight trials of the aircraft's coupled approach facility, which included the performance of the autopilot's duplex elevator channel, were initiated in March 1962 on a development aircraft leased back to Hawker Siddeley by B.E.A.

In February 1964 a certificate of airworthiness for coupled approaches to 300 ft was granted by the Directorate-General of Safety and Operations at the Board of Trade, on the recommendation of the A.R.B., after the results of a special certification test programme had been analysed. The first aircraft entered service with B.E.A. in April 1964. Although it was to have been delivered, according to contract, with a duplex autoflare capability, the development trials had indicated that many detailed changes were required to achieve acceptable performance. In fact it was not until June 1965 that certification was achieved for the automatic flare capability, after a test programme involving ninety-nine fully recorded autoflares flown by thirteen pilots – seven from B.E.A., four from Hawker Siddeley and two from the A.R.B. – over seven different I.L.S. installations. A week later, 10 June 1965, at London Airport on flight BE 343 from Paris with eighty-three passengers on board, B.E.A. were to make the world's first airline autoflare in scheduled service.[18]

Meanwhile, in 1963, a combination of factors had induced B.E.A. to upgrade their Trident procurement commitment from a duplex autoflare to a duplex autoland capability, with a strong presumption of subsequently proceeding to triplex level in pitch and roll channels. The decision had meant that the flight trials initiated in March 1964 for the purpose of autoflare certification could be simultaneously exploited for investigating the performance of duplex rudder and aileron channels, and they in fact began with a successful series of ten consecutive automatic landings. The airline's decision may be regarded as the outcome of a gradual focusing of their perception of the innovational environment with respect to five principal features.

First of these features was the prospective availability of adequate ground aids for terminal azimuth guidance. The results of the localiser feasibility study conducted by R.A.E. were not published until February 1965. But early in 1963 B.L.E.U. had made it known to all interested parties that '. . . the system was potentially viable provided that adequate monitoring devices were added'.[19]

Second of these features was the relationship between risk and visibility using current techniques for decision heights below 200 ft. An analysis of accident statistics, also conducted by R.A.E., had revealed that, in comparison to landing in clear visibility, the risk of accident on an automatic or instrument approach followed by manual take-over in runway visual ranges of 800 and 400 metres respectively was potentially multiplied by factors of ten and a hundred.[29]

Third of these features was the prospective definition of reasonable procedures for demonstrating compliance with the official A.R.B. safety requirement, agreed by the committee on landing and take-off aids, that an automatic landing system (comprising aircraft, ground equipment and pilot), when used in the worst permitted conditions, should not occasion an overall fatal accident risk greater than one in ten million per landing. With a triplex or any other kind of fail-operative system which was capable of surviving a single fault, it seemed that recordings of all relevant parameters showing no significant faults in rather more than three thousand landings – that is, the square root of ten million – were likely to be required. This implied the large-scale operation of a fleet of aircraft using fail-operative automatic systems in good visibility before civil blind landing could be introduced. It was estimated within B.E.A. that with a fleet of thirty-eight aircraft each fitted with a 64-channel flight recorder – that is, the twenty-four Trident 1's less the development aircraft leased back to Hawker Siddeley, and the fifteen Trident 2E's for which the contract was to be finalised in August 1965 – this recording programme should not take more than a year.[21]

Fourth of these features was the relationship between the rates of benefit to be secured from having a full blind landing capability on a fleet of aircraft (increased revenue-earning capability through a reduction in the number of delays and cancellations, and a notionally reduced accident rate) and the size of aircraft to which it was fitted. By 1963 it looked as if the post-Trident generation of aircraft could be carrying anything between 250 and 500 passengers. The cost of experience gained on the Trident might be offset by the competitive advantage B.E.A. might be able to secure over other airlines, in terms of pilot training and operational procedures, by the time these projected aircraft were introduced.[22]

Finally, there was the assurance of government financial

support, to a ceiling of £2 million, to offset the risk to industry of committing itself to the development of a category 3A automatic landing capability in both Trident and the VC-10 – that is, blind automatic touchdowns in operating limits of 200 metres runway visual range where the only limiting visibility factor was the pilot's ability safely to control the landing run – so long as any doubts could reasonably persist over the ultimate feasibility of using the I.L.S. localiser beam for terminal azimuth guidance.[23]

In May 1965, with the completion of the autoflare certification programme, flight trials directed towards the certification of the duplex automatic landing facility had begun. Twenty months' development, in the course of which over two thousand automatic landings at both duplex and triplex level were performed, culminated in January 1967 in an A.R.B. certification test programme of two hundred landings using a standard of equipment it was proposed ultimately to clear for category 3A conditions. In April 1967, after analysis and presentation to the A.R.B. of the landing performance data, certification was granted for the use of the system at duplex level in category 1 conditions. The following month, once clearance had been obtained by the airline's Flight Operations Department from the Board of Trade, in one week of scheduled operations a B.E.A. Trident equipped at duplex level made twenty-seven automatic landings at fifteen different airfields in nine European countries, carrying a total of 1,400 passengers.

Six months earlier, in December 1966, B.E.A. had finally and formally committed themselves to the procurement of a triplex fail-operative capability. The value of this decision to Smiths Industries – that is, upgrading the contract from duplex to triplex on twenty-three Trident 1's and fifteen Trident 2's – was of the order of £2 million.[24] Both in timing and substance, it may be regarded as the outcome of three parallel developments. Progress on Smiths Industries' triplex flight development programme had resulted in June 1966 in the first automatic landing of the 100-ton Belfast strategic freighter, and in a spectacular climax to the Trident programme in November 1966 when the triplex-fitted development aircraft made two series of automatic landings at a fogbound London Airport, with all other aircraft diverted owing to the weather. Sixteen months earlier, in August 1965, the airline had signed a contract with Hawker Siddeley Aviation for the delivery, from February 1968, of fifteen Trident 2E's, of similar capacity but longer range than the Trident 1's: clearly, it would be more economical to have these aircraft triplex-fitted during construction rather than to undergo the kind of retrospective modification programme – complicated, time-wasting and expensive – that had been imposed on the Trident 1 fleet. Finally, also in December 1966, the issue of an updated statement on certification for low visibilities had marked further progress in defining procedures for compliance with the highly exacting A.R.B. safety requirements.

In June 1967, at the Paris air show, it was announced by the Ministry of Technology that a further £2 million was to be made available for a development programme to extend system capabilities on the Trident and the VC-10 from category 3A to category 3B operating limits. Only at this point, it was stated, could the full benefits of a low visibility automatic landing capability be realised. Half the money was to go towards ground equipment development, and half towards the extended development of the Trident and VC-10 airborne systems to provide ground roll guidance and runway steering information for the pilot so that he could monitor the latter part of his landing run from his instruments without the need for external visual reference beyond the 50 metres runway visual range he would require for taxiing. The cost of airborne system development was to be shared with the avionics industry, the aircraft industry and the national airlines. The cost of ground system development, and subsequently procurement, would be borne almost entirely by the Government. The investment was justified on grounds of airline and airport operating efficiency, passenger convenience, and the potential export market for airborne and ground-based equipment.[25] In addition to the Tridents and the VC-10's, the eighteen B.A.C. 111-500's ordered by B.E.A. in November 1966 were also to be equipped with a fail-operative automatic landing capability (based on an anglicised Bendix PB-20 autopilot in the duplicate-monitored configuration designed by Elliotts).[26] So also would the twenty-six Trident 3B's (which the Government was then pressuring B.E.A. to buy in preference to the projected B.A.C. 2-11), the

Concorde supersonic airliner (with a system developed by the Elliotts–SFENA–Bendix consortium) on which B.O.A.C. had reserved delivery positions, and the projected European airbus A-300B which it then seemed a British Government would be sure to participate in and B.E.A. to acquire.

The following month, July 1967, the Trident 2E made its maiden flight from the Hawker Siddeley flight trials base at Hatfield and in February 1968 the first of these aircraft, triplex-fitted during construction, was delivered to B.E.A.

In July 1968 the flight trials stage of the Government-backed category 3B development programme for the Trident was initiated, culminating in May 1969 in a certification test programme for the A.R.B. of 120 landings, 120 take-offs and 24 overshoots in simulated 35 metres visibility. Twelve pilots took part – eight from B.E.A., two from the A.R.B., one from the Board of Trade and one from the Ministry of Technology.

Meanwhile, in September 1968 and February 1969 respectively, for both types of Trident, Hawker Siddeley Aviation had secured full category 2 certification (to 100-ft decision heights) from the A.R.B., and B.E.A. had obtained operating clearance from the Board of Trade to the intermediate limits of 150-ft decision heights and 500 metres runway visual range at the one runway at London Airport which was then equipped with category 2 ground aids (I.L.S. transmitters, lighting system and runway visual range measuring instruments).

In April 1969 B.E.A.'s Chief Engineer publicly expressed his regret that ground facilities in Britain should be lagging behind the rest of Europe when the aircraft programme was way ahead.[27]

(e) The avionics manufacturer – effort and impact

Majendie left Smiths Industries in 1967 to join the Molins Organisation as Managing Director, and at the same time he was appointed Chairman of the Civil Air Transport Industry Training Board. A year later, in November 1968, the Aviation Division carried out a detailed cost investigation into the development of the SEP-5 autopilot both for the Trident and the Belfast.

So far as the Trident was concerned, over the twelve-year

time scale 1960–72, expenditure on development, tooling and manufacturing costs exceeded revenue from sales, on an undiscounted basis, by something in the order of £1½ million.

The Trident venture could not, however, be isolated from the Belfast programme. It was this allocation of the military triplex autopilot development contract in 1960, on the understanding (justified or not) by Smiths Industries that the triplex autopilot was to form the basis of a standard flight control system for R.A.F. Transport Command, that induced the company to maintain the scale of its commitment to an ambitiously complex development programme whose costs could only be recouped according to the number of aircraft built to which the finished equipment was to be fitted. Unfortunately for the equipment manufacturer, it was subsequently decided within the Ministries of Defence and Aviation, first that only thirty-five Belfast aircraft would be required, and then only ten; finally, that Transport Command VC-10's should be fitted with an autopilot from another manufacturer. Moreover, the decision to accept the Belfast contract was in part the outcome of the De Havilland estimate in 1959, shared by Smiths Industries, of the number of Trident aircraft to which the developed triplex equipment would be fitted, on the twin presumptions that B.E.A. would eventually upgrade their procurement commitment to triplex level and that other airlines buying the Trident would follow suit. In 1959 it was thought reasonable to plan on the basis of eventually fitting triplex equipment to about two hundred Tridents. By 1969 it was thought unlikely that more than a hundred would be built.

If two hundred triplex-fitted Tridents and thirty-five Belfasts had in fact been ordered, the whole of the SEP-5 programme could have moved into a profitable situation after many years of adverse cash flow.

Ultimately, however, the impact of the project on divisional performance is unquantifiable. Perhaps the most serious of the unquantifiable costs would be the extent to which the concentration of the skills and energies of the Division on a single giant project might have impeded the development and diffusion of expertise in relevant advancing technologies, such as microelectronics. The unquantifiable benefits would include the boost given to sales of instruments and components improved or developed in the course of the multiplex project a virtually captive spares market for the Trident and a substantial amount of maintenance business; the timely introduction of improved divisional costing procedures; the international prestige derived by the company from their participation in a technically successful project (at least as far as the airborne system was concerned) of this magnitude and significance; and the consequent favourable impact on their bargaining position in the world avionics market.

References

1. This is an abridged version of a fuller account of the innovation, prepared by J. E. H. Hartland in collaboration with W. G. Evans and M. Gibbons, which it is hoped to publish in due course. We are grateful to Smiths Industries Aviation Division for making available internal reports, papers and memoranda.
2. O. B. St John, 'All-Weather Landing', *Shell Aviation News*, no. 364 (1968).
3. Smiths Industries Aviation Division, 'The Multiplex Philosophy', paper presented to the 15th International Air Transport Association Technical Conference, 1963.
4. K. Fearnside, 'Instrumental and Automatic Control for Approach and Landing', *Journal of the Institute of Navigation*, **12** 66 (1959).
5. W. J. Charnley, 'The Work of the Blind Landing Experimental Unit', paper presented to the Institute of Aeronautical Sciences (New York, 1959).
6. Flare: the gradual reduction of the aircraft's rate of descent to very nearly zero at touchdown, accomplished by raising the aircraft's nose.
7. Smiths Electric Pilot Mark 1.
8. F. W. Meredith, 'The Modern Autopilot', *Journal of the Royal Aeronautical Society*, **53** 409 (1949).
9. The preceding five paragraphs are a summary of material gathered in interviews with past and present members of B.L.E.U. in Oct and Nov 1969.
10. Interviews were conducted with A. M. A. Majendie on 24 Nov 1969 and with K. Fearnside on 17 May and 19 June 1968 and on 11 June and 2 Dec 1969.
11. A. M. A. Majendie, 'Automatic Landing: The Role of the Human Pilot', *Journal of the Institute of Navigation*, **16** 84 (1963).
12. £50,000 was allocated from the divisional budget for the feasibility study. For projects involving expenditure in excess of £250,000, reference had to be made to the main Board.
13. K. G. Wilkinson (Chief Engineer of B.E.A.), 'Automatic Landing in B.E.A.'s Trident Operations: A Review of Effort and Achievement', the De Havilland Memorial Lecture to the Royal Aeronautical Society, Apr 1969, printed in the *Aeronautical Journal*, **74** 187 (1970).
14. 'The Automatic Landing Era in Air Travel', *Smiths Industries Aviation Review*, no. 18 (June 1967).
15. Source: Ministry of Technology.
16. Source: Air Registration Board.
17. Source: Ministry of Technology.
18. F. J. Sullings and P. Waller, 'Automatic Landing Systems for All-Weather Operation by Civil Transport Aircraft', paper presented to the 10th Anglo-American Aeronautical Conference Organised by the American Institute of

Aeronautics and Astronautics and the Royal Aeronautical Society, Oct 1967.

19. Ibid.
20. St John, op. cit.
21. Wilkinson, op. cit.
22. Ibid., and private communication from B.E.A.
23. Source: Ministry of Technology.
24. *Smiths Industries Aviation Review*, no. 18 (June 1967).
25. Source: Ministry of Technology.
26. In the Elliott system, redundancy was achieved through having two completely independent channels, only one of which was actually controlling the aircraft at a time.
27. Wilkinson, op. cit.

35 SMITHS INDUSTRIES (KELVIN HUGHES): RADAR NAVIGATIONAL AID

A synthesis of developments from several areas of technology is involved in this case study. In addition, the experience gained from immediate post-war government contract work provided the basis, both financial and technical, for the later private venture work on marine radar. First-hand knowledge of operational requirements of the customers proved to be of great importance in the commercial success of these instruments.

(a) A new display for radar information

The display of radar information is, as is well known, achieved by varying the brightness of the time base of a cathode-ray tube which is rotating in synchronism with the aerial. Since by this means radar data are only renewed once in every revolution of the aerial, continuity of observation is achieved by using, on the face of the cathode-ray tube, a phosphor with long persistence characteristics. Presentation of radar information in this way suffers from several operational disadvantages, the chief ones being:

(i) In order to preserve contrast, the display has to be viewed in dim light.

(ii) The maximum available size of cathode-ray tube

restricts the number of people who can simultaneously view the display.

(iii) It is not easy to mark positions or directions on the display directly.

(iv) It is fatiguing for an observer to watch such a display for long periods of time, particularly because of the visible rotation of the time-base sweep.[1]

The Kelvin Hughes 'Photoplot' is designed to overcome these disadvantages, using photographic techniques to produce a large, bright, high-contrast, theatre-type presentation of radar data.

In the Photoplot, the radar information is first displayed in the normal way on a cathode-ray tube, then it is photographed on a specially made 16-mm film, processed in a few seconds and projected on to the underside of a translucent screen. The chemicals used for processing the film are blown on to it in a prearranged sequence by compressed air. If this process is continued film-frame by film-frame, a sequence of projected pictures of the radar display is produced, each picture being up to date, within a few seconds, with the latest radar data. This technique still allows the navigator to work on a chart-like presentation of the radar information.

The main line of development in the Photoplot was that surrounding the photographic process. The film processing takes place in a pot in which are arranged a number of jets which spray developer, fix and wash solutions as well as hot drying air on to the film sequentially. The jets are of the Venturi type (i.e. like a perfume spray) and are operated by compressed air so that the solutions are applied in the form of a fine atomised spray.

The use of compressed air is considered to be an important feature because it obviates the use of valves inl iquid lines. The processing solutions employed are highly concentrated and very corrosive. The use of valves for liquid switching would not be consistent with the requirement for long-term reliability. Switching the compressed air, however, allows processing to be fully automated and is suitable for computer-controlled operation.[2]

At the beginning of the project, consideration was given to

other techniques, such as projection cathode-ray tube systems. However, it was soon realised that such systems could not reach the requirements of screen size, brilliance and definition. Since storage tubes were not available at that time, production of flickerless displays could only be obtained by repeated generation of the signals, thus creating further problems of signal storage and the necessity for a very wide bandwidth electronic system.

(b) The Kelvin Hughes innovation[3,4]

The idea of trying to photograph cathode-tube displays seems to have originated in the Massachusetts Institute of Technology (M.I.T.) Radiation Laboratory during the Second World War. In 1946 the Navy, in consultation with engineers of the Royal Radar Establishment (R.R.E.), thought that high-speed photographic techniques might be applied in the Navy's military theatres and some preliminary work was begun at R.R.E. Malvern.

In 1949–50 Kelvin Hughes (K.H.) were in the process of diversifying. Radar was one of the fields into which they were moving. R.R.E. in 1950 asked K.H. to investigate the possibilities of a new photographic display for the proposed ROTA early warning defence system. The original contract from R.R.E. was intended for long-range (600 miles) detection of aircraft and eight models were requested.

The liaison was carried out between Mr G. Wikkenhauser and Mr E. R. Townley of K.H., and the Ministry of Supply. A small team was set up to study the problem. From the beginning, K.H. were in close collaboration with Ilford Photographic Company (I.P.) who developed what special films were required. The first few models were in the form of lash-ups and were 'big and a bit clumsy'. Around 1955 the contract with R.R.E. terminated and K.H. supplied sixteen display units to the Ministry.

All the work carried out by K.H. for the Ministry was top secret and the factory had been sectioned off to allow only those with clearance to work in the area. Somehow the Americans found out about the development and in 1955 the Hazeltine Company, who are subcontractors for the American

Government, placed a contract with K.H. for similar units to be fitted in the SAGE defence network. For the American contract, only projection and display equipment, 120-130 units, was required.

In the meantime, interest in the photographic display units was growing. R.R.E. placed a further contract with K.H. to extend the equipment from black-and-white to coloured displays, and the Race Finish Recording Company invited them to adapt their rapid photographic processes for photo-finishes at race meetings.

As K.H. are specialists in the field of marine instrumentation, particularly marine radar systems, the company had considerable experience of the operational problems that beset marine navigators. In 1959 some thought was given to the possibility of using the photographic display unit as an aid to navigation. The idea was pursued vigorously by Mr E. R. Townley, Mr S. R. Parsons and Mr C. Embling.

Various departments of the firm were frankly pessimistic and management was, in general, reluctant to commit any private venture capital until there was a guarantee of at least a small market. The possibility of the ship's personnel being able to maintain these instruments on board was seen as an important factor in the acceptance of such a display unit. Nevertheless, Townley, Parsons and Embling took one of the previously built models from the Radar Applications Department at Dagenham to the K.H. Radar Training School at Southend. There it was assembled as a working unit to be used for demonstration purposes. In addition to this, they installed a similar unit in a small ship to demonstrate the display under sea-going conditions. It was clear then that if the display were ever to be widely used, many modifications would need to be made.

Apparently, many people came to Southend to look at the system, but only one, Captain Washer of the P. & O. Line, showed an active interest. He wanted a Photoplot for the new ship, *Canberra*. This was a considerable incentive to the development of the Photoplot and, in addition, provided invaluable operational experience which could not really be gained until there was at least one working model in service.

The next contract came in 1960 from the Admiralty who

liked the Photoplot and ordered a number of them for an advanced type of warning system. The Navy provided their own radar. A survey was, it seems, carried out on both sides of the Atlantic and the Admiralty chose the K.H. system as the most realistic proposition for the Navy. The Navy, however, gave K.H. rather exacting specifications and, in order to comply with them, K.H. had to improve the detailed design of Photoplot. The successful outcome of the Navy contract and the operational experience gained on the *Canberra* encouraged K.H. to proceed with a commercial version of this equipment.

(c) Some key personnel

So far as the preparatory ground work of the innovation was concerned, E. R. Townley played a vital role. It was he who, in close collaboration with I.P., developed the chemical aspects of the rapid photographic process later to be used in the Photoplot. A graduate chemist from the University of London, Townley has now left K.H. and, after a short time with I.P., he emigrated to the United States where he now manages his own firm.

It was under S. R. Parsons and C. Embling that the rapid photographic process was applied to marine navigation. Parsons was the project engineer through the development of Photoplot. He originally joined K.H. before the Photoplot project, but he had previously been assigned to another division of the company. Since managing Photoplot, he has gone on to be Chief Engineer of K.H. where he is in charge of all R. & D. He holds a B.A. in physics from Oxford.

When Parsons was moved to the Photoplot project, he brought with him Mr C. Embling. It is to Embling that the final design acceptable to ship owners is due. Embling is one of those individuals who are able to get ideas into working hardware; he has no university training but he had acquired considerable experience of the operational problems associated with radar during the Second World War.

(d) Comments

Technologically, Photoplot represents an interesting combination of electronic, electromechanical, optical and photographic engineering. The *invention*, largely the work of Townley, lies in the development of rapid film-processing techniques for early warning systems. Radar technology was regarded as 'given' and the main effort went into development of both rapid-process photography and the compressed-air technique for spraying the film.

The *innovation* comes in the recognition by Wikkenhauser, Townley, Parsons and Embling among others of the possibility of using these previously developed techniques in another context – that of marine navigation. There was initially no clearly identifiable market, but one appeared in embryo when the P. & O. Line requested a Photoplot for the Canberra. Subsequently, the market began to grow.

Improvements in the design are continuing. Later versions of the Photoplot are fully transistorised and make use of the most recent advances in cathode-ray tube technology.

References

1. S. R. Parsons, 'The Application of Rapid Access Photographic Techniques to Radar Display Systems', *Journal of British Institution of Radar Engineers*, **24** 213 (1962).
2. G. Wikkenhauser, 'Applications of Rapid Photographic Processing to Electronic Displays', Congress of World Television Progress, 17–20 Oct 1965.
3. Interviews with C. Embling, Project Engineer, Kelvin Hughes, G. Smith, Public Relations Officer, Kelvin Hughes, and others.
4. Interview with S. Parsons, Chief Engineer, Kelvin Hughes.

36 THORIUM LTD:
RARE EARTH SEPARATION

A group of metals known somewhat misleadingly as the rare earths are found together in certain minerals and are difficult to separate from each other because of the similarity of their properties. They usually occur in association with thorium, which achieved commercial use in 1891 for the manufacture of gas mantles, and thorium production led to the availability of rare earth mixtures as by-products. Although chemically thorium is different from the rare earths, industrial usage has usually included thorium as a product of 'the rare earth industry'.[1]

The solvent extraction processes which gained Thorium Ltd a Queen's Award depend on the use of an organic solvent to extract an inorganic material from an aqueous phase. The pure science on which such processes can be said to be based is of nineteenth-century origin and in fact forms part of elementary courses in physical chemistry. Thorium Ltd's innovation is not, therefore, an application of a recent scientific discovery. Rather, it is an example of technology transfer; the solvent extraction systems derived from war-time research aimed at purifying uranium and then at separating nuclear fission products.

Unlike many sections of the chemical industry, the rare earth industry has been at the mercy of markets largely outside its own control. The basic attitude has been one of hope that other industries would discover uses for its products, and the development of improved separation techniques could only take place in an atmosphere of confidence in the development of new markets.

(a) The rare earth industry before the Second World War

Starting in 1794 with Gadolin's discovery of yttria, a considerable amount of scientific work was carried out on the separation and properties of the rare earths. Around the turn of the present century important centres of scientific research grew up in France, Austria and the United States. Two of these were connected with industrial applications. Carl Auer von Welsbach, an Austrian chemist, achieved a public statue and quite a fortune through developing a gas lighting mantle from thorium oxide containing 1 per cent of cerium oxide; it was first used in 1891. In the United States a small industry producing cerium and mixed rare earths was started in Chicago.[2]

The success of the Welsbach mantle caused a search for thorium-bearing deposits and provoked the interests of the governments of countries containing thorium-bearing ores, monazite being the most important. The world market in thorium was controlled by a German syndicate associated with Welsbach. In Britain an abortive attempt to compete with the Germans was made in 1903 by the South Metropolitan Gas Company, who formed the British Monazite Company to exploit some monazite deposits in North Carolina. The German syndicate reduced its prices and British Monazite ceased operations.[3]

Monazite contains more cerium, lanthanum and other rare earths than it does thorium. The development of the gas mantle industry provoked a search for applications for the rare earth by-products of the thorium extraction. A mixture of rare earths can in some contexts be regarded as one fictitious element sometimes referred to by the German name 'mischmetall'. It can be extracted and purified by standard chemical processes and compounds can be prepared. Mischmetall alloyed with iron found application as a 'flint' for gas lighters and later cigarette lighters. Mischmetall fluoride is readily volatilised in an electric arc and, being a mixture, it emits a spectrum with so many lines that intensity of emission is virtually uniform through the visible region. Mischmetall fluoride, therefore, found application as an additive to the carbon electrodes of arc lamps to improve the quality of the light.

Laboratory techniques for the separation of the individual rare earths existed and all the naturally occurring ones had been identified. The separation techniques were not suitable for industrial use except for lanthanum and cerium, which found minor applications, mainly in the glass industry. There was no stimulus to encourage the development of industrial separations.

In the rare earth field, as in synthetic dyestuffs, the First World War forced Britain to found an industry of its own. In 1914 Thorium Ltd was formed by Howard and Sons, Hopkin and Williams and the Volker Lighting Corporation Ltd. The war also released the monazite deposits in Travancore from German commercial control so that Thorium Ltd had a satisfactory supply of raw material and production was established at the Ilford site of the unsuccessful British Monazite Company.[1,3] The end of the First World War, however, left rare earth industries in Germany, Britain, America, France and

Brazil all facing a declining market as electric lighting began to reduce the need for gas mantles. Between the two wars the market did not improve and few new uses appeared, although research on rare earths in metal alloys was carried out. No new industrial separation techniques were developed and firms in Europe and America formed cartels, sharing out the small market so as to avoid bankruptcy.

In 1928 I.C.I. became an equal shareholder in Thorium Ltd with Howard and Sons Ltd of Ilford, and this was to have important effects after the Second World War.[3]

(b) The development of solvent extraction systems for inorganic materials

Solvent extraction, or more precisely liquid-liquid extraction, depends on the partition of a dissolved substance between two liquids. The partition law that the distribution of a dissolved substance between two solvents is a constant ratio independent of the amount of dissolved substance present was discovered empirically by Berthelot and Jungfleisch in 1872[4] and derived thermodynamically by Nernst in 1891.[5] The technique of extraction of organic substances from water by ether has been used for many years and sixth-form pupils do calculations showing, for example, that more material is extracted by two separate extractions using 50 ml of ether each time than by one extraction using 100 ml of ether.

That extraction of inorganic substances by partition between two liquids is possible has also been known for a long time. The partition of iodine between water and carbon disulphide has been in several generations of school textbooks and as long ago as 1842 Peligot used ether for extracting uranyl nitrate in his separation of uranium from pitchblende.[6] There are many interesting scientific problems involved in the partition of an inorganic substance between water and an organic solvent, but this area of research seems to have been somewhat neglected until the problems associated with atomic energy research focused attention on the need for new separation techniques. In the words of a review article published in 1951:[7]

Our knowledge of the extraction of inorganic substances (as opposed to inner complexes) is largely based on empirical

observations and until recently few attempts have been made
to interpret the data on any rational theoretical basis.

R. J. Callow, formerly Research Manager, Thorium Ltd, has
claimed that he was responsible for the first full-scale commercial use of solvent extraction of an inorganic material. In his
own words:

> Solvent extraction purification relies on the fact frequently
> obscured by chemistry courses divided too rigidly into
> 'organic' and 'inorganic' that many 'inorganic' compounds
> are readily soluble in 'organic' liquids. When the solubility
> of a compound in a liquid immiscible with water is reasonably
> high and differs from those of associated impurities, it is
> possible to exploit the differences in an elegant, continuous
> purification process.
>
> The first major use of the technique for purifying inorganic
> materials was in obtaining uranyl nitrate of adequate purity
> for atomic energy purposes. Later it was used on a semi-
> industrial scale for separating niobium and tantalum and
> hafnium and zirconium. The first full-scale commercial use,
> however, was for purifying thorium nitrate. This process was
> first established in Britain in 1957 under the author's direc-
> tion and has since been adopted elsewhere.[1]

The uranium extraction process was developed by I.C.I. and
used ether as the solvent. It was, in fact, a larger-scale process
than Callow's thorium extraction.[16] To Callow's summary it
should be added that it took the discovery of n-tributyl phosphate (TBP) as a specially useful solvent for rare earths before
solvent extraction could become a commercial proposition in
competition with precipitation techniques and ion exchange
separation. The effectiveness of TBP as a solvent for cerium was
discovered by J. C. Warf at Ames (Iowa State College) working
under contract for the Manhattan District (atomic bomb)
Project. At the same time he discovered that TBP could be used
for the extraction of thorium and uranium. The results were
classified (i.e. kept secret) until 1948 and published in 1949.[8]
There are apparently only two recorded publications on
liquid-liquid extraction of rare earths in the inter-war years.
Imre in 1927 used ether to separate cerium from concentrated

nitric acid,[9] and Fischer, Dietz and Jübermann claimed in a very short article in 1937 to have studied the partition of rare earths between water and ethers, alcohols and ketones. A detailed report of their conclusions was promised but seems not to have appeared.[10] Imre had to go back to 1884 to find previous work on the ether extraction of cerium. Imre's results showed that the solvent was oxidised by the cerium (IV), the cerium (IV) becoming cerium (III) which is not extractable by ether. Warf took up this work and carried out a search for solvents which would not be oxidised. He finally selected nitromethane and TBP, the latter proving to be a useful solvent.

Atomic energy research resulted in advances in many fields of analytical and separative chemistry. This work was not confined to the United States. As mentioned above, a solvent extraction process for the purification of uranium was developed by I.C.I. in 1941.[11] The radiochemical team at Montreal under Goldschmidt (who had worked for Seaborg, the leader of the radiochemical group at Chicago) included British and French workers and developed a solvent system for separating plutonium from irradiated uranium obtained from Chicago.[12] The Montreal work provided the basis for a United Kingdom separation plant at Windscale. American research concentrated on ion exchange systems for the separation of individual fission products, rare earths, etc., and it is perhaps unusual that Warf's discovery should have been exploited outside the United States, with American firms now paying royalties to Thorium Ltd.

War-time research had another important effect so far as the rare earth industry is concerned, namely a renewed interest in thorium. Bretscher and Feather[11] suggested in 1940 that neutron bombardment of thorium would give the fissile isotope uranium 233. This was confirmed by Seaborg in 1942 and by 1944 the Chicago group believed that, after a reactor containing thorium had been started with uranium, it might produce enough U-233 to permit the reaction to sustain itself without the addition of anything but more thorium. Thus thorium, which is much more plentiful than uranium, became a strategic material.[12] It was included with uranium in the Declaration of Trust signed by Roosevelt and Churchill in 1944 in an attempt to control world supplies of these materials. A search for sources of thorium followed and by 1959 thorium was being produced from the Ontario uranium deposits. This marked the end of the dependence of the production of thorium and associated rare earths on monazite.

Visions of large-scale thorium production resulted in renewed interest in the rare earths which would be available as by-products from thorium extraction. Atomic energy agencies gave contracts to people looking for rare earth applications. Some of the results suggested that large quantities might be used in ferrous metallurgy and this resulted in the exploitation of bastnasite deposits in California. Bastnasite contains little thorium, being mainly a rare earth fluorocarbonate, and its exploitation was an unwelcome development for those who were regarding the rare earths as a by-product of thorium production.

Another area of research stimulated by the war was the use of mischmetall in magnesium alloys to give higher strength and better creep resistance at elevated temperatures than any other alloying ingredient available at the time of the early development of the jet engine.

Thus, as a result of the war, the rare earth industry changed from an impoverished industry with little innovative activity to an industry 'having promise' and beginning to look at new extraction techniques.

(c) Thorium Ltd since the Second World War

After the war, Thorium Ltd had three I.C.I. nominees as directors. The Technical Director was A. M. Roberts, the Deputy Research Manager of I.C.I.'s General Chemical Division at Widnes. Being aware of the possible development of new outlets for thorium and the rare earths, the Thorium board sanctioned an expansion of the small research team at Ilford. R. J. Callow was recruited from Pilkington to be Research Manager and during the fifties there were about six graduates and twelve assistants carrying out research and development work. About 20–30 per cent of the R. & D. effort was devoted to the development of solvent extraction systems, and the total R. & D. expenditure was of the order of 7 per cent of turnover[13] (i.e. about twice the national average for the chemical industry).

It was A. M. Roberts who suggested in 1950 that solvent

systems should be examined. At that time it was not obvious that solvent extraction would be better than ion exchange or other methods. In fact, one publication[14] in 1950 claimed that the future for solvent extraction of rare earths was pessimistic and this comment might have been true had it not been for TBP. Roberts was able to suggest the use of solvent extraction because of his previous experience. He had worked on the alternative method of ion exchange separation and during the war had been in charge of the chemical engineering work for the ether extraction of uranium. In 1946 classified documents had been exchanged with the United States during a visit and Warf's discovery of the effectiveness of TBP was known before its publication in 1949.[16]

During 1950–1 Callow talked about solvent extraction, read the literature and selected methyl isobutyl ketone (MIBK) and TBP as being potentially the most useful solvents for thorium. (At that time thorium was still a potential alternative to uranium as a nuclear fuel.) Laboratory work was carried out using micro-scale mixer settler units developed by war-time research and loaned by I.C.I. MIBK was found to be unsatisfactory for several reasons including its comparatively high solubility in water and its volatility which resulted in solvent losses and the risk of an explosion.[13] Two years of laboratory work with TBP by Kemp and Ponting between 1951 and 1953 showed the commercial promise of a TBP-based system for the production of pure thorium nitrate and a semi-pilot plant was set up consisting of six stages of a continuous manufacturing unit. There were a lot of teething troubles but in 1957 a 24-stage plant produced a product of acceptable quality continuously and Callow celebrated by buying champagne for the people who had worked on the plant.

The expected requirements for thorium as a nuclear fuel did not develop (though some experimental work on thorium reactors is still carried out), but the possibility of metallurgical developments requiring individual rare earths for alloys led Thorium Ltd in 1956 to start work on a TBP system for the separation of lanthanum. A plant was operating successfully by 1959. It should perhaps be mentioned that in the rare earth industry a 'plant' is a collection of quite small separating units. A plant capable of the continuous production of 10 lb per hour of pure lanthanum occupies a space smaller than a normal living room.[13]

In 1959 the promise of successful large new markets for rare earths seemed as far away as ever and I.C.I. and Howard and Sons Ltd, who at that time jointly owned the shares in Thorium Ltd, sold out to Rio Tinto Corporation Ltd and Dow Chemie International A.G., Zürich.[3] (Rio Tinto and Dow of Canada supply the thorium concentrates which are the starting point of the extraction process.) The main reason for the sale was Howard's need of capital for a new process.

One outlet that had expanded during the war was the use of cerium in arc-lamp carbons for searchlights. This expansion had resulted in a new factory being built with Government assistance at Widnes. Following the take-over by Rio and Dow, the company's Research Department had to be removed. Only three members of the research team were prepared to move to Widnes; these included Kemp and Copper, an engineer who had developed the original plant. R. J. Callow and the others left Thorium Ltd. Callow's deputy, L. G. Sherrington, took charge of the research laboratory at Widnes and new staff were recruited.[13]

Other phosphate esters were investigated following the discovery in Illinois by Peppard et al.[15] that di(2-ethylhexyl) phosphoric acid gave better distribution ratios than TBP when used to extract rare earths from dilute nitric acid. This compound was the first of a group of esters known by the generic term HDGP or hydrogen di(generalised organic group) phosphate. The use of novel esters in commercial extraction depends on their availability at a reasonable price and, although American firms carried out research into the many permutations of HDGP esters, Thorium concentrated on looking at readily available solvents. As a result, Thorium have been able to develop new extraction processes using versatic acid, a Shell product of mixed naturally-derived carboxylic acids.

Apart from research into polishing powders for telescope glass, Thorium Ltd has not carried out applications research, i.e. it has not attempted to find applications for rare earths. Instead it has concentrated on making available samples of compounds for a wide range of industrial research in the hope that unpredictable new markets would develop. Prior to the late

1950s, the individual rare earths were not freely available except at high prices, so industrial research ignored them as possible ingredients in 'bottle off the shelf' research, otherwise known as 'try a bit of this'. As there was no obvious market there was no incentive for rare earth producers to develop the extraction systems necessary for making samples available. In a vicious-circle context like this, there might be a case for Government-sponsored research to start things moving and in fact some work was financed by the Department of Scientific and Industrial Research in the now defunct Chemical Research Laboratory. This work, however, had no application and it was really an act of faith in the future that led Thorium Ltd to develop extraction systems for individual rare earths.

In the 1960s markets began to develop as research workers looking for novel effects included rare earths in their investigations. One such development has been the discovery that yttrium orthovanadate activated by europium helps to make a superior red phosphor for the colour television industry.[17,18] The publicity surrounding the new phosphor since its announcement in a letter headed, 'A New Highly Efficient Red-Emitting Cathodoluminescent Phosphor (YVO_4:Eu) for Colour Television'[18] has led to other applications for yttrium oxide including the manufacture of yttrium iron garnets (YIGS) for microwave devices and yttrium aluminium garnets (YAGS) for lasers. Several other potential applications have also been described. Thorium's extraction process for yttrium, using versatic acid as well as TBP and other solvents, has been very successful and has resulted in substantial royalty earnings. The older markets in special glasses and ceramic glazes have continued to expand as new outlets have been discovered. For example, praseodymium oxide is used in the preparation of a clear brilliant yellow ceramic stain.[19]

In the 1950s Thorium's sales fluctuated between about a quarter and half a million pounds per year. In the 1960s sales increased, being about £700,000 in 1966 and over £1 million in 1967. Export sales climbed from £69,000 in 1961 to £490,000 in 1967, earning a second Queen's Award (for exports) in 1968.[20, 21] Part of the 'sales' income is, in fact, from licensing arrangements with American and Japanese companies who have found Thorium's extraction processes to be the best available.

In 1967 new research laboratories were opened at the Widnes site and by 1968 the company was employing 110 people, including 21 professionally qualified in science or engineering.[21] In November 1968 a new £500,000 plant was opened with a potential capacity four times greater than existed before. This plant was claimed to be the largest rare earth separating plant in the world.[22]

To some extent the expansion was based on the same faith in the future that typified Thorium Ltd in the 1950s, but a 1968 return on capital of over 20 per cent[22] demonstrates that this is one example of innovation that has been financially successful. The company is not restricting itself just to separating rare earth compounds by solvent extraction systems. Two logical diversifications have been started with the formation of new companies, Solvent Extraction Processes Ltd to apply Thorium Ltd's knowledge in other areas, and Rare Earth Products Ltd, a joint company formed with Johnson Matthey to use a different extraction technique, ion exchange.

(d) Comments

The Thorium case illustrates several factors that can be of importance in the process of innovation. The transfer of information by 'agents rather than agencies' (see p. 44) is clearly shown by Roberts bringing his war-time experience and knowledge of American work to Thorium Ltd. The importance of research in new areas stimulated by the war is also shown. Another interesting point is the long time-lag from the laboratory extraction of cerium by an organic solvent in 1884 to the commercial use of solvent techniques. The long time-lag is primarily due to the lack of any commercial need for new extraction methods under stagnant or declining market conditions for which the older precipitation techniques were adequate.

Another point of general interest from the Thorium case concerns the difficulty of defining a 'failed' innovation. 'Research into research' has now reached the point where workers in this area are attempting to compare successful research projects with commercially unsuccessful ones. This is, of course, a laudable aim but it is beset with the difficulty that success is time-dependent. At any given point in time what seems to be a 'failed' innovation may turn out to be highly successful in the future and vice versa.

In the Thorium case, six years' work went into the production of a new commercial process for thorium extraction. The possible nuclear fuel market for thorium did not develop and the small amounts of thorium required for other applications could be obtained more economically by the existing precipitation process. At that point in time the TBP extraction of thorium could have been described as a 'failed' innovation, in that its major objective had not been realised. However, the objectives and results of actual research projects are more complex than the simple models produced by some theoreticians. Callow was not too sorry that the nuclear fuel market for thorium did not develop; if it had, the process would probably have been taken over by the Government, as had happened with processes for the extraction of radioactive materials started by Thorium Ltd at the Amersham site which became the Radio-chemical Centre.

The engineering and other know-how acquired by the development of a process that was not used turned out to be extremely valuable in the development of other processes that were used. If the objective of the early research is described as 'the acquisition of useful knowledge about the industrial solvent extraction of rare earths', then it can be said to have been successful because the knowledge gained was useful – eventually. The results of almost any research may turn out to be useful at some time in the future, so it is not possible at a given point in time finally to write off any particular research project as a 'failure'.

Another point of interest in the Thorium case is the disruption of its research team with the move to Widnes. Research managers are often reluctant to disrupt a successful research team with specialised expertise. In the Thorium case, however, several people including Roberts and Callow left and fresh people were brought in to form a new research team. The new group was responsible for the development of processes using versatic acid. Thus the change in research staff clearly did not prevent a major discovery from being made.

References

1. R. J. Callow, *The Rare Earth Industry* (Pergamon, Oxford, 1966).
2. F. H. Spedding and A. H. Daane (eds), *The Rare Earths* (Wiley, London, 1961).
3. D. W. F. Hardie and J. D. Pratt, *A History of the Modern British Chemical Industry* (Pergamon, Oxford, 1966) p. 322.
4. Berthelot and Jungfleisch, *Ann. Chim. Phys.*, **26**(4) 396 (1872).
5. W. Nernst, *Z. phys. Chem.*, **8** 110 (1891).
6. Peligot, *Ann. Chim. Phys.*, **5**(7) 42 (1842).
7. H. M. Irving, *Quarterly Reviews*, **5** 200 (1951).
8. J. C. Warf, *J. Amer. Chem. Soc.*, **71** 3257 (1949).
9. Imre, *Z. inorg. u. allgem. Chem.*, **164** 214 (1927).
10. Fischer, Dietz and Jübermann, *Naturwissenschaften*, **25** 348 (1937).
11. M. Gowing, *Britain and Atomic Energy 1939–1945* (Macmillan, London, 1965) pp. 69, 300.
12. Hewlett and Anderson, *The New World 1939–1946* (Pennsylvania State U.P., University Park, (1962).
13. R. J. Callow, private communications, 1968 and 1969.
14. R. Bock, *Angew. Chem.*, **62A** 375 (1950).
15. Peppard *et al.*, *J. Inorg. & Nuclear Chem.*, **4** 334 (1957).
16. A. M. Roberts, interview, 1969.
17. Chang, *J. Appl. Phys.*, **34** 3500 (1963).
18. Levine and Palilla, *Appl. Phys. Letters*, **5** 118 (1964).
19. Ceramic pigments, British Patent 965,863 (1963).
20. *Financial Times*, 22 Apr 1968.
21. Thorium Ltd, Press handout on the occasion of the opening of new research laboratories, 11 Apr (1967).
22. *Sunday Times*, 13 Nov 1968.

Index